Impact of Societal Development and Infrastructure on Biodiversity Decline

Ashok Kumar Rathoure
Saraca Research Inc., India

A volume in the Practice, Progress, and Proficiency in Sustainability (PPPS) Book Series

IGI Global
PUBLISHER of TIMELY KNOWLEDGE

Published in the United States of America by
IGI Global
Engineering Science Reference (an imprint of IGI Global)
701 E. Chocolate Avenue
Hershey PA, USA 17033
Tel: 717-533-8845
Fax: 717-533-8661
E-mail: cust@igi-global.com
Web site: http://www.igi-global.com

Library of Congress Cataloging-in-Publication Data

CIP DATA PROCESSING

2024 Engineering Science Reference
ISBN(hc) 9798369369500 | ISBN(sc) 9798369369517 | eISBN 9798369369524

This book is published in the IGI Global book series Practice, Progress, and Proficiency in Sustainability (PPPS) (ISSN: 2330-3271; eISSN: 2330-328X)

British Cataloguing in Publication Data
A Cataloguing in Publication record for this book is available from the British Library.

Practice, Progress, and Proficiency in Sustainability (PPPS) Book Series

ISSN:2330-3271
EISSN:2330-328X

MISSION

In a world where traditional business practices are reconsidered and economic activity is performed in a global context, new areas of economic developments are recognized as the key enablers of wealth and income production. This knowledge of information technologies provides infrastructures, systems, and services towards sustainable development.

The **Practices, Progress, and Proficiency in Sustainability (PPPS) Book Series** focuses on the local and global challenges, business opportunities, and societal needs surrounding international collaboration and sustainable development of technology. This series brings together academics, researchers, entrepreneurs, policy makers and government officers aiming to contribute to the progress and proficiency in sustainability.

COVERAGE

- Socio-Economic
- Eco-Innovation
- Technological learning
- Strategic Management of IT
- Global Business
- Environmental informatics
- Innovation Networks
- Global Content and Knowledge Repositories
- Outsourcing
- Green Technology

IGI Global is currently accepting manuscripts for publication within this series. To submit a proposal for a volume in this series, please contact our Acquisition Editors at Acquisitions@igi-global.com or visit: http://www.igi-global.com/publish/.

The Practice, Progress, and Proficiency in Sustainability (PPPS) Book Series (ISSN 2330-3271) is published by IGI Global, 701 E. Chocolate Avenue, Hershey, PA 17033-1240, USA, www.igi-global.com. This series is composed of titles available for purchase individually; each title is edited to be contextually exclusive from any other title within the series. For pricing and ordering information please visit http://www.igi-global.com/book-series/practice-progress-proficiency-sustainability/73810. Postmaster: Send all address changes to above address.

Titles in this Series

For a list of additional titles in this series, please visit:
ttp://www.igi-global.com/book-series/practice-progress-proficiency-sustainability/73810

Promoting Multi-Sector Sustainability With Policy and Innovation
Aftab Ara (University of Hail, Saudi Arabia) and Renuka Thakore (University of Central Lancashire, UK)
Engineering Science Reference • copyright 2024 • 286pp • H/C (ISBN: 9798369321133) • US $290.00 (our price)

The Role of Female Leaders in Achieving the Sustainable Development Goals
Mercia Selva Malar Justin (Xavier Institute of Management and Entrepreneurship, India) and Joycia Thorat (Churches Auxiliary for Social Action, India)
Business Science Reference • copyright 2024 • 341pp • H/C (ISBN: 9798369318348) • US $285.00 (our price)

Emerging Technologies and Marketing Strategies for Sustainable Agriculture
Jabulani Garwi (University of the Free State, South Africa) Reason Masengu (Middle East College, Oman) and Option Takunda Chiwaridzo (University of Science and Technology Beijing, Beijing, China)
Engineering Science Reference • copyright 2024 • 315pp • H/C (ISBN: 9798369348642) • US $255.00 (our price)

Quality of Life and Climate Change Impacts, Sustainable Adaptation, and Social-Ecological Resilience
Kasturi Shukla (Symbiosis Institute of Health Sciences, Symbiosis International University (Deemed), Pune, India) Yogesh B. Patil (Symbiosis Centre for Research and Innovation, Symbiosis International University (Deemed), Pune, India) Ronald C. Estoque (Center for Biodiversity and Climate Change, Forestry and Forest Products Research Institute, Tsukuba, Japan) and Pedro Antonio López de Haro (Universidad Autónoma Indígena de México, Mexico)
Engineering Science Reference • copyright 2024 • 499pp • H/C (ISBN: 9781668498637) • US $250.00 (our price)

IGI Global
PUBLISHER of TIMELY KNOWLEDGE

701 East Chocolate Avenue, Hershey, PA 17033, USA
Tel: 717-533-8845 x100 • Fax: 717-533-8661
E-Mail: cust@igi-global.com • www.igi-global.com

The book is dedicated to
Anika A. Rathoure
“ELA”

Editorial Advisory Board

Table of Contents

Detailed Table of Contents

In the era of rapid global development, the delicate balance between progress and environmental preservation is strained. "Vanishing Harmony" addresses biodiversity loss, revealing the link between development and Earth's ecosystems decline. The book examines causes, consequences, and solutions against urbanization, industrialization, and infrastructure expansion. It unravels interconnected factors like deforestation, pollution, climate change, and resource exploitation. Biodiversity loss affects ecological stability and human well-being, impacting food security, medicine, and ecosystem resilience. Addressing root causes is vital for sustainability. Policy, international collaboration, and community engagement are key. The book advocates for a shift in societal attitudes and economic models to prioritize development and ecological preservation. It aims to inform policymakers, environmentalists, researchers, and the public, fostering understanding and inspiring collective commitment to preserve our planet's biological heritage.

Today, the global decline of biological diversity, or biodiversity, is one of the key environmental problems. Keeping in the view of day by day reduction of biodiversity, Convention on Biological Diversity advised a comprehensive analysis that how developmental activity specially, road widening and new highway affect biodiversity must be included in the environmental impact assessment (EIA) process: "Each Contracting Party shall introduce appropriate procedures requiring environmental impact assessment of its proposed projects that are likely to have significant adverse impact on biological diversity." In response to this suggestion, a number of governmental organizations have released guidelines on biodiversity and impact assessments and several non-governmental organizations, like The World Conservation Union and the International Association for Impact Assessment, are also working in this field. As a result, a new field of study called biodiversity impact assessment (BIA) was created.

Stream biodiversity loss is a major environmental issue that directly affects the delicate ecological balance of these aquatic habitats. The stability and health of the surrounding environment are supported by the wide variety of plants, fish, and invertebrates that may be found in streams. These habitats are becoming less healthy largely because of human activity, which may lead to a decline in biodiversity and the extinction of many species. Understanding the origins, impacts, and challenges involved, as well as increasing awareness of the issue, are crucial goals for the preservation of stream ecosystems.

Due to the growing need for freshwater resources, water diversion projects are becoming an essential part of international water management plans. While hydropower is considered renewable, its environmental impact includes altered river flows, disrupted fish migration, and sediment retention affecting ecosystems. Water diversion projects impact aquatic flora and fauna and their habitats. Sustainable water management is crucial, considering the potential effects such as variations in temperature, sediment movement, and flow patterns caused by dam construction. In aquatic habitats building hydropower projects reduced connectivity, altered nutrient dynamics, and habitat fragmentation can decrease organism diversity and numbers, posing a significant threat to biodiversity in affected areas. A comprehensive approach that includes adaptive management and environmental impact evaluations is crucial for water diversion projects, as this abstract highlights. Addressing water diversion impacts on biodiversity is essential for long-term environmental sustainability.

The variety among living things from various environments, such as desert, marine, and terrestrial ecosystems, as well as the ecological complexes they are a part of, is known as biodiversity. The wellbeing of the earth and all of its people depends on the preservation of biodiversity. Human-induced long-term changes in temperatures and weather patterns disrupt the delicate life-supporting equilibrium, causing climate change. Pollution of land, water, and air jeopardizes ecosystems and species globally. According to the World Health Organization, air pollution affects 92% of the population and contributes to 11.6% of global fatalities. Climate change impacts biodiversity through altered breeding and migration behaviors, shifting distribution patterns, invasive species incursions, increased insect attacks, and heightened forest fire risks. India can enhance coordination by enacting eco-friendly laws, integrating these issues into national plans, and adopting biodiversity-based climate change mitigation strategies.

The world is continuously progressing towards technological advancement. Eventually, global biodiversity is taking a consistent downfall. This is not only disrupting the ecosystem services and livelihood spatially but also impacts carbon offsetting globally in the present context. In the arena of biodiversity loss in the last three decades the problem remains unresolved, despite environmental regulations present at national and international levels to protect biodiversity. Forest diversion has been observed to be one of the major root causes for these losses which is further resulting in habitat deterioration and fragmentation. Numerous case studies have been reported that support biodiversity loss due to forest diversions. The conversion of forest land to commercial purposes such as plantations, infrastructural developments, and residential projects has shown a major impact on biodiversity loss, especially in the tropics.

The mining sector, despite its significant role in the climate and biodiversity crises, has not garnered substantial attention in global discussions on these matters. COP15 presents an additional chance to integrate biodiversity and forests into standard practices within the mining sector. In 2019, a UN report highlighted a global biodiversity crisis characterized by an alarming pace of animal and plant extinctions, putting one million species at risk of vanishing. Surprisingly, the biodiversity crisis doesn't garner the same level of media and political focus as the climate crisis, despite the imperative to address both simultaneously. Preserving habitats such as forests serves a dual purpose: safeguarding species diversity while also serving as a solution to mitigate greenhouse gas emissions. Forests act as vital carbon sinks, absorbing and storing atmospheric carbon as biomass. Mining activities, essential for meeting global resource demands, have become a pervasive threat to biodiversity, leading to widespread ecological disruption and biodiversity loss.

Biodiversity loss is one of the most critical environmental problems nowadays. Several natural and anthropogenic causes are known to trigger it. Environmental pollution, including land degradation, has been long considered a significant contributor to biodiversity loss. However, the role of plastic pollution, especially microplastics, in biodiversity loss has been recognised relatively recently. This chapter delves into the complex relationship between biodiversity and plastic pollution, assessing how plastic degrades natural habitats, distorts the ecosystem, and threatens the survival of countless species, eventually disrupting the ecosystem's delicate balance. Through a review of scientific literature, the chapter probes the sources of plastic pollution, its various forms and the impacts of plastic on marine and terrestrial ecosystems. It examines how plastic waste ingestion, entanglement, and chemical contamination affect individual species, populations, and ecological communities. Lastly, the chapter discusses measures to manage plastic pollution and safeguard biodiversity.

The concentration of atmospheric CO2 is currently at around 415 ppm and is increasing by approximately 2-3 ppm annually. This rise is mainly caused by human activities, particularly the burning of fossil fuels for energy production, industrial processes, and changes in land use. Although annual emission rates have fluctuated, they generally remain high, with recent estimates indicating around 40-45 GtCO2 being emitted each year. This chapter examines the crucial involvement of terrestrial, aquatic, and marine ecosystems in carbon sequestration, a key process in contending climate change. By investigating various mechanisms such as terrestrial vegetation absorption of carbon dioxide through photosynthesis and the deposition of organic material in marine sediments, the chapter elucidates the intricate connections between these ecosystems and their impact on the global carbon cycle and regulating climate and stresses the vital requirement for cooperative efforts to ensure their ability to sequester carbon for the benefit of future generations.

Vishal M. Makwana, Zoology Department, Sir P. P. Institute of Science, M. K. Bhavnagar University, India
Foram P. Patel, D. K. V. Arts and Science College, Saurashtra University, Rajkot, India
Pravinsang P. Dodia, Zoology Department, Sir P. P. Institute of Science, M. K. Bhavnagar University, India

In this book chapter, the authors explore the intricate relationship between urbanization and biodiversity conservation. Despite urban environments being traditionally associated with habitat loss and fragmentation, they hold immense potential as critical arenas for safeguarding biodiversity. The authors delve into innovative conservation strategies that capitalize on the unique opportunities offered by urban settings, including the utilization of unconventional spaces and the integration of Indigenous perspectives. Through case studies and examples, they highlight the importance of community involvement and citizen science initiatives in fostering urban biodiversity. By reconceptualizing urban areas as vital components of conservation efforts, the authors aim to ensure the survival and recovery of threatened species within their natural habitats. This chapter offers insights into how urban landscapes can be transformed into resilient ecosystems that coexist harmoniously with human communities.

"Harmony in Nature" offers a thorough examination of biodiversity conservation methods. It serves as a guide for scientists, policymakers, and enthusiasts. Beginning with the urgent need for conservation, it emphasizes ecosystems' role in sustaining life. The book covers traditional and modern assessment methods, including field studies and advanced technologies like satellite imagery. It stresses integrating indigenous knowledge, citizen science, and community-based approaches. Adaptive management, policy frameworks, and global collaborations are discussed. Case studies illustrate successful initiatives and lessons learned. Ethical considerations and sustainability are integrated throughout. The book concludes with a call to action for global biodiversity preservation. It aims to be a valuable resource for academia, practitioners, and policymakers, fostering understanding of essential methodologies for successful conservation amidst environmental challenges.

Plantation methodologies and strategies play a crucial role in addressing environmental concerns and promoting sustainable development. This article focuses on two specific approaches – avenue plantation and compensatory forestry. Avenue plantation involves the strategic planting of trees along roadways, providing various environmental, social, and aesthetic benefits. Compensatory forestry, on the other hand, is an essential strategy for mitigating the environmental impact of development projects. This chapter explores the methods, advantages, challenges, and environmental implications of these plantation strategies.

Foreword

In the relentless march of human progress, we have reshaped the world in ways both profound and far-reaching. Our cities reach ever skyward, our industries churn ceaselessly, and our technologies push the boundaries of what was once thought possible. Yet, amidst this whirlwind of advancement, there lies a silent casualty: the intricate web of life that sustains us all. *Impact of Societal Development and Infrastructure on Biodiversity Decline* is a timely and sobering exploration of the consequences of our collective actions on the natural world. In its pages, authors are confronted with the stark realities of biodiversity loss—the silent disappearance of species, the unravelling of ecosystems, and the erosion of the very foundation of life on Earth.

Authored by experts in the fields of ecology, environmental science, and sustainable development, this book offers a comprehensive analysis of the multifaceted challenges we face. From the sprawling expanses of agricultural monocultures to the concrete jungles of urbanization, from the construction of highways and dams to the insidious creep of pollution and climate change, each chapter delves deep into the complex interplay between societal development, infrastructure expansion, and biodiversity decline.

Yet, amidst the sombre narrative, there exists a glimmer of hope—a recognition that while we have been architects of destruction, we can also be agents of restoration. Through innovative methodologies, visionary policies, and collective action, we have the power to reverse the tide of biodiversity loss and forge a path towards a more sustainable future. As we navigate through these pages, let us heed the call to action that resonates from every chapter. Let us recognize that the fate of biodiversity is not a distant concern but a fundamental imperative for the well-being of humanity and the health of our planet. Let us stand together, as stewards of the Earth, and work tirelessly to preserve the rich tapestry of life that sustains us all.

Impact of Societal Development and Infrastructure on Biodiversity Decline is more than just a book—it is a rallying cry, a call to arms, and a roadmap for action. May it inspire and empower all who read it to join in the urgent task of safeguarding the precious diversity of life on our planet, for the sake of generations yet unborn.

Vishnu D Rajput
Southern Federal University, Russia

Preface

In the intricate tapestry of life on Earth, biodiversity forms the very threads that bind ecosystems together. It encompasses the variety of life forms, from the smallest microorganisms to the largest mammals, and the intricate webs of interactions between them. However, in recent times, this delicate balance has been under severe threat.

Welcome to this comprehensive exploration of the *Impact of Societal Development and Infrastructure on Biodiversity Decline*. This book delves deep into the multifaceted issues surrounding biodiversity loss and its repercussions on ecosystems and human societies alike. Each chapter within this volume is meticulously crafted to unravel the complexities of biodiversity decline from various angles. We embark on a journey that traverses the vast landscapes of modern agriculture, infrastructure development, climate change, pollution, and urbanization, among other factors, all of which contribute to the erosion of our planet's rich biological heritage.

Chapter by chapter, we examine the root causes and consequences of biodiversity loss, shedding light on the intricacies of each challenge. From the impact of road widening to the insidious effects of plastic pollution, from the ecological toll of urbanization to the consequences of mining activities, no stone is left unturned in our quest to understand the myriad ways in which human activities threaten the diversity of life on Earth. But this book is not merely a chronicle of despair. Within these pages lie seeds of hope and resilience, as we explore methodologies and approaches for biodiversity conservation. We delve into the potential of carbon sequestration by terrestrial, aquatic, and marine biodiversity, and we uncover the promise of plantation methodologies and strategies in mitigating biodiversity loss.

As we navigate through these pages, we invite you to embark on a journey of discovery and reflection. Let us confront the challenges ahead with courage and determination, recognizing that the fate of biodiversity is intricately intertwined with our own. Together, let us strive to safeguard the rich tapestry of life that sustains us all.

INSIDE THIS ISSUE

The first chapter focuses on rapid advancement of human society has brought about significant changes to the global landscape, with infrastructure development often at the forefront of these transformations. While such developments have undoubtedly improved human quality of life, they have also had profound implications for the natural world, particularly in terms of biodiversity decline. This article explores the complex interplay between societal development, infrastructure expansion, and the degradation of biodiversity, examining both the direct and indirect impacts of these factors on ecosystems worldwide. Additionally, authors highlighted the importance of sustainable development practices and proactive conservation efforts in mitigating the adverse effects of societal growth on the planet's biological diversity.

Chapter 2, concentrating on biodiversity, or the diversity of life forms on Earth, has declined considerably as a result of the shift towards monoculture farming and hybrid crops. Genetic uniformity in hybrid crops, which is frequently caused by clonal propagation, jeopardizes crop species' genetic diversity. High-input agriculture also causes chemical runoff, which degrades soil quality and has an impact on soil microbes. When agricultural land use is unsustainable and causes disruption of local ecosystems, it can lead to mass extinctions. Hybrid crops can cause unintentional gene flow and potential genetic pollution, leading to the emergence of "superweeds." Agrochemical runoff degrades water quality, putting aquatic biodiversity in danger. Chemical runoff endangers non-target animals, aquatic habitats, and water quality, creating pollution and changing aquatic ecosystems. These findings underscore the necessity of sustainable agriculture methods to mitigate biodiversity loss, highlighting the need for collaboration among farmers, policymakers, and researchers for long-term agricultural profitability.

Chapter 3 speaks about Biodiversity, the variety of life on Earth, is fundamental to the functioning of ecosystems and the well-being of humanity. This article explores the intricate connections between biodiversity and ecosystem services, examining their ecological, economic, and cultural significance. Through a multidisciplinary lens, the author elucidate the role of biodiversity in providing essential ecosystem services such as pollination, soil fertility, water purification, and climate regulation. Moreover, they discussed the implications of biodiversity loss for human societies and the urgent need for conservation efforts to safeguard both biodiversity and the services it provides.

Chapter 4 highlights the reasons for biodiversity loss. The worldwide threat to human welfare that is expanding the quickest right now is climate change. Since the current generation is thought to be the last to be able to take meaningful action to lessen its effects, the world has understood how responsible it is. In the event that it fails, human society is probably destined to never recover. Many organizations have stressed that the climate problem is fundamentally unjust, with poorer nations bearing a heavier burden than others. India is one of the nine nations where climate change is said to be having a significant impact. A WHO estimate from 2016 suggested that by 2050, India would account for over 25% of all climate-related deaths worldwide because of dwindling food supplies. China is predicted to see the greatest per-capita mortality rate from food insecurity.

Chapter 5 focuses on the current scenario of biodiversity loss due to developmental activities. In the era of rapid global development, the delicate balance between progress and environmental preservation is strained. Vanishing Harmony addresses biodiversity loss, revealing the link between development and Earth's ecosystem decline. The book examines causes, consequences, and solutions against urbanization, industrialization, and infrastructure expansion. It unravels interconnected factors like deforestation, pollution, climate change, and resource exploitation. Biodiversity loss affects ecological stability and human well-being, impacting food security, medicine, and ecosystem resilience. Addressing root causes is vital for sustainability. Policy, international collaboration, and community engagement are key. The article advocates for a shift in societal attitudes and economic models to prioritize development and ecological preservation. It aims to inform policymakers, environmentalists, researchers, and the public, fostering understanding and inspiring collective commitment to preserve our planet's biological heritage.

Chapter 6 provides the impact of road widening & new highway construction on biodiversity and its solution. Today global decline of biological diversity, or biodiversity is one of the key environmental problems. Keeping in the view of day by day reduction of biodiversity, Convention on Biological Diversity advised a comprehensive analysis that how developmental activity specially, road widening and new highway affect biodiversity must be included in the Environmental Impact Assessment (EIA) process: "Each Contracting Party shall introduce appropriate procedures requiring environmental impact assessment of its proposed projects that are likely to have significant adverse impact on biological diversity." In response to this suggestion, a number of governmental organizations have released guidelines on biodiversity and impact assessments and several non-governmental organizations, like The World Conservation Union and the International Association for Impact Assessment, are also working in this field. As a result, a new field of study called Biodiversity Impact Assessment (BIA) was created.

Chapter 7 elaborates biodiversity loss in streams due to human diversion activities. Stream biodiversity loss is a major environmental issue that directly affects the delicate ecological balance of these aquatic habitats. The stability and health of the surrounding environment are supported by the wide variety of plants, fish, and invertebrates that may be found in streams. These habitats are becoming less healthy largely because of human activity, which may lead to a decline in biodiversity and the extinction of many species. Understanding the origins, impacts, and challenges involved, as well as increasing awareness of the issue, are crucial for the preservation of stream ecosystems.

Chapter 8 explains about biodiversity loss due to construction of dams and hydropower projects and its impact on aquatic ecosystem. Due to the growing need for freshwater resources, water diversion projects are becoming an essential part of international water management plans. While hydropower is considered renewable, its environmental impact includes altered river flows, disrupted fish migration, and sediment retention affecting ecosystems. Water diversion projects impact aquatic flora and fauna and their habitats. Sustainable water management is crucial, considering the potential effects such as variations in temperature, sediment movement, and flow patterns caused by dam construction. In aquatic habitats building hydropower projects reduced connectivity, altered nutrient dynamics, and habitat fragmentation can decrease organism diversity and numbers, posing a significant threat to biodiversity in affected areas. A comprehensive approach that includes adaptive management and environmental impact evaluations is crucial for water diversion projects, as this abstract highlights. Addressing water diversion impacts on biodiversity is essential for long-term environmental sustainability.

Chapter 9 speaks about Biodiversity loss due to climate change and pollution. The variety among living things from various environments, such as desert, marine, and terrestrial ecosystems, as well as the ecological complexes they are a part of, is known as biodiversity. The wellbeing of the earth and all of its people depends on the preservation of biodiversity. Human-induced long-term changes in temperatures and weather patterns disrupt the delicate life-supporting equilibrium, causing climate change. Pollution of land, water, and air jeopardizes ecosystems and species globally. According to the World Health Organization, air pollution affects 92% of the population and contributes to 11.6% of global fatalities. Climate change impacts biodiversity through altered breeding and migration behaviors, shifting distribution patterns, invasive species incursions, increased insect attacks, and heightened forest fire risks. India can enhance coordination by enacting eco-friendly laws, integrating these issues into national plans, and adopting biodiversity-based climate change mitigation strategies.

Chapter 10 highlights the biodiversity loss due to forest diversion. The world is continuously progressing towards technological advancement. Eventually, global biodiversity is taking a consistent downfall. This is not only disrupting the ecosystem services and livelihood spatially but also impacts carbon offsetting globally in the present context. In the arena of biodiversity loss in the last three decades the problem remains unresolved, despite environmental regulations present at national and international levels to protect biodiversity. Forest diversion has been observed to be one of the major root causes for these losses which is further resulting in habitat deterioration and fragmentation. Numerous case studies have been reported that support the biodiversity loss due to forest diversions. The conversion of forest land to commercial purposes such as plantations, infrastructural developments, and residential projects has shown a major impact on biodiversity loss, especially in the tropics.

Chapter 11 speaks about biodiversity loss due to mining activities. The mining sector, despite its significant role in the climate and biodiversity crises, has not garnered substantial attention in global discussions on these matters. COP15 presents an additional chance to integrate biodiversity and forests into standard practices within the mining sector. In 2019, a UN report highlighted a global biodiversity crisis characterized by an alarming pace of animal and plant extinctions, putting one million species at risk of vanishing. Surprisingly, the biodiversity crisis doesn't garner the same level of media and political focus as the climate crisis, despite the imperative to address both simultaneously. Preserving habitats such as forests serves a dual purpose: safeguarding species diversity while also serving as a solution to mitigate greenhouse gas emissions. Forests act as vital carbon sinks, absorbing and storing atmospheric carbon as biomass. Mining activities, essential for meeting global resource demands, have become a pervasive threat to biodiversity, leading to widespread ecological disruption and biodiversity loss.

Chapter 12 focuses on plastic pollution and its impact on biodiversity. Biodiversity loss is one of the most critical environmental problems nowadays. Several natural and anthropogenic causes are known to trigger it. Environmental pollution, including land degradation, has been long considered a significant contributor to biodiversity loss. However, the role of plastic pollution, especially microplastics, in biodiversity loss has been recognized relatively recently. This chapter delves into the complex relationship between biodiversity and plastic pollution, assessing how plastic degrades natural habitats, distorts the ecosystem, and threatens the survival of countless species, eventually disrupting the ecosystem's delicate balance. Through a review of scientific literature, the chapter probes the sources of plastic pollution, its various forms and the impacts of plastic on marine and terrestrial ecosystems. It examines how plastic waste ingestion, entanglement, and chemical contamination affect individual species, populations, and ecological communities. Lastly, the chapter discusses measures to manage plastic pollution and safeguard biodiversity.

Chapter 13 provides the information of carbon sequestration by terrestrial, aquatic and marine biodiversity. The concentration of atmospheric CO2 is currently at around 415 ppm and is increasing by approximately 2-3 ppm annually. This rise is mainly caused by human activities, particularly the burning of fossil fuels for energy production, industrial processes, and changes in land use. Although annual emission rates have fluctuated, they generally remain high, with recent estimates indicating around 40-45 GtCO2 being emitted each year. This chapter examines the crucial involvement of terrestrial, aquatic, and marine ecosystems in carbon sequestration, a key process in contending climate change. By investigating various mechanisms such as terrestrial vegetation absorption of carbon dioxide through photosynthesis and the deposition of organic material in marine sediments, the chapter elucidates the intricate connections between these ecosystems and their impact on the global carbon cycle and regulating climate and stresses the vital requirement for cooperative efforts to ensure their ability to sequester carbon for the benefit of future generations.

Chapter 14 expresses the urbanization's ecological toll: exploring its role in biodiversity decline. Despite urban environments being traditionally associated with habitat loss and fragmentation, they hold immense potential as critical arenas for safeguarding biodiversity. Authors delve into innovative conservation strategies that capitalize on the unique opportunities offered by urban settings, including the utilization of unconventional spaces and the integration of Indigenous perspectives. Through case studies and examples, authors highlighted the importance of community involvement and citizen science initiatives in fostering urban biodiversity. By reconceptualizing urban areas as vital components of conservation efforts, authors aim to ensure the survival and recovery of threatened species within their natural habitats. This chapter offers insights into how urban landscapes can be transformed into resilient ecosystems that coexist harmoniously with human communities.

Chapter 15 speaks about methodology and approaches for biodiversity conservation. Harmony in Nature offers a thorough examination of biodiversity conservation methods. It serves as a guide for scientists, policymakers, and enthusiasts. Beginning with the urgent need for conservation, it emphasizes ecosystems' role in sustaining life. The article covers traditional and modern assessment methods, including field studies and advanced technologies like satellite imagery. It stresses integrating indigenous knowledge, citizen science, and community-based approaches. Adaptive management, policy frameworks, and global collaborations are discussed.

The last chapter 16 highlights the plantation methodologies & strategies: avenue plantation & compensatory forestry to conserve the biodiversity. Plantation methodologies and strategies play a crucial role in addressing environmental concerns and promoting sustainable development. This article focuses on two specific approaches – Avenue Plantation and Compensatory Forestry. Avenue Plantation involves the strategic planting of trees along roadways, providing various environmental, social, and aesthetic benefits. Compensatory Forestry, on the other hand, is an essential strategy for mitigating the environmental impact of development projects. This article explores the methods, advantages, challenges, and environmental implications of these plantation strategies.

This book delves deep into the complex relationship between societal development, infrastructure expansion, and the alarming decline of biodiversity. Across its sixteen chapters, it explores the myriad ways in which human activities, from modern agriculture to urbanization, from pollution to climate change, threaten the rich tapestry of life on Earth. Each chapter examines a different facet of biodiversity loss, from the impact of road widening and construction projects to the consequences of plastic pollution and mining activities. Through meticulous analysis, it highlights the root causes and far-reaching consequences of these challenges, underscoring the urgent need for action.

However, amidst the challenges, the book also offers glimpses of hope and resilience. It explores methodologies and approaches for biodiversity conservation, from carbon sequestration to sustainable plantation strategies. By recognizing the interconnectedness of all life forms and embracing innovative solutions, it paints a vision of a future where biodiversity thrives and ecosystems flourish. Ultimately, this book serves as a call to action, urging readers to confront the realities of biodiversity loss and to work towards a more sustainable and harmonious relationship with the natural world. It reminds us that the fate of biodiversity is intertwined with our own, and that only through collective effort can we ensure a thriving planet for future generations.

Ashok K. Rathoure
Saraca Research Inc. India

Acknowledgment

This book is the result of the dedicated effort of numerous individuals. We would like to acknowledge the help of all the people involved in this project and, more specifically, to the authors and reviewers that took part in the review process. Without their support, this book would not have become a reality.

First, we would like to thank each one of the authors for their contributions. Our sincere gratitude goes to the chapter's authors who contributed their time and expertise to this book.

Second, we wish to acknowledge the valuable contributions of the reviewers regarding the improvement of quality, coherence and content presentation of chapters. Most of the authors also served as referees; we highly appreciate their double task.

Additionally, we would like to thank our Editorial Advisory Board.

Our Production Coordinator skillfully managed the production of the textbook. We wish to thank the publisher for the sustained interest shown by him during the entire work. A sincere expression of the thanks goes to the entire staff of IGI Global.

Generous participation of faculties and students to eliminate errors in the text and to eliminate errors in the text and to refine the presentation will be greatly acknowledged.

Last and not least, we beg forgiveness of all those who have been with us over the course of the years and whose names we have failed to mention.

Ashok K. Rathoure
Saraca Research Inc. India

Introduction

Welcome to *Impact of Societal Development and Infrastructure on Biodiversity Decline*. In the pages of this book, we embark on a journey to explore the intricate interplay between human progress, infrastructure expansion, and the alarming decline of biodiversity. Through sixteen meticulously crafted chapters, we delve into the myriad ways in which our societal activities shape and, oftentimes, imperil the rich tapestry of life on Earth.

Chapter by chapter, we uncover the complex web of factors driving biodiversity loss, from the relentless march of urbanization to the relentless pursuit of economic development. We examine the impact of modern agricultural practices on ecosystems, the consequences of infrastructure projects like road widening and dam construction, and the insidious effects of pollution and climate change on biodiversity.

Each chapter serves as a window into a different dimension of the challenge at hand, shedding light on the root causes and far-reaching consequences of our actions. We confront the stark realities of biodiversity loss in streams diverted by human activities, in forests cleared for development, and in oceans choked by plastic pollution.

Yet, amidst the gloom, there are rays of hope. Throughout these pages, we explore methodologies and approaches for biodiversity conservation, from innovative carbon sequestration strategies to sustainable plantation methodologies. We highlight the resilience of nature and the potential for human ingenuity to chart a course towards a more sustainable future.

As we navigate through these pages, we are reminded of the urgent need for action. The fate of biodiversity is not merely an abstract concern but a matter of survival for countless species, including our own. It is a reminder that we are but one thread in the intricate tapestry of life, and that our choices today will shape the world for generations to come.

In closing, let this book serve as a call to arms—a call to embrace our role as stewards of the planet and to work tirelessly towards a future where biodiversity thrives, ecosystems flourish, and all life is cherished and respected. Together, let us embark on this journey of discovery and action, knowing that the fate of biodiversity is in our hands.

Ashok K. Rathoure
Saraca Research Inc., India

Chapter 1
Interplay Between Societal Development, Infrastructure, and Biodiversity Decline

Ashok K. Rathoure
https://orcid.org/0000-0001-9131-1346
Saraca Research Inc., India

ABSTRACT

The rapid advancement of human society has brought about significant changes to the global landscape, with infrastructure development often at the forefront of these transformations. While such developments have undoubtedly improved human quality of life, they have also had profound implications for the natural world, particularly in terms of biodiversity decline. This chapter explores the complex interplay between societal development, infrastructure expansion, and the degradation of biodiversity, examining both the direct and indirect impacts of these factors on ecosystems worldwide. Through a multidisciplinary approach encompassing ecological, sociological, and economic perspectives, the authors seek to elucidate the mechanisms driving biodiversity loss in the face of human progress. Additionally, they highlight the importance of sustainable development practices and proactive conservation efforts in mitigating the adverse effects of societal growth on the planet's biological diversity.

INTRODUCTION

The relationship between human society and the natural environment is multifaceted, characterized by a dynamic interplay of socio-economic activities and ecological

DOI: 10.4018/979-8-3693-6950-0.ch001

processes. As societies evolve and expand, they invariably require infrastructure to support their burgeoning populations and burgeoning economies. However, the construction of roads, buildings, dams, and other forms of infrastructure often comes at a cost to biodiversity, leading to habitat destruction, fragmentation, and pollution. Moreover, the secondary effects of infrastructure development, such as urbanization, industrialization, and resource extraction, further exacerbate the decline of biodiversity on a global scale (Biodiversity International, 2017).

Societal development refers to the overall progress and advancement of human society across various dimensions, including economic, social, cultural, and technological aspects. It encompasses improvements in living standards, education, healthcare, infrastructure, governance, and human rights, among other factors. Societal development is often measured using indicators such as GDP per capita, literacy rates, life expectancy, access to basic services, and political stability. It involves the continuous evolution and transformation of societies over time, driven by factors such as innovation, globalization, demographic changes, and public policies. Sustainable societal development seeks to balance economic growth with social equity and environmental stewardship, ensuring the well-being of current and future generations (Cook 2018).

BACKGROUND

The interplay between societal development, infrastructure, and biodiversity decline is a complex and dynamic relationship shaped by numerous socio-economic and environmental factors. Understanding this interplay requires examining the historical context, current trends, and future implications of human activities on natural ecosystems and species diversity. Historically, human societies have evolved in close connection with the natural world, relying on ecosystems for food, water, shelter, and other essential resources (Dudley et al.,2018). However, the industrial revolution and subsequent waves of urbanization, technological advancement, and globalization have led to unprecedented levels of societal development and infrastructure expansion. These processes have transformed landscapes, altered ecosystems, and exerted significant pressures on biodiversity. Infrastructure development, including transportation networks, energy systems, and urban infrastructure, is a cornerstone of modern societal development. While infrastructure projects contribute to economic growth, improved living standards, and enhanced connectivity, they often come at the expense of natural habitats and biodiversity. Roads, highways, dams, and other infrastructure projects fragment landscapes, disrupt wildlife corridors, and degrade ecosystems, leading to habitat loss and species decline (FAO, 2018; 2019).

Moreover, societal development drives changes in land use, agriculture, and resource extraction practices, further exacerbating biodiversity decline. Agricultural expansion, deforestation, mining, and industrial activities contribute to habitat destruction, pollution, and the depletion of natural resources, threatening the survival of many plant and animal species. Climate change adds another layer of complexity to the interplay between societal development, infrastructure, and biodiversity decline. Rising temperatures, changing precipitation patterns, and extreme weather events exacerbate existing threats to biodiversity, altering habitats, and disrupting ecological processes (Grooten & Almond, 2018).

Despite these challenges, there are opportunities for reconciling societal development with biodiversity conservation. Sustainable development practices, conservation strategies, and policy interventions can help mitigate the negative impacts of infrastructure development on ecosystems and species. By integrating biodiversity considerations into land-use planning, infrastructure design, and resource management, societies can promote sustainable development pathways that prioritize both human well-being and environmental conservation. The interplay between societal development, infrastructure, and biodiversity decline is a critical issue that requires comprehensive understanding and concerted action. By recognizing the complex interactions between human activities and natural ecosystems, societies can work towards sustainable solutions that support biodiversity conservation while meeting the needs of present and future generations (IPBES, 2019).

The Arcus Foundation, in collaboration with Cambridge University Press, has released a comprehensive report addressing the complex interplay between ape conservation efforts and infrastructure development. The report highlights the pressing need to minimize and mitigate the adverse effects of expanding infrastructure on both biodiversity and local communities. Examining the current environmental and social standards upheld by financing institutions for development projects, including infrastructure expansion, the report underscores their inadequacy in safeguarding biodiversity and critical habitats. Despite the noble intentions of many infrastructure projects to spur economic growth and alleviate poverty, they often fall short of achieving these goals (Lanjouw et al., 2018).

Annette Lanjouw, co-editor of the 'State of the Ape' series, emphasizes that a significant portion of global infrastructure expansion is planned for developing nations, particularly in regions boasting exceptional biodiversity and crucial ecosystem services. She warns that the repercussions of such growth could be catastrophic for both natural systems and the communities reliant on them for sustenance. Disturbingly, the report predicts that by 2030, industrial activities will encroach upon an estimated 90 percent of ape habitats in Africa and 99 percent in Asia. In sub-Saharan Africa alone, more than 30 "development corridors" are slated to traverse over 400 protected areas and degrade over 1,800 nature reserves, resulting in the erosion of ecological integrity and wildlife loss. Of notable concern is China's Belt and Road Initiative, which is expected to span over 70 countries across Asia, Europe, and Africa. The report projects that an additional 25 million kilometers of paved roads will crisscross the globe by 2050, with nearly 90 percent of them being constructed in developing nations. Many of these roads will traverse areas rich in biodiversity and ecosystem services, exacerbating the threat to fragile ecosystems (Lanjouw et al., 2018).

COMPARATIVE ANALYSIS OF BIODIVERSITY DECLINE: INFRASTRUCTURE VS. SOCIETAL DEVELOPMENT

Both infrastructure development and societal development can significantly contribute to biodiversity decline. Here's a comparative analysis exploring the specific impacts and potential mitigation strategies:

Impact on Biodiversity

Infrastructure Development

- **Habitat Fragmentation:** Construction of roads, railways, and dams physically divides habitats, isolating populations and hindering movement of animals.
- **Habitat Loss:** Clearing land for infrastructure projects destroys existing ecosystems, eliminating resources for various species.
- **Pollution:** Construction activities generate dust, noise, and water pollution, impacting nearby ecosystems.
- **Invasive Species:** Infrastructure projects can act as corridors for invasive species, disrupting native ecosystems.

Societal Development

- **Urbanization:** Expanding cities and towns lead to habitat loss and fragmentation, displacing wildlife.
- **Agriculture:** Intensive farming practices with monoculture crops and pesticide use reduce biodiversity by eliminating natural habitats and harming pollinators.
- **Resource Extraction:** Mining, logging, and oil drilling destroy natural habitats and pollute the environment, leading to biodiversity loss.

- **Climate Change:** Emissions from energy production and industrial processes contribute to global warming, altering ecosystems and threatening species survival.

Similarities

Both infrastructure and societal development often prioritize short-term economic gains over long-term environmental consequences. They can lead to habitat destruction, pollution, and climate change, all of which harm biodiversity.

Differences

- **Scale:** Infrastructure projects typically have a more localized impact, while societal development can have a broader and more long-term effect on a regional or even global scale.
- **Planning:** Infrastructure projects can be planned with more mitigation measures in place, such as building wildlife corridors or using sustainable construction practices. Societal development, on the other hand, often involves complex economic and social factors that make incorporating environmental considerations more challenging.

Mitigation Strategies

- **Environmental Impact Assessments (EIAs):** Conducting thorough EIAs for infrastructure projects can identify potential biodiversity impacts and suggest mitigation measures like habitat restoration or wildlife crossings.
- **Sustainable Resource Management:** Implementing sustainable agricultural practices, responsible forestry, and renewable energy sources can minimize environmental impact and protect biodiversity.
- **Protected Areas:** Establishing and maintaining protected areas like national parks and wildlife sanctuaries provides critical refuges for biodiversity.
- **Green Infrastructure:** Integrating green spaces, such as urban parks and green roofs, into infrastructure and development projects can provide habitat for some species and promote connectivity within urban landscapes.
- **Public Awareness and Education:** Raising public awareness about the importance of biodiversity and encouraging sustainable consumption patterns can drive societal change towards more responsible development.

Both infrastructure and societal development are essential for human progress, but they need to be balanced with protecting biodiversity. By implementing strategic planning, mitigation measures, and promoting sustainable practices, we can minimize the negative impacts on our planet's rich tapestry of life. Collaboration between governments, businesses, and communities is crucial to achieve a future where development and biodiversity can coexist (Lanjouw, 2018).

DRIVERS OF BIODIVERSITY DECLINE

Several interconnected factors contribute to biodiversity decline in the context of societal development and infrastructure expansion (Monastersky, 2014). Land-use change, driven primarily by agriculture, urban sprawl, and industrial activities, is a leading cause of habitat destruction and fragmentation, which disrupts ecosystems and displaces native species. Infrastructure projects, such as roads, highways, and pipelines, not only directly encroach upon natural habitats but also create barriers to wildlife movement, leading to genetic isolation and reduced population viability. Additionally, the extraction and exploitation of natural resources, including timber, minerals, and fossil fuels, degrade ecosystems and threaten species survival (Millennium Ecosystem Assessment, 2005; Watts, 2018).

The drivers of biodiversity decline due to societal development are multifaceted and interconnected, with various socio-economic activities contributing to habitat degradation, species loss, and ecosystem disruption. Some of the key drivers include:

Land-use Change: Societal development often necessitates the conversion of natural habitats, such as forests, grasslands, and wetlands, into urban areas, agricultural fields, and industrial sites. This land-use change results in habitat fragmentation and loss, depriving many species of their homes and reducing overall biodiversity.

Urbanization: Rapid urbanization, driven by population growth and economic development, leads to the expansion of cities and infrastructure networks. Urban areas consume vast amounts of land and resources, displacing native vegetation and wildlife. Moreover, urban environments are often characterized by pollution, habitat fragmentation, and the introduction of invasive species, further threatening biodiversity (Monastersky, 2014; Pimm, 2014).

Infrastructure Development: The construction of roads, highways, railways, dams, and other infrastructure projects alters natural landscapes and disrupts ecological processes. Infrastructure development fragments habitats, creates barriers to wildlife movement, and facilitates the spread of invasive species. Additionally, large-scale infrastructure projects can lead to habitat destruction and pollution, directly impacting local biodiversity (Sandbrook et al., 2019).

Industrial Activities: Industrialization is a major driver of biodiversity decline, as it involves the extraction of natural resources, pollution emissions, and habitat destruction. Industries such as mining, logging, manufacturing, and energy production contribute to air and water pollution, soil degradation, and the loss of biodiversity. Chemical pollutants, heavy metals, and other industrial wastes can contaminate ecosystems, harming both terrestrial and aquatic species (Pimm, 2014).

Agricultural Expansion: Agriculture is a significant driver of biodiversity loss, as it often involves the clearing of forests and other natural habitats to make way for croplands and pastures. Intensive agricultural practices, such as monoculture farming, agrochemical use, and irrigation, degrade soil fertility, water quality, and biodiversity. Moreover, the expansion of agricultural land leads to the loss of critical habitats for many plant and animal species (Monastersky, 2014).

Overexploitation of Natural Resources: Societal development increases demand for natural resources, leading to overexploitation and depletion of ecosystems. Overfishing, hunting, logging, and poaching threaten the survival of many species, including commercially valuable ones. Unsustainable resource extraction practices can lead to population declines, ecosystem degradation, and loss of biodiversity (IPBES, 2019).

Climate Change: While not directly linked to societal development, climate change exacerbates the impacts of other drivers of biodiversity decline. Rising temperatures, changing precipitation patterns, and extreme weather events alter habitats and disrupt ecological relationships, leading to shifts in species distributions, phenology, and community dynamics. Climate change also exacerbates existing stressors, such as habitat loss, pollution, and invasive species, further threatening biodiversity (IPBES, 2019).

These drivers of biodiversity decline interact in complex ways, often exacerbating each other's effects and creating synergistic impacts on ecosystems and species. Addressing the root causes of biodiversity loss requires concerted efforts to promote sustainable development, conserve natural habitats, and mitigate anthropogenic impacts on the environment.

CONSEQUENCES FOR ECOSYSTEMS AND SPECIES

The consequences of biodiversity decline are far-reaching and profound, affecting ecosystems, species, and the services they provide to humanity. Ecosystem degradation and loss diminish the resilience of natural systems, making them more susceptible to environmental stressors such as climate change, invasive species, and disease outbreaks. Furthermore, the loss of biodiversity undermines essential ecosystem functions, such as pollination, water purification, and carbon sequestration, with detrimental implications for human well-being and socio-economic stability.

The consequences for ecosystems and species resulting from societal development are profound and far-reaching, with impacts ranging from habitat degradation to species extinction. Here are some of the key consequences:

Habitat Loss and Fragmentation: Societal development often leads to the conversion of natural habitats into urban areas, agricultural lands, and industrial zones. This habitat loss and fragmentation disrupt ecological processes and diminish the availability of suitable habitats for many species. Fragmented habitats isolate populations, restrict gene flow, and increase the vulnerability of species to environmental stressors and population declines (IPBES, 2019).

Loss of Biodiversity: Habitat loss, fragmentation, and degradation are major drivers of biodiversity loss. As natural habitats are converted or destroyed, species that depend on those habitats for food, shelter, and reproduction are displaced or pushed towards extinction. The loss of biodiversity weakens ecosystem resilience, reduces ecosystem services, and diminishes the overall health and functioning of ecosystems (FAO, 2018).

Species Extinctions: Societal development accelerates the rate of species extinctions through habitat destruction, pollution, overexploitation, and other anthropogenic activities. Species unable to adapt to rapidly changing environments or compete with human activities are at heightened risk of extinction. The loss of species has cascading effects on ecosystem dynamics, disrupting food webs, altering nutrient cycling, and reducing ecosystem stability (IPBES, 2019).

Ecosystem Degradation: Societal development contributes to the degradation of ecosystems through pollution, habitat destruction, and resource exploitation. Industrial activities, urbanization, and agricultural practices release pollutants into the air, water, and soil, contaminating ecosystems and harming wildlife. Ecosystem degradation compromises the ability of ecosystems to provide essential services, such as clean water, carbon sequestration, and climate regulation.

Disruption of Ecological Processes: Societal development disrupts natural ecological processes, such as nutrient cycling, pollination, and seed dispersal, which are essential for ecosystem functioning. Habitat fragmentation, pollution, and invasive species can alter ecological interactions and lead to imbalances in ecosystems. Disruptions to ecological processes can have cascading effects on ecosystem structure and function, impacting the abundance and distribution of species (FAO, 2018; Watts, 2018).

Loss of Ecosystem Services: Ecosystem services, including provisioning, regulating, supporting, and cultural services, are vital for human well-being and socio-economic development. Societal development undermines ecosystem services by degrading natural habitats, reducing biodiversity, and compromising ecosystem resilience. The loss of ecosystem services has direct and indirect impacts on human health, livelihoods, and quality of life (IPBES, 2019).

Increased Vulnerability to Environmental Stressors: Societal development increases the vulnerability of ecosystems and species to environmental stressors, such as climate change, pollution, and invasive species. Habitat loss and degradation diminish the resilience of ecosystems, making them more susceptible to disturbances and less capable of adapting to changing conditions. Species already facing multiple threats are further challenged by the cumulative impacts of societal development (FAO, 2018).

The consequences of societal development for ecosystems and species underscore the urgent need for sustainable development practices, conservation efforts, and integrated land-use planning to minimize negative impacts and promote the coexistence of humans and nature.

MITIGATION STRATEGIES AND CONSERVATION SOLUTIONS

Addressing the complex challenges posed by biodiversity decline requires a multifaceted approach that integrates conservation, sustainable development, and stakeholder engagement. Policy interventions, such as land-use planning, protected area designation, and environmental regulations, can help minimize the negative impacts of infrastructure development on biodiversity. Similarly, innovative conservation strategies, including habitat restoration, species reintroduction, and community-based management, can enhance ecosystem resilience and promote coexistence between humans and wildlife. Moreover, fostering public awareness and fostering a culture of environmental stewardship are essential for promoting sustainable lifestyles and fostering a sense of responsibility towards the natural world (United Nations, 2015).

Mitigation strategies and conservation solutions for biodiversity conservation encompass a range of approaches aimed at minimizing human impacts on natural ecosystems, preserving species diversity, and promoting sustainable coexistence between humans and wildlife (United Nations, 2015). Here are some key strategies:

Protected Areas and Habitat Conservation: Establishing and effectively managing protected areas, including national parks, wildlife reserves, and marine protected areas, is crucial for conserving biodiversity. These areas provide sanctuary for wildlife, protect critical habitats, and safeguard vulnerable species from human disturbances and habitat loss (FAO, 2018).

Sustainable Land Use Planning: Implementing land-use planning measures that prioritize conservation and sustainable development can help minimize habitat fragmentation, protect biodiversity hotspots, and maintain ecological connectivity across landscapes. Incorporating biodiversity considerations into land-use policies, zoning regulations, and infrastructure development plans is essential for mitigating the impacts of societal development on ecosystems and species (United Nations, 2015).

Habitat Restoration and Connectivity: Restoring degraded habitats, reforesting deforested areas, and creating wildlife corridors can enhance habitat quality and connectivity, facilitating the movement of species and promoting genetic diversity. Restorative actions such as reforestation, wetland restoration, and shoreline stabilization contribute to ecosystem resilience and support the recovery of threatened species.

Species Conservation and Recovery Programs: Implementing species-specific conservation and recovery programs is essential for protecting endangered species and preventing extinctions. These programs may include captive breeding and reintroduction efforts, habitat restoration, population monitoring, and anti-poaching measures targeted at species at risk (United Nations, 2015).

Sustainable Resource Management: Adopting sustainable resource management practices in forestry, fisheries, agriculture, and other sectors helps reduce pressure on natural ecosystems and mitigate biodiversity loss. Sustainable harvesting techniques, habitat-friendly farming practices, and ecosystem-based fisheries management promote the long-term viability of natural resources while safeguarding biodiversity (IPBES, 2019).

Community-Based Conservation: Engaging local communities in conservation efforts fosters stewardship of natural resources and promotes sustainable livelihoods. Community-based conservation initiatives empower local stakeholders to participate in decision-making processes, implement conservation projects, and benefit from ecotourism and sustainable resource management activities (Sandbrook et al., 2019).

Policy Interventions and Legal Protections: Enacting and enforcing policies, laws, and regulations that promote biodiversity conservation is essential for addressing systemic threats to ecosystems and species. Governments can establish protected areas, regulate land use and development, implement wildlife protection laws, and enforce regulations aimed at combating illegal wildlife trade and habitat destruction (Secretariat of the Convention on Biological Diversity, 2010).

Education and Awareness: Raising public awareness about the importance of biodiversity conservation and the consequences of human activities on ecosystems and species is critical for fostering a culture of environmental stewardship. Educational programs, outreach campaigns, and community engagement initiatives can empower individuals to take action to protect biodiversity in their daily lives and communities (IPBES, 2019).

International Collaboration and Partnerships: Collaborative efforts among governments, non-governmental organizations, academia, and the private sector are essential for addressing transboundary conservation challenges and implementing coordinated conservation strategies at regional and global scales. International agreements, such as the Convention on Biological Diversity (CBD) and the United Nations Sustainable Development Goals (SDGs), provide frameworks for international cooperation on biodiversity conservation and sustainable development (United Nations Environment Programme, 2019; IPBES, 2019).

By implementing a combination of these mitigation strategies and conservation solutions, stakeholders can work together to address the root causes of biodiversity decline, preserve ecosystems, and safeguard the rich diversity of life on Earth for future generations.

The ongoing decline of biodiversity represents a critical challenge for humanity in the 21st century, exacerbated by the relentless march of societal development and infrastructure expansion. However, by recognizing the intricate links between human activities and ecological systems, we can adopt more sustainable practices that safeguard biodiversity while meeting the needs of present and future generations. Through concerted efforts at local, national, and global levels, we can strive towards a harmonious coexistence between human society and the natural world, preserving the rich tapestry of life that sustains us all.

CONCLUSION

The impact of societal development on biodiversity decline is undeniable and multifaceted. As human societies advance and expand, they exert increasing pressure on natural ecosystems, leading to habitat destruction, species loss, and ecosystem degradation. Drivers such as land-use change, urbanization, infrastructure development, industrial activities, agricultural expansion, and climate change collectively contribute to the unprecedented rates of biodiversity decline observed globally. The consequences for ecosystems and species are severe, with far-reaching implications for ecological integrity, ecosystem services, and human well-being. Habitat loss and fragmentation disrupt ecological processes, reduce species diversity, and compromise ecosystem resilience. Species extinctions, habitat degradation, and the loss of ecosystem services further undermine the stability and functioning of natural ecosystems, exacerbating environmental degradation and threatening the survival of countless species.

However, amidst these challenges, there is hope. Conservation efforts, sustainable development practices, and international cooperation offer pathways towards mitigating the negative impacts of societal development on biodiversity. Protected areas, habitat restoration, sustainable resource management, community-based conservation, and policy interventions are critical tools for preserving biodiversity and promoting sustainable coexistence between humans and the natural world. By recognizing the intricate links between societal development and biodiversity decline and adopting holistic approaches to conservation and sustainable development, we can work towards a future where human societies thrive in harmony with diverse and resilient ecosystems. It is imperative that we act decisively to address the root causes of biodiversity loss and safeguard the irreplaceable wealth of life on Earth for present and future generations.

REFERENCES

Biodiversity International. (2017). *Mainstreaming agrobiodiversity in sustainable food systems: scientific foundations for an agrobiodiversity index*. Biodiversity International.

Cook, S. (2018). *The spice of life: the fundamental role of diversity on the farm and on the plate*. (Discussion Paper, IIED and Hivos. London and The Hague).

Dudley, N., Jonas, H., Nelson, F., Parrish, J., Pyhälä, A., Stolton, S., & Watson, J. (2018). The essential role of other effective area-based conservation measures in achieving big bold conservation targets. *Global Ecology and Conservation, 15*, e00424. doi:10.1016/j.gecco.2018.e00424

FAO. (2018). *State of the world's forests 2018: Forest pathways to sustainable development*. Food and Agriculture Organization of the United Nations.

FAO. (2019). *The state of the world's biodiversity for food and agriculture*. FAO. http://www.fao.org/3/CA3129EN/ ca3129en.pdf

Grooten, M., & Almond, R. E. A. (Eds.). (2018). WWF Living Planet Report – 2018: Aiming higher.

IPBES. (2019). *Global assessment report on biodiversity and ecosystem services of the Intergovernmental Science-Policy Platform on Biodiversity and Ecosystem Services*. IPBES Secretariat.

Lanjouw, A., Rainer, H., & White, A. (2018). *Infrastructure Development and Ape Conservation. Vol III. Arcus Foundation New York*. Great Apps & Gibbons program. https://www.stateoftheapes.com/volume-3-infrastructure-development/

Millennium Ecosystem Assessment. (2005). *Ecosystems and human well-being: Synthesis*. Island Press.

Monastersky, R. (2014). Life – A status report. *Nature, 516*(7530), 159–161. doi:10.1038/516158a

Pimm, S. L., Jenkins, C. N., Abell, R., Brooks, T. M., Gittleman, J. L., Joppa, L. N., Raven, P. H., Roberts, C. M., & Sexton, J. O. (2014). The biodiversity of species and their rates of extinction, distribution, and protection. *Science*, *344*(6187), 1246752. doi:10.1126/science.1246752 PMID:24876501

Sandbrook, C., Fisher, J. A., Holmes, G., Luque-Lora, R., & Keane, A. (2019). The global conservation movement is diverse but not divided. *Nature Sustainability*, *2*(4), 316–323. doi:10.1038/s41893-019-0267-5

Secretariat of the Convention on Biological Diversity. (2010). *Global biodiversity outlook 3*. Secretariat of the Convention on Biological Diversity.

United Nations. (2015). *Transforming our world: The 2030 agenda for sustainable development*. United Nations.

United Nations Environment Programme. (2019). *Global environment outlook: Summary for policymakers*. United Nations Environment Programme.

Watts, J. (2018). Destruction of nature as dangerous as climate change, scientists warn. *The Guardian*. http://tinyurl.com/y56erzml

Chapter 2
Betting on Biodiversity: Rethinking Agriculture Beyond Hybrid Horizons

Anakha Mariya Jacob
Sree Narayana Training College, Poochakkal, India

ABSTRACT

Biodiversity, or the diversity of life forms on Earth, has declined considerably as a result of the shift towards monoculture farming and hybrid crops. Genetic uniformity in hybrid crops, which is frequently caused by clonal propagation, jeopardizes crop species' genetic diversity. High-input agriculture also causes chemical runoff, which degrades soil quality and has an impact on soil microbes. When agricultural land use is unsustainable and causes disruption of local ecosystems, it can lead to mass extinctions. Hybrid crops can cause unintentional gene flow and potential genetic pollution, leading to the emergence of "superweeds." Agrochemical runoff degrades water quality, putting aquatic biodiversity in danger. Chemical runoff endangers non-target animals, aquatic habitats, and water quality, creating pollution and changing aquatic ecosystems. These findings underscore the necessity of sustainable agriculture methods to mitigate biodiversity loss, highlighting the need for collaboration among farmers, policymakers, and researchers for long-term agricultural profitability.

INTRODUCTION

Biodiversity encompasses the diversity of life forms on Earth at different levels of taxonomic hierarchy viz., species level, genetic level, and within different levels of ecological groups like populations, communities, ecosystems, biomes, etc. These

DOI: 10.4018/979-8-3693-6950-0.ch002

organisms interact among themselves and with the environment in which they dwell bringing about different complex food webs contributing to the well-being of each individual and thus maintaining the balance of the biome in which they exist. The extinction of even a single species in an ecosystem sets off a chain reaction. It's like taking out a piece from a puzzle — everything starts to fall apart. Over time, this disruption spreads, causing more and more problems for the plants and animals that depend on each other. Each species, no matter how seemingly insignificant, plays a unique role in the functioning of ecosystem of the nature. Their loss creates a void that can't easily be filled, leading to a gradual constriction of the once-vibrant community. Eventually, the whole ecosystem can suffer, losing its balance and becoming less healthy. The extinction of one species signals not only the end of a lineage but also the onset of a process that threatens the resilience and richness of the environment in which we live. Thus, the loss of even a single species can impede the balance of the ecosystem and, with time can lead to community closure. In recent years, to boost yields and profitability, the agricultural sector has shifted towards monoculture farming and hybrid crops, but this has had unfavourable effects as well, such as soil degradation, insect vulnerability, and biodiversity loss. A promising strategy that is gaining acceptance is incorporating biodiversity into agricultural methods. This movement aims to balance agricultural practices in agricultural areas with biodiversity preservation and promotion. It seeks to improve the resilience, productivity, and sustainability of agricultural systems for the benefit of people and the environment by promoting a symbiotic interaction between farming activities and various ecosystems.

Genetic Uniformity Inherent in Hybrid Crops

The term "genetic uniformity" in hybrid crops describes the low level of genetic variability found in a given crop variety as a result of the particular breeding techniques utilized to create hybrids. Hybridization involves the crossing of genetically dissimilar parents to capitalize on desirable traits from each. It may be intraspecific, interspecific, or even intergeneric. Though beneficial in some respects such as improved yield, and abilities of disease resistance, this can pose a threat to the genetic diversity of the crop species. The genetic uniformity of hybrids produced in F1 generation is focused intentionally, to achieve the goals of hybridization initially set forth. This is done to maintain consistency in the performance of the hybrid variety developed (Goswami Yashwant Singh Parmar, n.d.).This can lead to reduced genetic variability. These hybrids are often subjected to clonal propagation to ensure that they are genetically similar to the original hybrid developed. Genetic uniformity can compromise the adaptive potential of a crop population and can make them more susceptible to certain diseases or environmental stressors. Due to the lack of genetic diversity, the entire population may be at risk if a specific pest or disease arises to target the unique properties of a hybrid crop. In contrast to this vulnerability, varied communities may contain individuals who possess innate resistance. In contrast to the vulnerability posed by the loss of biodiversity, diverse communities often harbour individuals with innate resistance to various threats. Within populations, genetic diversity can produce individuals with unique traits that provide natural defences against diseases, pests, and external stresses. This innate resistance may be caused by genetic mutations, adaptations, or even symbiotic connections with other organisms. These resilient individuals are an important resource for their societies' survival and adaptation to changing environments. By preserving biodiversity, we create a store of genetic variation that has the potential to yield useful qualities for agricultural and ecological resilience. Harnessing the intrinsic resistance present in many populations can be crucial in improving the sustainability and productivity of agricultural systems while minimizing the effects of environmental concerns. Thus, Genetic uniformity in hybrid crops calls into question the long-term sustainability and resilience of the crop species.

Loss of Traditional Crop Varieties

In recent years, the agriculture sector has experienced a paradigm shift in favour of monoculture farming methods and hybrid crop adoption. The desire to pursue higher agricultural yields and greater revenues has caused this shift (Wale et al., 2012). However, this approach has led to many negative consequences, including the loss of biodiversity, especially with regard to traditional crop types. Demands from both local and international markets frequently favour crop varieties that are deemed to be more financially viable. Traditional crops may become less popular in the market and eventually be abandoned, even though they may have special flavours, nutritional benefits, or cultural importance. In order to increase yields and reap financial rewards, farmers frequently switch to these cultivars. Local agricultural techniques and traditional knowledge, handed down through generations, often intertwine with the cultivation of traditional crop varieties. These time-honoured practices reflect a deep understanding of local ecosystems and the cultivation of crops best suited to their specific conditions, preserving both cultural heritage and biodiversity. There's a chance that farmers switching to more contemporary hybrid crops would lose the knowledge and methods related to traditional crops, such as cultivation techniques and seed saving. The technique of growing a single crop over a vast area—and frequently switching out the crop in following seasons—is known as monoculture. It is associated with large-scale farming operations and contemporary industrial agriculture. Large-scale production and uniformity are frequently given priority in monoculture, which may neglect the cultivation of a variety of traditional crops. Moreover, the impact of climate change can disrupt the suitability of traditional crop varieties in certain regions, prompting farmers to abandon them in favour of more resilient alternatives better adapted to evolving environmental conditions. This transition underscores the dynamic relationship between agricultural practices, biodiversity conservation, and the imperative to adapt to changing climates. Government policies and subsidies often favour high-yielding crops over traditional ones, potentially deterring farmers from cultivating and preserving traditional crops. Traditional crops, which rely on diverse pollinator species for reproduction, are at risk of disappearing due to changes in agricultural practices. This could disrupt ecological interactions and affect plant species' reproduction. Additionally, the elimination of traditional crop varieties may increase dependency on synthetic pesticides, disrupting natural pest management processes and inflicting collateral damage to non-target organisms. This shift emphasizes the interdependence of agricultural biodiversity loss, dependency on chemicals, and the possibility of unintended environmental impacts.

Dependency on High-Input Agriculture

High-input agriculture is a modern farming system that heavily relies on external inputs like synthetic fertilizers, pesticides, and irrigation to increase crop yields (Shankar et al., 2021). Concern over high-input agriculture is growing since it brings with it higher yields but also hazards including nutrient runoff, soil erosion, and water pollution. To keep diseases and pests under control, it uses chemical pesticides, which can damage species that are not intended targets and breed pesticide-resistant bugs. Aquatic habitats can suffer from excessive irrigation systems' effects on soil salinity and water depletion. Utilizing machines and cutting-edge technologies, modern high-input farming has the potential to destroy the environment and displace conventional methods. Even though they are advantageous, genetically modified crops raise long-term environmental and health issues. The widespread adoption of monoculture and high-yielding crop types compounds habitat destruction, soil erosion, and biodiversity loss due to their uniform and intensive cultivation methods. The energy-intensive aspect of high-input agriculture not only elevates greenhouse gas emissions but also raises concerns about the effects of climate change, jeopardizing biodiversity on our planet. Addressing these issues demands the switch to more sustainable agriculture practices that prioritize the conservation of biodiversity, good soil health, and resilience to climate change.

Disruption of Local Ecosystems

Disruption of local ecosystems refers to significant alterations, disturbances, or changes in the structure, composition, and functioning of a specific ecological community within a defined geographical area. Agriculture and urbanization fragment natural habitats, leading to a decline in biodiversity. This fragmentation disrupts the continuity of ecosystems, making it difficult for species to migrate and maintain genetic diversity. The loss of native plant and animal species can result in local extinctions, reducing the diversity and abundance of species adapted to the original ecosystem. Introduced non-native species, whether purposefully or accidentally, have the potential to seriously threaten biodiversity and ecological stability by outcompeting native species, upsetting natural food webs, and drastically changing ecosystem dynamics. (Prakash & Verma, 2022). Agricultural runoff, carrying pesticides, fertilizers, and sediment, can contaminate water bodies, negatively impacting aquatic ecosystems. Unsustainable land-use practices, such as deforestation and intensive agriculture, contribute to soil erosion, reducing fertile topsoil and plant growth, and causing a decline in biodiversity. Changes in land use and human activities can also disrupt natural fire regimes, affecting species adapted to specific fire conditions. Climate change, with rising temperatures, altered precipitation patterns, and extreme weather events, exacerbates the disruption of local ecosystems. Disruptions, such as the removal or decline of keystone species, can have cascading effects, leading to imbalances in populations and a loss of biodiversity. Pollinator populations, including bee and butterfly populations, can be greatly impacted by changes in land use and pesticide application. This can reduce their reproductive success and pose a significant risk to efforts to conserve biodiversity. Disruptions can result in a disconnect between people and their natural environment, leading to decreased awareness and appreciation for biodiversity and ecosystem health.

Pollination Challenges and Genetic Pollution

Hybrid crops with agriculturally engineered traits can crossbreed with wild relatives, potentially transferring genetically modified traits to wild populations, and potentially altering native plant genetic makeup. In the case of hybrid crops, the pollen containing modified genes if carried by pollinators to nearby wild or non-GMO crops, can lead to unintended gene flow and potential genetic pollution (Saba & Balwan, 2023). Gene flow from hybrid crops to wild relatives or non-GMO crops can lead to the emergence of "superweeds" that possess both hybrid traits and wild plant ecological resilience. These superweeds can be more difficult to control and can outcompete native vegetation, impacting natural ecosystems. Continuous gene flow from genetically modified crops to wild relatives can lead to homogenization of plant populations, reducing genetic diversity and making them more vulnerable to environmental changes. Genetic pollution can affect not only plants but also non-target organisms in the ecosystem, such as beneficial insects and soil organisms. The introduction of genetically modified traits into wild populations can disrupt natural ecological balances, potentially increasing pest populations. Over time, continuous exposure to genetically modified traits in wild populations may lead to the evolution of resistance in target pests, reducing the effectiveness of engineered resistance traits and necessitating the development of new genetic modifications or increased reliance on chemical pesticides. The unintentional spread of genetically modified traits puts the maintenance of distinctive and regionally adapted genetic variations—which are crucial for preserving ecosystem resilience—in jeopardy, which is concerning for the conservation of native plant species. This unintentional spread serves as a reminder of how crucial it is to protect biodiversity in the face of developing agricultural technologies by implementing strict regulations and thorough risk assessments.

Impact of Hybrid Crops on Soil Microbial Diversity

Hybrid crops, bred for traits like increased yield or pest resistance, release different types and quantities of root exudates, which can affect the composition and activity of soil microorganisms. These exudates can selectively promote certain microbes while inhibiting others (Chauhan et al., 2023). The introduction of hybrid crops, especially those with enhanced nutrient uptake capabilities, can influence nutrient cycling in the soil, impacting the abundance and diversity of microbial communities involved in nutrient transformations. Genetic modifications of hybrid crops can result in different interactions with soil microorganisms, resulting in shifts in the structure and composition of microbial communities. High-input agriculture releases agrochemicals, which can have direct and indirect effects on soil microbial diversity, negatively impacting non-target microbes and altering nutrient availability, affecting microbial communities. Soil disturbance during planting and harvesting can affect soil structure and microbial habitat, potentially leading to changes in microbial diversity and community composition. Hybrid crops can be modified to alter the organic matter returned to the soil, influencing carbon inputs that sustain soil microbial communities.

Water Resource Management Issues

Hybrid crop cultivation intensifies water resource management challenges, owing to increased water demand and intensive irrigation practices. The pursuit of increased yields frequently results in more competition for water resources, which negatively impacts natural ecosystems as well as agriculture. Over-extraction of water affects riparian zones, river dynamics, and aquatic habitats by changing hydrological patterns. Conversion of wetlands for hybrid crops interferes with essential processes like water filtration and adds to the loss of biodiversity. Aquatic biodiversity is at risk due to additional water quality degradation caused by agrochemical discharge (Bashir et al., 2020). Together, these problems make the water shortage in the impacted areas worse, which harms ecosystems and puts the existence of species that depend on steady water sources in jeopardy. Implementing precision irrigation techniques, such as drip irrigation, to optimize water usage and minimize waste is one solution to these difficulties. Furthermore, encouraging the adoption of drought-resistant crop types and agroforestry methods can improve soil moisture retention and minimize the need for irrigation. Moreover, developing watershed management plans and rehabilitating degraded riparian zones can assist in offsetting the effects of water extraction on aquatic habitats and biodiversity (Borah & Borah, n.d.). Collaborative initiatives including stakeholders, policymakers, and local communities are critical for adopting sustainable water management techniques that guarantee ecosystem health and support the livelihoods of those who rely on water.

Pesticide Resistance and Chemical Runoff

Hybrid crops are frequently cultivated with greater pesticide use, which breeds pesticide-resistant pests over time. The emergence of resistance complicates pest control, limiting the efficacy of chemical interventions. Furthermore, the extensive use of pesticides in hybrid crop production contributes to chemical runoff. This pesticide-laden runoff contaminates surrounding bodies of water, causing harm to the environment. As these compounds harm non-target animals, aquatic habitats, and water quality, they can cause water pollution, disturb aquatic ecosystems, and pose potential risks to biodiversity (Devi et al., 2018).

CONCLUSION

Agricultural activities have a huge impact on biodiversity loss, accounting for an estimated 80% of worldwide deforestation. According to the Food and Agriculture Organisation (FAO), around 24% of the world's land surface is being degraded, mostly as a result of unsustainable farming methods. The use of chemical fertilizers and pesticides in agriculture also creates risks, with agricultural runoff accounting for up to 60% of total global freshwater contamination. Furthermore, monoculture methods and the widespread cultivation of high-yield variants contribute to the decline and extinction of traditional crop varieties, resulting in a significant reduction in overall agricultural biodiversity. These frightening findings underline the critical need for sustainable agriculture methods to help minimize biodiversity loss. Agroecological farming approaches are one way to reimagine agriculture outside the boundaries of hybridity (Scherr & McNeely, 2008). The goal of agroecology is to increase biodiversity in agricultural systems by simulating natural ecosystems. Numerous techniques, including crop rotation, intercropping, and the use of cover crops, can help achieve this. To reduce biodiversity loss due to hybrid crops and modern agriculture, sustainable techniques such as crop diversification and integrated pest management are critical. Land-use planning contributes to the conservation of natural habitats but promoting genetic diversity through traditional and regionally adapted crop varieties increases resilience. Responsible use of agrochemicals and reduced water use protect ecosystems. Small-scale farmers and local communities benefit from equal involvement in modern agricultural changes. Collaboration among farmers, policymakers, and researchers is crucial for integrated strategies. Beyond the confines of monoculture farming, we may build a more resilient and sustainable food system by incorporating biodiversity into agricultural methods. Using agroecological principles and embracing a variety of crop varieties improves soil health and encourages natural pest management and nutrient cycling, which lowers the need for chemical inputs. The use of biodiverse farming techniques is safeguarding the long-term viability of agricultural landscapes, improving food security, and fostering ecosystem resilience. Changing policies, investing more in R&D, and implementing on-farm methods are just a few of the many components that this will take. Nonetheless, there are several possible advantages to such a change that might enhance food security, preserve the environment, and guarantee the long-term profitability of agriculture.

REFERENCES

Bashir, I., Lone, F. A., Bhat, R. A., Mir, S. A., Dar, Z. A., & Dar, S. A. (2020). Concerns and threats of contamination on aquatic ecosystems. In Bioremediation and Biotechnology: Sustainable Approaches to Pollution Degradation (pp. 1–26). Springer International Publishing. doi:10.1007/978-3-030-35691-0_1

Borah, I. D., & Borah, S. R. (n.d.). *Precision Irrigation Systems: Enhancing Water Efficiency and Sustainability in Agriculture.* www.vigyanvarta.com

Chauhan, P., Sharma, N., Tapwal, A., Kumar, A., Verma, G. S., Meena, M., Seth, C. S., & Swapnil, P. (2023). Soil Microbiome: Diversity, Benefits and Interactions with Plants. In Sustainability (Switzerland) (Vol. 15, Issue 19). Multidisciplinary Digital Publishing Institute (MDPI). doi:10.3390/su151914643

Devi, Y. B., Meetei, T. T., & Kumari, N. (2018). Impact of Pesticides on Soil Microbial Diversity and Enzymes: A Review. *International Journal of Current Microbiology and Applied Sciences*, *7*(06), 952–958. doi:10.20546/ijcmas.2018.706.113

Goswami Yashwant Singh Parmar, M. (n.d.). Hybridization: Importance, Techniques and Consequences. *Recent Trends in Agriculture*. https://doi.org/

Prakash, S., & Verma, A. K. (2022). ANTHROPOGENIC ACTIVITIES AND BIODIVERSITY THREATS. *International Journal of Biological Innovations*, *04*(01), 94–103. doi:10.46505/IJBI.2022.4110

Saba, N., & Balwan, W. K. (2023). Genetic Pollution: A Safe or Risky Bet? *Scholars Academic Journal of Biosciences*, *11*(04), 159–162. doi:10.36347/sajb.2023.v11i04.004

Scherr, S. J., & McNeely, J. A. (2008). Biodiversity conservation and agricultural sustainability: Towards a new paradigm of "ecoagriculture" landscapes. *Philosophical Transactions of the Royal Society of London. Series B, Biological Sciences*, *363*(1491), 477–494. doi:10.1098/rstb.2007.2165 PMID:17652072

Shankar, T., Sahoo, U., & Maitra, S. (2021). *Intensive Farming: It's Effect on the Environment*. Research Gate. https://www.researchgate.net/publication/357423421

FAO. (2019). *The State Of The World's Biodiversity For Food And Agriculture*. FAO.

Wale, E., Drucker, A. G., & Zander, K. K. (2012). The economics of managing crop diversity On-Farm: Case studies from the genetic resources policy initiative. In The Economics of Managing Crop Diversity On-Farm: Case Studies from the Genetic Resources Policy Initiative. Taylor and Francis. doi:10.4324/9781849775649

Chapter 3
Biodiversity and Ecosystem Services:
An Interconnected Web

Ashok K. Rathoure
https://orcid.org/0000-0001-9131-1346
Saraca Research Inc., India

Vineet Kumar
https://orcid.org/0000-0002-4415-4971
Dr. Yashwant Singh Parmar Government PG College, Nahan, India

ABSTRACT

Biodiversity, the variety of life on Earth, is fundamental to the functioning of ecosystems and the well-being of humanity. This chapter explores the intricate connections between biodiversity and ecosystem services, examining their ecological, economic, and cultural significance. Through a multidisciplinary lens, the authors elucidate the role of biodiversity in providing essential ecosystem services such as pollination, soil fertility, water purification, and climate regulation. Moreover, the authors discuss the implications of biodiversity loss for human societies and the urgent need for conservation efforts to safeguard both biodiversity and the services it provides.

INTRODUCTION

Biodiversity is the cornerstone of life-supporting systems on Earth, encompassing a rich tapestry of species, genes, and ecosystems. The interdependence between

DOI: 10.4018/979-8-3693-6950-0.ch003

biodiversity and ecosystem services is profound, with biodiversity underpinning the delivery of essential services that sustain human well-being and socio-economic development. Understanding the complex relationships between biodiversity and ecosystem services is essential for effective conservation and sustainable management of natural resources (Berkes, 2012). Biodiversity, the intricate tapestry of life encompassing a vast array of species, genes, and ecosystems, forms the bedrock of our planet's ecological richness and resilience. Ecosystem services, the myriad benefits that humans derive from nature's functioning, are deeply intertwined with biodiversity, providing essential resources and conditions for human well-being and socio-economic development (Boyd & Banzhaf, 2007). The interplay between biodiversity and ecosystem services is complex and multifaceted, shaping the intricate web of life that sustains our existence on Earth.

In this paper, we delve into the profound significance of biodiversity and ecosystem services, exploring their ecological, economic, and socio-cultural dimensions. We embark on a journey to unravel the interconnectedness between biodiversity and the essential services it provides, from pollination and nutrient cycling to climate regulation and cultural enrichment. Through a multidisciplinary lens, we seek to elucidate the critical role of biodiversity in supporting ecosystem functioning and human prosperity, while also addressing the pressing challenges posed by biodiversity loss and environmental degradation.

As we navigate through the intricate web of biodiversity and ecosystem services, we uncover the profound implications of their decline for ecosystems, species, and human societies. From the loss of species diversity to the degradation of ecosystem functions, the consequences of biodiversity loss reverberate across landscapes, impacting food security, livelihoods, and cultural heritage. Yet, amidst these challenges, there is hope. Through concerted conservation efforts, sustainable management practices, and policy interventions, we can strive to safeguard biodiversity and restore ecosystems, ensuring a sustainable future for generations to come.

BACKGROUND

Biodiversity and ecosystem services are of paramount importance for the functioning of ecosystems, the well-being of humanity, and the sustainability of our planet (Cardinale et al., 2012). Here's why they are crucial:

Ecological Stability: Biodiversity is essential for maintaining the balance and stability of ecosystems. Different species play unique roles in ecosystems, such as pollination, nutrient cycling, and pest control, which contribute to ecosystem resilience and the ability to adapt to environmental changes.

Essential Services: Ecosystem services, including provisioning (e.g., food, water), regulating (e.g., climate regulation, water purification), supporting (e.g., nutrient cycling, soil formation), and cultural (e.g., recreation, spiritual enrichment) services, are fundamental for human well-being. These services provide the resources and conditions necessary for human survival, economic development, and cultural fulfillment.

Economic Value: Biodiversity and ecosystem services have significant economic value. They contribute to various sectors of the economy, such as agriculture, fisheries, tourism, and pharmaceuticals, generating income and employment opportunities. Moreover, healthy ecosystems provide cost-effective solutions for addressing environmental challenges, such as flood control, erosion prevention, and carbon sequestration.

Food Security: Biodiversity is critical for food security and nutrition. Genetic diversity within species ensures resilience to pests, diseases, and environmental stressors, which is essential for maintaining agricultural productivity and food availability. Additionally, diverse ecosystems provide a variety of wild foods and non-timber forest products that support dietary diversity and livelihoods for millions of people.

Climate Regulation: Ecosystems play a vital role in regulating the Earth's climate by sequestering carbon dioxide, stabilizing temperatures, and moderating extreme weather events. Forests, wetlands, and oceans act as carbon sinks, absorbing and storing large amounts of carbon, thus mitigating climate change and its impacts on both natural and human systems.

Cultural and Spiritual Significance: Biodiversity and ecosystems hold cultural and spiritual significance for communities around the world. They provide inspiration for art, literature, music, and traditional knowledge systems, enriching human experiences and fostering a sense of connection with nature. Indigenous and local communities often depend on biodiversity for their cultural identity, rituals, and traditional practices.

Health and Well-being: Biodiversity and ecosystems contribute to human health and well-being in various ways. Access to green spaces and natural environments promotes physical and mental health, reduces stress, and enhances overall quality of life. Moreover, many medicines and pharmaceuticals are derived from natural sources, highlighting the importance of biodiversity for medical research and drug discovery.

The biodiversity and ecosystem services are essential for sustaining life on Earth, supporting human societies, and maintaining the health and integrity of the planet's ecosystems. Recognizing their importance and implementing measures to conserve and sustainably manage them are critical for achieving global sustainability goals and ensuring a prosperous and resilient future for all.

ECOLOGICAL SIGNIFICANCE OF BIODIVERSITY

Biodiversity plays a critical role in maintaining ecosystem structure and function, enhancing ecosystem resilience, and promoting stability in the face of environmental change. Species diversity contributes to ecosystem productivity, nutrient cycling, and energy flow, ensuring the provision of vital ecosystem services that support life on Earth (CBD, 2005). Moreover, genetic diversity within species confers resilience to environmental stressors and facilitates adaptation to changing conditions, thereby enhancing ecosystem capacity to withstand disturbances.

The ecological significance of biodiversity lies in its crucial role in maintaining the balance, stability, and functioning of ecosystems (Costanza et al., 1997). Here are key points outlining its ecological significance:

Ecosystem Stability: Biodiversity contributes to the stability of ecosystems by increasing their resilience to environmental changes and disturbances. Diverse ecosystems are better able to withstand and recover from disturbances such as natural disasters, climate variability, and invasive species infestations.

Species Interactions: Biodiversity facilitates intricate networks of species interactions within ecosystems, including predation, competition, mutualism, and symbiosis. These interactions regulate population dynamics, maintain ecosystem structure, and influence nutrient cycling, energy flow, and ecosystem processes.

Resilience to Environmental Changes: Genetic diversity within species enables populations to adapt to changing environmental conditions, such as temperature fluctuations, precipitation patterns, and habitat alterations. This genetic variability enhances the resilience of ecosystems to environmental stressors and promotes their long-term survival.

Ecosystem Services: Biodiversity underpins the provision of ecosystem services, including provisioning (e.g., food, water), regulating (e.g., climate regulation, water purification), supporting (e.g., nutrient cycling, soil formation), and cultural (e.g., recreation, aesthetic enjoyment) services. These services are essential for human well-being, supporting livelihoods, food security, clean water, and cultural heritage.

Nutrient Cycling: Biodiversity plays a critical role in nutrient cycling and energy flow within ecosystems. Different species contribute to nutrient uptake, recycling, and redistribution, maintaining soil fertility, plant growth, and ecosystem productivity. Microorganisms, fungi, and decomposers break down organic matter, releasing nutrients back into the soil and facilitating nutrient cycling.

Habitat Provision: Biodiversity provides a variety of habitats and microhabitats for species, offering shelter, nesting sites, and foraging opportunities. Different species have specific habitat requirements, and the presence of diverse habitats supports a wide range of species, enhancing biodiversity and ecosystem resilience.

Ecosystem Functioning: Biodiversity influences ecosystem functioning by affecting processes such as primary productivity, decomposition, and nutrient cycling. Species diversity enhances ecosystem functioning, increasing productivity, stability, and the efficiency of resource use within ecosystems.

The biodiversity is essential for the functioning, resilience, and sustainability of ecosystems. Recognizing and conserving biodiversity is critical for maintaining ecosystem health, supporting human well-being, and ensuring the provision of essential ecosystem services for present and future generations.

ECONOMIC VALUE OF ECOSYSTEM SERVICES

Ecosystem services provided by biodiversity have significant economic value, contributing to human well-being and socio-economic development. Pollination services by insects, birds, and bats are essential for agricultural productivity and food security, with an estimated global economic value of billions of dollars annually (Costanza et al., 2014). Soil fertility, water purification, climate regulation, and pest control services provided by ecosystems are invaluable for agricultural production, water management, climate mitigation, and disaster risk reduction (Daily, 1997).

The economic value of ecosystem services refers to the tangible and intangible benefits that ecosystems provide to humans, which can be quantified in economic terms (Daily et al., 1997). Here are key points outlining the economic value of ecosystem services:

Provisioning Services: Ecosystems provide a wide range of goods and resources that are directly harvested or extracted by humans for consumption or commercial use. Examples include food (e.g., crops, seafood), freshwater, timber, fiber, fuelwood, medicinal plants, and genetic resources for pharmaceuticals and biotechnology. The economic value of provisioning services is often measured in terms of market prices and the contribution of these goods to national economies.

Regulating Services: Ecosystems regulate various environmental processes and provide services that mitigate natural hazards, regulate climate, purify air and water, control pests and diseases, and detoxify pollutants. Examples include climate regulation (e.g., carbon sequestration, climate regulation), water purification (e.g., wetlands filtering pollutants), pollination (enhancing agricultural productivity), and natural hazard mitigation (e.g., coastal protection by mangroves). The economic value of regulating services is often assessed by estimating the costs of human-made alternatives or the economic benefits of avoided damages.

Supporting Services: Ecosystems provide supporting services that are essential for maintaining the structure, function, and productivity of ecosystems. Examples include nutrient cycling, soil formation, photosynthesis, habitat provision, and biodiversity maintenance. While these services may not have direct market value, they underpin the provision of other ecosystem services and support human well-being in indirect ways. The economic value of supporting services is often inferred from the economic value of the ecosystem services they support.

Cultural Services: Ecosystems provide cultural and recreational benefits that contribute to human well-being, spiritual fulfillment, and quality of life. Examples include aesthetic enjoyment, cultural heritage, spiritual significance, recreational activities (e.g., hiking, birdwatching, ecotourism), and educational opportunities. The economic value of cultural services is challenging to quantify but can be estimated using methods such as travel cost analysis, contingent valuation, or willingness-to-pay surveys.

Non-Use Values: Ecosystems also have non-use values that reflect people's preferences for the existence or preservation of ecosystems, regardless of any direct use or benefit. These non-use values include intrinsic value, existence value (the value of knowing that ecosystems exist), and bequest value (the value of preserving ecosystems for future generations). Non-use values are often assessed using stated preference methods, such as contingent valuation or choice experiments.

The economic value of ecosystem services provides important insights into the contributions of ecosystems to human well-being, the costs of ecosystem degradation and loss, and the potential benefits of ecosystem conservation and restoration. Recognizing and quantifying the economic value of ecosystem services is essential for informing policy decisions, land-use planning, and natural resource management strategies that promote sustainable development and the conservation of natural capital.

CULTURAL IMPORTANCE OF BIODIVERSITY

Biodiversity is deeply intertwined with human cultures, traditions, and identities, providing spiritual, aesthetic, and recreational benefits to societies around the world (de Groot et al., 2010). Indigenous and local communities rely on biodiversity for food, medicine, shelter, and cultural practices, maintaining traditional knowledge and cultural heritage (Díaz et al., 2018). Biodiversity also inspires art, literature, music, and tourism, enriching human experiences and fostering connections with the natural world.

The cultural importance of biodiversity refers to the ways in which biodiversity and natural ecosystems contribute to human cultures, traditions, identities, and well-being (Huntington, 2000). Here are key points outlining the cultural importance of biodiversity:

Spiritual and Religious Significance: Biodiversity and natural landscapes hold spiritual and religious significance for many cultures and indigenous communities around the world. Sacred sites, such as mountains, forests, rivers, and groves, are revered as manifestations of the divine, ancestral spirits, or sacred beings. These places serve as centers for rituals, ceremonies, and prayers, connecting people with the natural world and reinforcing cultural beliefs and traditions.

Cultural Identity: Biodiversity is deeply intertwined with the cultural identity and heritage of communities, shaping their customs, languages, arts, and worldviews. Indigenous and local cultures often have profound connections to specific ecosystems, species, and landscapes, which are integral to their sense of place, belonging, and self-expression. Traditional knowledge systems, passed down through generations, contain invaluable insights into the uses of biodiversity for food, medicine, shelter, and livelihoods.

Traditional Practices and Knowledge: Biodiversity sustains a wealth of traditional practices, knowledge, and skills that have been developed over centuries in harmony with nature. Indigenous and local communities possess deep ecological knowledge about the behavior, ecology, and medicinal properties of plants, animals, and ecosystems. This traditional knowledge is essential for biodiversity conservation, sustainable resource management, and adaptation to environmental changes.

Aesthetic and Recreational Enjoyment: Biodiversity provides aesthetic enjoyment and recreational opportunities for people to connect with nature, experience beauty, and find inspiration. Natural landscapes, wildlife, and biodiversity hotspots attract tourists, artists, photographers, and nature enthusiasts seeking tranquility, adventure, and cultural immersion. Protected areas, such as national parks, wildlife reserves, and cultural landscapes, serve as havens for biodiversity and cultural heritage, offering opportunities for ecotourism, recreation, and education.

Cultural Expressions and Creativity: Biodiversity inspires cultural expressions, creativity, and artistic endeavors across various forms of art, literature, music, dance, storytelling, and folklore. Traditional ecological knowledge is often conveyed through oral traditions, songs, dances, and artistic representations, preserving cultural heritage and fostering a deeper connection with nature. Indigenous art, crafts, and artifacts reflect the diversity of local ecosystems, species, and landscapes, celebrating the richness of biodiversity and cultural diversity.

Sense of Place and Belonging: Biodiversity shapes people's sense of place, belonging, and attachment to their homelands and ancestral territories. Landscapes, ecosystems, and species are integral parts of cultural narratives, histories, and memories, shaping collective identities and fostering social cohesion. Cultural landscapes, shaped by human interactions with nature over time, are valued for their aesthetic, historical, and symbolic significance, serving as repositories of cultural and ecological heritage.

The cultural importance of biodiversity highlights the profound connections between humans and the natural world, emphasizing the intrinsic value of biodiversity for cultural diversity, spiritual well-being, and sustainable development. Recognizing and respecting the cultural significance of biodiversity is essential for promoting cultural diversity, social justice, and inclusive conservation efforts that honor the rights, values, and aspirations of indigenous and local communities.

IMPLICATIONS OF BIODIVERSITY LOSS

The loss of biodiversity poses significant risks to ecosystems, species, and the services they provide to humanity. Habitat destruction, pollution, climate change, invasive species, and overexploitation threaten biodiversity at unprecedented rates, leading to species extinctions, ecosystem degradation, and loss of ecosystem services (IPBES, 2019). The consequences of biodiversity loss include reduced agricultural productivity, compromised water quality, increased vulnerability to natural disasters, and diminished cultural heritage (Millennium Ecosystem Assessment, 2003).

The implications of biodiversity loss are far-reaching and can have significant impacts on ecosystems, species, human societies, and the planet as a whole. Here are key points outlining the implications of biodiversity loss:

Ecosystem Degradation: Biodiversity loss can lead to the degradation and collapse of ecosystems, compromising their structure, function, and resilience. Reduced biodiversity can disrupt key ecological processes such as nutrient cycling, pollination, and pest control, resulting in decreased ecosystem productivity, stability, and the ability to provide essential ecosystem services.

Species Extinctions: Biodiversity loss increases the risk of species extinctions, threatening the survival of plants, animals, and microorganisms around the world. Species extinctions can disrupt food webs, alter ecosystem dynamics, and diminish genetic diversity, weakening the capacity of ecosystems to adapt to environmental changes and recover from disturbances.

Decline in Ecosystem Services: Biodiversity loss diminishes the capacity of ecosystems to provide essential services that support human well-being, including food production, water purification, climate regulation, and disease regulation. The decline in ecosystem services can exacerbate food insecurity, water scarcity, climate change impacts, and the spread of infectious diseases, particularly in vulnerable communities.

Loss of Genetic Resources: Biodiversity loss reduces the pool of genetic resources available for breeding and genetic improvement of crops, livestock, and medicinal plants. The loss of genetic diversity diminishes resilience to pests, diseases, and environmental stressors, increasing the vulnerability of agricultural systems and threatening global food security.

Disruption of Cultural Practices: Biodiversity loss undermines the cultural practices, traditions, and livelihoods of indigenous and local communities that depend on biodiversity for food, medicine, shelter, and cultural identity. Disruptions to traditional knowledge systems, cultural landscapes, and sacred sites erode cultural diversity and threaten the cultural survival of indigenous peoples and local communities.

Economic Costs: Biodiversity loss imposes significant economic costs on society, including losses in ecosystem services, declines in agricultural productivity, increased healthcare expenditures, and the loss of potential economic opportunities derived from biodiversity-based industries such as ecotourism, bioprospecting, and pharmaceuticals.

Global Environmental Challenges: Biodiversity loss exacerbates other global environmental challenges, including climate change, deforestation, habitat destruction, pollution, and invasive species. These interconnected threats can create feedback loops and synergistic effects that amplify the impacts on ecosystems, species, and human societies, leading to cascading ecological and socio-economic consequences.

Loss of Cultural and Aesthetic Values: Biodiversity loss diminishes the cultural and aesthetic values associated with diverse ecosystems, species, and landscapes. The disappearance of iconic species, natural habitats, and cultural landscapes erodes the beauty, inspiration, and sense of wonder derived from nature, affecting human well-being and quality of life.

The implications of biodiversity loss underscore the urgent need for concerted action to conserve and restore biodiversity, protect ecosystems, and promote sustainable development pathways that prioritize both human well-being and environmental integrity. Recognizing the interconnectedness of biodiversity with ecological, social, economic, and cultural systems is essential for addressing the root causes of biodiversity loss and fostering a more harmonious relationship between humans and the natural world.

CONSERVATION IMPERATIVE

Addressing the challenges of biodiversity loss requires urgent and concerted action to conserve and restore ecosystems, protect species, and promote sustainable use of natural resources (Maffi, 2001). Conservation strategies such as habitat preservation, species reintroduction, protected area management, and sustainable resource management are essential for safeguarding biodiversity and the ecosystem services it provides (Millennium Ecosystem Assessment, 2005). Moreover, fostering public awareness, engaging stakeholders, and strengthening policy frameworks are critical for mainstreaming biodiversity conservation into decision-making processes and promoting sustainable development pathways that prioritize both human well-being and environmental integrity.

The conservation imperative for biodiversity and ecosystem services is grounded in the recognition of their fundamental importance for the health of the planet and the well-being of humanity (Posey 1999). Here are key points outlining the conservation imperative for biodiversity and ecosystem services:

Preservation of Life Support Systems: Biodiversity and ecosystem services are the foundation of life on Earth, providing essential resources, regulating environmental processes, and supporting human societies. Conservation efforts are necessary to preserve the functioning of ecosystems as life support systems, ensuring the provision of clean air, fresh water, fertile soils, and a stable climate for present and future generations.

Protection of Species and Genetic Diversity: Conservation efforts aim to protect species from extinction and preserve genetic diversity within populations. Species conservation is critical for maintaining ecosystem resilience, supporting ecosystem services, and preventing the loss of valuable genetic resources for agriculture, medicine, and biotechnology. Protected areas, species reintroduction programs, and habitat restoration initiatives are essential tools for safeguarding biodiversity.

Sustainable Use of Natural Resources: Conservation involves promoting the sustainable use of natural resources to ensure their long-term availability and avoid overexploitation. Sustainable resource management practices, such as sustainable forestry, fisheries management, and agroecology, aim to balance human needs with the capacity of ecosystems to regenerate and replenish natural resources. Sustainable development approaches prioritize the integration of environmental, social, and economic considerations to achieve long-term sustainability.

Restoration and Rehabilitation of Degraded Ecosystems: Conservation efforts include the restoration and rehabilitation of degraded ecosystems to enhance their ecological function, resilience, and biodiversity. Ecosystem restoration initiatives aim to reverse habitat loss, degradation, and fragmentation by restoring natural habitats, reintroducing native species, and implementing ecosystem-based management practices. Restoration projects contribute to biodiversity conservation, climate change mitigation, and the provision of ecosystem services.

Promotion of Connectivity and Landscape Connectivity: Conservation strategies promote connectivity and landscape connectivity to facilitate the movement of species, maintain genetic diversity, and enhance ecosystem resilience. Connectivity corridors, green infrastructure, and ecological networks provide pathways for species dispersal, migration, and adaptation to changing environmental conditions. Protecting and restoring habitat connectivity is essential for conserving biodiversity, reducing fragmentation, and promoting ecosystem resilience in the face of climate change and habitat loss.

Integration of Traditional Knowledge and Indigenous Perspectives: Conservation efforts recognize the value of traditional knowledge, indigenous perspectives, and community-based approaches to biodiversity conservation. Indigenous and local communities possess valuable insights, practices, and stewardship traditions that contribute to biodiversity conservation, sustainable resource management, and cultural resilience. Collaborative conservation initiatives that integrate traditional knowledge with scientific expertise can enhance the effectiveness and legitimacy of conservation interventions.

Policy Support and Institutional Frameworks: Conservation requires strong policy support, legal frameworks, and institutional mechanisms to promote biodiversity conservation and sustainable development. National and international policies, such as protected area designations, biodiversity conservation strategies, and environmental regulations, provide legal protection for biodiversity and support conservation efforts. Multilateral agreements, such as the Convention on Biological Diversity (CBD) and the Sustainable Development Goals (SDGs), provide frameworks for global cooperation and collective action on biodiversity conservation and sustainable development (Sternberg, 2000; Secretariat of the Convention on Biological Diversity, 2010; United Nations, 2015).

The conservation imperative for biodiversity and ecosystem services underscores the urgent need for concerted action to protect, restore, and sustainably manage natural ecosystems for the benefit of present and future generations (Sala et al., 2000; Tilman et al., 2001). By recognizing the interconnectedness of biodiversity with human well-being, promoting sustainable practices, and fostering collaboration among diverse stakeholders, we can work towards a more resilient, equitable, and sustainable future for all.

CONCLUSION

The interconnected web of biodiversity and ecosystem services underscores the critical importance of conservation efforts to safeguard the health of our planet and the well-being of humanity. Biodiversity, the intricate tapestry of life, and ecosystem services, the benefits derived from nature's functioning, are essential for supporting life on Earth, regulating environmental processes, and sustaining human societies. However, biodiversity loss and ecosystem degradation pose significant threats to ecosystems, species, and human well-being, with far-reaching implications for ecological integrity, economic stability, and cultural diversity. The conservation imperative for biodiversity and ecosystem services requires urgent and concerted action to address the root causes of biodiversity loss, promote sustainable development pathways, and foster harmonious relationships between humans and the natural world. Conservation efforts must prioritize the preservation of life support systems, protection of species and genetic diversity, sustainable use of natural resources, restoration of degraded ecosystems, promotion of connectivity and landscape connectivity, integration of traditional knowledge and indigenous perspectives, and support for policy frameworks and institutional mechanisms.

By recognizing the intrinsic value of biodiversity, respecting indigenous knowledge, and embracing sustainable practices, we can work towards a future where ecosystems thrive, species flourish, and human societies prosper in harmony with nature. Through collaborative conservation initiatives, informed decision-making, and collective action at local, national, and global levels, we can strive to conserve biodiversity, restore ecosystem health, and secure a sustainable future for generations to come. Ultimately, the conservation of biodiversity and ecosystem services is not only a moral imperative but also a fundamental necessity for ensuring the resilience, vitality, and beauty of our planet for all living beings.

REFERENCES

Berkes, F. (2012). *Sacred ecology* (3rd ed.). Routledge. doi:10.4324/9780203123843

Boyd, J., & Banzhaf, S. (2007). What are ecosystem services? The need for standardized environmental accounting units. *Ecological Economics*, *63*(2-3), 616–626. doi:10.1016/j.ecolecon.2007.01.002

Cardinale, B. J., Duffy, J. E., Gonzalez, A., Hooper, D. U., Perrings, C., Venail, P., & Naeem, S. (2012). Biodiversity loss and its impact on humanity. *Nature*, *486*(7401), 59–67. doi:10.1038/nature11148 PMID:22678280

CBD (Convention on Biological Diversity). (2005). *Biodiversity: Importance and Benefits*. CBD. https://www.cbd.int/biodiversity

Costanza, R., d'Arge, R., de Groot, R., Farber, S., Grasso, M., Hannon, B., & van den Belt, M. (1997). The value of the world's ecosystem services and natural capital. *Nature*, *387*(6630), 253–260. doi:10.1038/387253a0

Costanza, R., de Groot, R., Sutton, P., van der Ploeg, S., Anderson, S. J., Kubiszewski, I., & Turner, R. K. (2014). Changes in the global value of ecosystem services. *Global Environmental Change*, *26*, 152–158. doi:10.1016/j.gloenvcha.2014.04.002

Daily, G. C. (1997). *Nature's Services: Societal Dependence on Natural Ecosystems*. Island Press.

Daily, G. C., Alexander, S., Ehrlich, P. R., Goulder, L., Lubchenco, J., Matson, P. A., & Woodwell, G. M. (1997). Ecosystem services: Benefits supplied to human societies by natural ecosystems. *Issues in Ecology*, *2*, 1–16.

de Groot, R. S., Alkemade, R., Braat, L., Hein, L., & Willemen, L. (2010). Challenges in integrating the concept of ecosystem services and values in landscape planning, management and decision making. *Ecological Complexity*, *7*(3), 260–272. doi:10.1016/j.ecocom.2009.10.006

Díaz, S., Pascual, U., Stenseke, M., Martín-López, B., Watson, R. T., Molnár, Z., & Shirayama, Y. (2018). Assessing nature's contributions to people. *Science*, *359*(6373), 270–272. doi:10.1126/science.aap8826 PMID:29348221

Huntington, H. P. (2000). Using traditional ecological knowledge in science: Methods and applications. *Ecological Applications*, *10*(5), 1270–1274. doi:10.1890/1051-0761(2000)010[1270:UTEKIS]2.0.CO;2

IPBES (Intergovernmental Science-Policy Platform on Biodiversity and Ecosystem Services). (2019). *Summary for policymakers of the global assessment report on biodiversity and ecosystem services*. IPBES Secretariat.

MA (Millennium Ecosystem Assessment). (2003). *Ecosystems and Human Well-being: A Framework for Assessment*. Island Press.

Maffi, L. (Ed.). (2001). *On biocultural diversity: Linking language, knowledge, and the environment*. Smithsonian Institution Press.

Millennium Ecosystem Assessment. (2005). *Ecosystems and Human Well-being: Biodiversity Synthesis*. World Resources Institute.

Posey, D. A. (Ed.). (1999). *Cultural and spiritual values of biodiversity*. United Nations Environment Programme.

Sala, O. E., Chapin, F. S. III, Armesto, J. J., Berlow, E., Bloomfield, J., Dirzo, R., & Wall, D. H. (2000). Global biodiversity scenarios for the year 2100. *Science*, *287*(5459), 1770–1774. doi:10.1126/science.287.5459.1770 PMID:10710299

Secretariat of the Convention on Biological Diversity. (2010). *Global biodiversity outlook 3*. Secretariat of the Convention on Biological Diversity.

Sternberg, R. (2000). *Our cultural heritage: The role of biodiversity in preserving cultural landscapes. United Nations Educational, Scientific and Cultural Organization*. UNESCO.

Tilman, D., Fargione, J., Wolff, B., D'Antonio, C., Dobson, A., Howarth, R., & Swackhamer, D. (2001). Forecasting agriculturally driven global environmental change. *Science*, *292*(5515), 281–284. doi:10.1126/science.1057544 PMID:11303102

United Nations. (2015). *Transforming our world: The 2030 agenda for sustainable development*. United Nations.

Chapter 4
Reasons for Biodiversity Loss

P. Selvakumar
https://orcid.org/0000-0002-3650-4548
Nehru Institute of Technology, India

ABSTRACT

The worldwide threat to human welfare that is expanding the quickest right now is climate change. Since the current generation is thought to be the last to be able to take meaningful action to lessen its effects, the world must understood how responsible it is. In the event that it fails, human society is probably destined to never recover. Many organizations have stressed that the climate problem is fundamentally unjust, with poorer nations bearing a heavier burden than others. India is one of the nine nations where climate change is said to be having a significant impact. A WHO estimate from 2016 suggested that by 2050, India would account for over 25% of all climate-related deaths worldwide because of dwindling food supplies. China is predicted to see the greatest per-capita mortality rate from food insecurity.

INTRODUCTION

A natural area's richness is derived from the mosaic of plants, animals, and microorganisms that make up its biodiversity, as well as from the genetic variability exhibited in these species' populations and habitats. Humans are sustained physically, aesthetically, medically, and spiritually by biodiversity, or biological resources (Allan et al. 2015). For its riches and other advantages, biodiversity in the living world is essential to human survival. However, overcrowding, overexploitation, pollution, habitat destruction, invasive species introduction, and climate change are all contributing factors to human activity's detrimental effects on biodiversity. The primary causes of biodiversity loss include improper development, overuse of natural

DOI: 10.4018/979-8-3693-6950-0.ch004

resources, pollution, overharvesting, and habitat degradation. The effects of human activity on the environment have been profound. Globally, the quantity of natural vegetation decreases by half between the 1700s and 2000s, while the area used for agriculture quintupled (Arya S. 2021). Moreover, anthropogenic performance is the primary cause of the topical increase in atmospheric CO_2 levels, according to numerous scientists. Increased CO_2 is mostly caused by burning fossil fuels like coal and petroleum. For all of biota, including humans, climate change and global warming present an overwhelming challenge.

GLOBAL WARMING

Global warming is the primary source of climate change, and increased greenhouse effect brought on by excessive human activity is the primary cause of global warming. These negative alterations are mostly the result of non-systematic human activity (Ashok K.V., 2016). Numerous human actions, including as habitat change and destruction, overexploitation of species, disjointed pockets of native vegetation, and pollution of the air and water, can lead to a loss of biodiversity (Kalal V., et.al, 2021). Nearly every region of our planet faces threats to biodiversity due to human activity and population expansion. Saving of biodiversity is a functional approach to environmental and natural resource management.

HABITAT LOSS

Natural habitat is being lost as a result of the growing usage of it for metropolitan areas, industrial infrastructure, agriculture, and mining. The world's population is expanding, per capita consumption is rising more quickly, and consumption patterns are changing, all of which contribute to the growing demand for natural resources (Kumar Ajay and Verma A.K. (2017). Natural habitat loss occurs when an area is changed from one that is suitable to one that is useless. The main drivers of habitat loss include mining, deforestation, aquaculture, industrial operations, and water exploitation. Deforestation to provide wood for cooking is one example of this. Loss of habitat due to resource exploitation, agricultural conversion, and urbanization is the primary driver of biodiversity loss. One of the largest terrorizations to biodiversity is habitat fragmentation, which is the loss of large ecological units. Due to human activity, biodiversity in both plants and animals is decreasing globally. A major issue for preserving the diversity of life on land and in the ocean is habitat conservation. The urgent ecological issues allow research on the processes, and effects of marine biodiversity to be prioritized.

Consequently, a thorough understanding of how human disturbances impact biodiversity eventually leads to a better understanding of the signs of alterations in the sea brought on by human activity. While coral reefs are diminishing as a result of increased pollution and overexploitation, tropical forest are mostly endangered by land conversion to other purposes. In the absence of human deforestation, the pace of decline would be 1000–10,000 times faster than the predicted rate of extinction. In the next 200–300 years, up to half of all animal and bird variety may turn into extinct, according to some studies. Overharvesting (20%) for commodities like food, medicine, and lumber is also more challenging in areas with less biodiversity. Therefore, the key factors contributing to the loss of biodiversity are human activities that disrupt essential ecological services like pollination, the cycling of carbon and other nutrients, flood control, soil fertility maintenance, water purification, and garbage assimilation.

OVEREXPLOITATION

If a species is taken from the wild more quickly than its natural populations can recover, it is being overharvested or overexploited. Genetic diversity decline is caused by overharvesting. Genetic diversity serves as a safeguard for biodiversity. Disturbances in age distribution and gender composition lead to a reduction in population size. Of the elements of biodiversity, humans are one. On the early earth, biodiversity and humanity coexisted peacefully, i.e., there was a dynamic equilibrium between the two. Subsequently, due to the following significant changes, humans have placed a great deal of pressure on biodiversity: (1) rapid population growth; (2) industrialization and animal husbandry; (3) animal domestication; (4) increased human capability with technological development to exploit biodiversity; and (5) overexploitation of innate resources. There are numerous instances of legally protected commercial fisheries that were under the observation of fisheries scientists but ultimately failed. The most notable recent decline is in the cod fishery in the western Atlantic. Political and economic factors are the root reasons of the decline of fisheries. Even when a country's territorial seas contain fishing territory, most fisheries are administered as common (shared) resources. Overexploitation of shared resources is an inevitable result of resource harvesting. This pressure still exists, even if big fisheries are managed in an effort to mitigate it. Technology that allows fishermen to overfish and unrestricted access to the fisheries both worsen this overexploitation. There are certain fisheries where the organic growth of the resource is smaller than the possible expansion of fishing earnings if the time and money were put to better use.

Figure 1. Activities of biodiversity loss

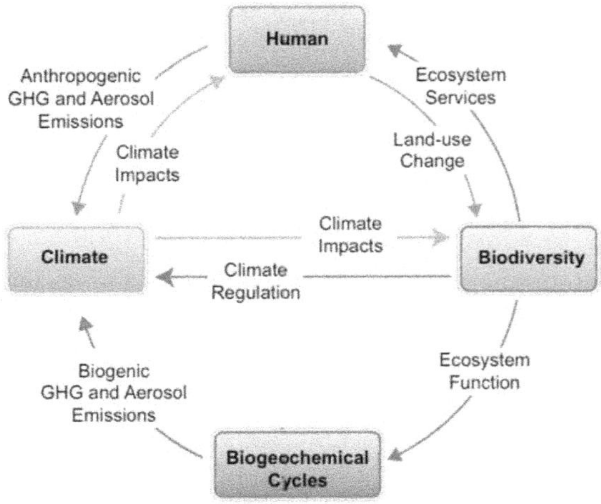

DEFORESTATION

The loss of priceless plant and animal species is a direct consequence of deforestation. The devastation of forest lands and inappropriate agriculture methods have resulted in severe degradation of native forest. One major factor contributing to the loss of biodiversity is deforestation, which is strongly impacted by the expansion of agriculture. One of the main causes of soil erosion and runoff creation is extremely poor management (Peters G.P., et.al, 2011). Land structure changes, the depletion of natural resources, and urbanization have always been associated with human activity. Tropical forests have experienced extraordinary rates of change during the past century due to human activity-induced degradation or destruction. Regrettably, the pace of deforestation and the use of marginal areas for cultivation are disrupted due to the daily increase in population, aberrant and erroneous use of natural resources, and inadequate agricultural measurement procedures. Reduced water infiltration, fertility, and the soil's capacity to hold onto nutrients are frequently signs of soil degradation, which is always easily identified by declining soil organic matter. Additionally, plowing exposes soils to erosion from wind and water, which severely contaminates freshwater supplies. Deforestation is known to be the second main factor contributing to rising CO levels.

INCREASING WILDLIFE TRADE

Nowadays, hunting is one of the main causes of biotic population decreases worldwide. In hunted forests, hunting has resulted in declines in bird and mammal populations of 58% and 83%, respectively (Abhilasha, et.al, 2022). The demise of iconic animals like tigers, elephants, rhinoceroses, and so on is closely linked to the unlawful killing of wildlife for food, pets, and internationally traded goods. In the worst-case scenario, overhunting may lead to the extinction of large-bodied animals in otherwise intact, healthy habitat, which would alter the composition of the forest.

INVASIVE SPECIES

Non-native species that have spread beyond their natural range are considered invasive, whereas imported species have been brought into an ecosystem where they are not native due to human activity. Both introduced and invasive species have the ability to bring about diseases, change abiotic environments, become pests, or cause extinctions. Species that are more vulnerable to competition or predators or have a reduced reproductive capacity are often the targets of invasive species. According to numerous studies, invasive species can modify ecosystem functioning and biodiversity, making them important drivers of global change. Invasion-related changes to disturbance patterns or nutrient levels may have a direct impact on ecosystem functioning (Bradshaw et.al, 2009). The indirect effects of biodiversity loss will only become important in relation to other factors when invaders significantly destroy biodiversity and when biodiversity has a large impact on functioning. The indirect effects of diversity loss on ecosystem functioning, mediated by plant diversity, to be just as strong as the direct effects of land-use intensification, despite a study showing that these effects were much smaller than the direct effects of nitrogen addition.

The native species may be impacted by the introduction of a new animal, plant, or microbe in a number of ways. New species have the potential to cause problems for existing ones by interbreeding, disrupting vital interactions, bringing unknown diseases, hybridizing with them, or parasitizing or preying upon them (Cai, et.al., 2016). For instance, it is believed that the arrival of the brown snake to the island of Guam is what led to the loss of 12 out of the 18 native birds that once lived there. These species have the potential to have catastrophic effects on native biota, leading to extinctions and altering both wild and farmed ecosystems. The reason most alien species introductions fail is most likely due to either a small number of individuals introduced or inadequate adaptability to the new ecology (Poudel et.al, 2017). However, there are instances when these alien species proliferate too quickly in an environment, which poses a number of issues for the local native species and even contributes to the spread of some native species.

POLLUTION

Pollution is the most detrimental effect of anthropogenic activity. Anthropogenic pollutants-such as those from homes, cities, and industries-have a significant impact on the quality of water. These pollutants find their way into bodies of water through human activity (NAS (1995). Cleaning, cooking, and bathing are examples of household tasks that alter the quality of water by adding chemical residue. When food is cooked with firewood, smoke may be released into the atmosphere. Pesticides and fertilizers may leak into the environment as a result of agricultural activity (Chakraborty B.K., et.al., 2021). Every human and industrial activity releases certain undesirable compounds into the environment. Pollution is the term used to describe the concentration of undesired elements that might negatively impact the environment and living things. Pollution is the term for anything introduced into the environment that is extremely dangerous, poisonous, or lethal to nearby animals, people, or other living things (Isbell,et.al, 2013). To put it simply, pollution is when an area of the environment gets contaminated by a chemical or another pollutant and becomes unsuitable for its intended or desired usage.

Poor water quality and biodiversity loss are caused by a variety of factors, including the removal of riparian vegetation, residential, and agricultural pollutants into water resources Industrial pollution has spread around the world and grown to be a significant socioeconomic issue in the most industrialized regions of the planet (Prakash, 2014). Industrial pollution is the result of the release of various pollutants in the form of gases, liquids, and solids that have an adverse effect on the physical, chemical, and biological conditions of the environment as well as the health of people, animals, plants, and soil properties (Dayton, et.al, 1995). Various enterprises are releasing various kinds of chemicals into the environment. These substances range widely in nature and have a variety of quantitative and qualitative effects on plants and animals (Selvakumar P., et al.,2024). The ecology is negatively impacted by industrial waste effluents, which can also cause species extinction in a given ecosystem and instability in plant communities. Pollution of the air, land, and water has detrimental consequences on biodiversity (Goudie., 2013). The biota of the industrial areas is negatively impacted by the highest concentration of pollutants in the soil.

Pollutants emitted from a variety of sources have altered the makeup, structure, and character of biological communities and may be responsible for the extinction of certain significant species (Khaleghi, 2017). Plants can absorb ions of heavy metals such as Pb, Cd, Ni, Ti, Cu, Hg, Cr, Al, and Zn through their roots. High concentrations of heavy metals were examined in soil samples taken from different parts of contaminated cities. Furthermore, the physiology of aquatic species, including fish, is impacted by chemical pollutants like pesticides and herbicides that seep into soils and water bodies (Masih S.C. (2021). Fish populations eventually decline as a result of fish abandoning their native spawning grounds due to pollution in streams (Pejchar L. and Mooney H.A. (2009). Pollution sensitivity varies amongst species. However, because of the accumulation of hazardous chemicals in the top predators of food chains and the disruption of predator-prey interactions, a large number of species are susceptible to the indirect impacts of pollution (Ghosh P. and Brand W.A., 2003). Ecological balances have been altered locally and in distant ecosystems due to human-caused reactive nitrogen synthesis, which primarily comes from the development of synthetic fertilizer to boost agricultural output.

Reactive nitrogen produced by humans causes atmospheric nitrogen molecules to be released, which are then deposited on the biosphere. A certain amount of phytoplankton growth and primary production is stimulated by the addition of nutrients like phosphate and nitrate. Through eutrophication, the influx of nutrient-rich sewage alters the algal community (Huey R.B., et.al, 2009). Nitrogen levels in ecosystems rise as a result of atmospheric deposition. Furthermore, soluble nitrogen from soils seeps into groundwater, causing water bodies to become more eutrophic. This leads to anoxic (oxygen-free) zones forming in inshore marine areas, as well as excessive plant growth and algal blooms.

CLIMATE CHANGE

Climate change as climate variability, it is still a daily occurrenceLand-use change is anticipated to lose ground to climate change as the factor most affecting the biodiversity of tropical forests. The physiological processes of plants and the makeup of species are impacted by the rise in CO levels in the atmosphere. Climate change has an impact on sustainable development, biodiversity, ecosystem balance, and environmental ethics (Goudie A., 2013). Greenhouse gases, which keep heat from exiting the surface of the earth, are produced by burning biomass and fossil fuels, deforestation, and agricultural practices. These gases also contain carbon dioxide (Pounds J.A. (2001). In addition to increased air pollution and increased weather variability, rising temperatures brought on by rising greenhouse gas concentrations will also cause changes in the distribution of flora over the terrain. Certain species will go extinct because they are unable to adapt to these environmental changes.

As the temperature rises, it is anticipated that a large number of plant and animal species would try to move to higher latitudes and elevations (Ramakrishnan T.,et.

al, 2022). Therefore, biodiversity will be lost as a result of any physical obstacles in the landscape, like as cities and highways, that restrict migration to more hospitable locations (Dueck T.A. and Endendijk G.J., 1987). Ocean levels will rise as a result of melting glaciers and more warmer water entering the ocean, reducing the extent of islands and potentially causing the extinction of some species. In the upcoming decades, climate change brought on by humans will likely play a significant role in the decline of biological diversity (Mahadev Madgule, et.al. 2023). The demands associated with economic development, such as the rising demand for biological resources, are largely responsible for the stresses placed on biodiversity.

Climate change is altering the customary ranges of species by modifying the temperature and precipitation patterns to which they have grown accustomed. This drives certain animals to either relocate in search of more favorable living environments or to adjust to their new environment. Climate change will bring about changes that some species can adapt to, while others won't be able to. The development of different species depends on the crucial ecosystem services that coastal areas supply (Froehner S., et.al, 2010). Climate change and the availability of commodities and services that sustain the communities along the shore are both impacted by human activity. Numerous issues lead to different dynamics in these ecosystems, particularly the fast urbanization and growth of social and cultural activities. The numerous additional factors at play make it challenging to establish a direct correlation between changes in species richness and climate change.

ECOSYSTEM FUNCTIONING

Human activity has always resulted in altered land use, depletion of natural resources, and urbanization, all of which have contributed to the decline in biodiversity Not only does a species disappear from an environment when it becomes extinct, but also its interactions and the ecological services they provide, including seed dispersion (GBS, 1992). The ecosystem's overall health or the survival of another species may depend on these relationships. Many species can perform the same task in an environment, creating ecological redundancy (Pongratz J., et.al, 2008). For instance, numerous insects may be involved in the pollination of a single plant species. Although this might mitigate the effects of a single species' extinction, many species loss is acceptable before it begins to negatively impact ecosystem functioning.

Biodiversity is significantly impacted by industrialization. The industrialization-related rise in human population and economic expansion puts non-human species in jeopardy and drives them out of competition (Nijmegen R.U. (2017). Additionally, the loss of genetic variety and vulnerable organisms in agricultural settings is a result of industrial sectors that rely on biodiversity (Froehner S., et.al, 2010). Around the

Reasons for Biodiversity Loss

Figure 2. Reasons for biodiversity loss

world, losses in biodiversity and ecosystem services are caused by changes in land use brought about by industrial operations like forestry and agriculture. The strain on biodiversity resources is further increased by the growth of trade, which is made easier by industrialization, especially in parts of the global South that are rich in biodiversity (Memmott J., et.al, 2007). In order to tackle the crisis of biodiversity, it is imperative that governments acknowledge their remote responsibilities and encourage economic development initiatives that have minimal impact on biodiversity.

Threats to biodiversity rank among the top issues on the minds of intellectuals today. These hazards result from environmental issues that are brought about by both manmade and natural causes. In actuality, important environmental challenges are those caused by human activity that impact biodiversity (Deutsch C.A., et.al, 2008). These have a negative impact on all living things, not just humans. In human affairs, biodiversity is not a bonus; rather, it is the basis of life and essential to human survival and long-term growth (Kearney M., et.al, 2009). Therefore, protecting biodiversity is not only essential to preserving species, but it also aids in habitat protection and may even slow down climate change. It is almost obvious that factors such as overexploitation, invasive species, habitat destruction and fragmentation, and climate change could seriously harm biodiversity loss. Human activity, particularly overcrowding and other unsustainable practices, plays a significant role in this.

CONCLUSION

Animal life is disappearing at a startling rate due to human activity. Resources are being taken out of the environment in an unsustainable way because they are needed by the growing human population. Population expansion and unsustainable resource usage are the main threats to biodiversity. As of right now, the introduction of foreign species, deforestation, habitat loss, overharvesting, and climate change are the main drivers of extinctions. It is anticipated that human activity will significantly contribute to biodiversity extinctions in the next century through climate change. Reclamation of forests, overfishing, river damming, and other human activities are the main causes of habitat loss. Aquatic species are particularly vulnerable to overharvesting. Numerous native species have gone extinct as a result of exotic species, which are particularly harmful to lakes and islands. The introduction of exotic species is rising as a result of expanding international trade, transportation, and population movement among humans. As a result, we must safeguard our ecological capital, which includes soils, wetlands, grasslands, forests, and wildlife. Enforcing laws is also necessary to prevent environmental degradation and biodiversity loss.

REFERENCES

Abhilasha, A., Sreenivasulu, A., Manimozhi, T., Satheesh Kumar, P., Selvakumar, P., & Singh, P. (2022). *The Model of Smart Sensing Device For Sensitive Nanoclusters Modification in Sensing Properties.* 2022 2nd International Conference on Advance Computing and Innovative Technologies in Engineering (ICACITE), Greater Noida, India. 10.1109/ICACITE53722.2022.9823846

Allan, E., Manning, P., Alt, F., Binkenstein, J., Blaser, S., Blüthgen, N., Böhm, S., Grassein, F., Hölzel, N., Klaus, V. H., Kleinebecker, T., Morris, E. K., Oelmann, Y., Prati, D., Renner, S. C., Rillig, M. C., Schaefer, M., Schloter, M., Schmitt, B., & Fischer, M. (2015). Land use intensification alters ecosystem multifunctionality via loss of biodiversity and changes to functional composition. *Ecology Letters, 18*(8), 834–843. doi:10.1111/ele.12469 PMID:26096863

Arya S. (2021). Freshwater Biodiversity and Conservation Challenges: A Review. *International Journal of Biological Innovations, 3*(1), 74-78. . doi:10.46505/IJBI.2021.3106

Ashok, K. V. (2017). Genetic Diversity as Buffer in Biodiversity. *Indian Journal of Biology., 4*(1), 61–63. doi:10.21088/ijb.2394.1391.4117.9

Bradshaw, C. J. A., Sodhi, N. S., & Brook, B. W. (2009). Tropical turmoil: A biodiversity tragedy in progress. *Frontiers in Ecology and the Environment*, *7*(2), 79–87. doi:10.1890/070193

Cai M., Liu Y., Chen K., Huang D., & Yang S. (2016). Quantitative analysis of anthropogenic influences on coastal water: A new perspective. *Ecological indicators*. *67*(3).

Chakraborty, B. K., Bhattacharjee, S., & Muniya, S. (2021). A Study on Aquatic Biodiversity of Shuthi-Shaiduli River of Bangladesh. *International Journal of Biological Innovations.*, *3*(1), 58–67. doi:10.46505/IJBI.2021.3104

Dayton, P. K., Simon, F., Thrush, M., Tundi, A., & Hofman, R. J. (1995). Environmental E f f e c t s o f Ma r i n e Fi s h i n g. A q u a t i c Conservation. *Marine and Freshwater Ecosystems.*, *5*(3), 205–232. doi:10.1002/aqc.3270050305

Deutsch C.A., Tewksbury J. J., Huey R.B., Sheldon K.S., Ghalambor C.K., Haak D.C. and Martin P.R. (2008). Impacts of climate warming on terrestrial ectotherms across latitude. *Proc. Natl Acad. Sci., 105*(18).

Dueck, T. A., & Endendijk, G. J. (1987). Soil pollution and changes in vegetation. *Chemosphere*, *16*(5), 1021–1030. doi:10.1016/0045-6535(87)90039-7

Froehner, S., Souza, D. B., Machado, K. S., & Rosa, E. C. (2010). Tracking anthropogenic inputs in Barigui River, Brazil using biomarkers. *Water, Air, and Soil Pollution*, *201*(1-4), 33–41. doi:10.1007/s11270-009-0220-8

GBS. (1992). *Global Biodiversity Strategy: Guidelines for Action to Save, Study and Use Earth's Biotic Wealth sustainable and equitably. World Resources Institute (WRI), World Conservation Union (IUCN), United Nations Environmental programme (UNEP)*. IUCN.

Ghosh P. and Brand W.A. (2003). Stable isotope ratio mass spectrometry in global climate change research. Int. *Journal of Mass Spec., 3806*(03).

Goudie, A. (2013). The human impact on the natural environment: past, present and future. (7th ed). A John Wiley & Sons.

Isbell, F., Reich, P. B., Tilman, D., Hobbie, S. E., Polasky, S., Binder, S., & Bren, D. (2013). Nutrient enrichment, biodiversity loss, and c o n s e q u e n t d e c l i n e s i n e c o s y s t e m productivity. *Proceedings of the National Academy of Sciences of the United States of America*, *110*(29), 11911–11916. doi:10.1073/pnas.1310880110 PMID:23818582

Kalal, V., Giri, M. J., Baskar, S., & Kuba, R. (2021). Detection of Pollutants present in the surface water of Ayad River, Udaipur, Rajasthan. *International Journal of Biological Innovations.*, *3*(1), 212–220. doi:10.46505/IJBI.2021.3123

Kearney M., Shine R. and Porter W.P. (2009). The potential for behavioral hermoregulation to buffer 'cold-blooded' animals against climate warming. *Proc. Natl Acad. Sci. U S A., 106* (10). /pnas.0808913106.

Khaleghi M.R. (2017). The influence of deforestation and anthropogenic activities on runoff generation. *Journal of Forest Science*, *63*(6), 245 - 253 . . doi:10.17221/130/2016-JFS

Khalid, F., Iqbal, M. Z., & Qureshi, M. S. (1996). Concentration of heavy metals determined in leaves and soil from various areas of Karachi, city. *Environmental Sciences*, *4*, 213–219.

Kumar, A. & Verma, A.K. (2017). Biodiversity loss and its Ecological impact in India. *International Journal of Biological Sciences*, *8*(2), 156–160.

Mahadev Madgule, N. (2023). Mechanical properties and microstructure of activated TIG welded similar joints of Inconel alloys by desirability approaches. *Materials Today: Proceedings*, *77*, 528–533. doi:10.1016/j.matpr.2022.12.250

Masih, S. C. (2021). Impact of Monocrotophos pesticide on serum biochemical profile in freshwater fish, Cirrhinus mrigala (Hamilton, 1822). *International Journal of Biological Innovations.*, *3*(2), 402–406. doi:10.46505/IJBI.2021.3222

Memmott, J., Gibson, R., Gigante, C. L., Henson, K., Huttel, H. R., Lopezaraiza, M. M., & Pearce, S. (2007). *The conservation of ecological interactions.* doi:10.1079/9781845932541.0226

NAS. (1995). *National Academies of Sciences, Engineering, and Medicine. Understanding Marine Biodiversity*. The National Academies Press. doi:10.17226/4923

Nijmegen, R. U. (2017). Hunting accounts for 83 and 58 percent declines in tropical mammal and bird populations. *Science Daily*. www.sciencedaily.com/releases/2017/04/170413141120.htm

Pejchar, L., & Mooney, H. A. (2009). Invasive species, ecosystem services and human wellbeing. *Trends in Ecology & Evolution*, *24*(9), 497–504. doi:10.1016/j.tree.2009.03.016 PMID:19577817

Peters, G. P., Marland, G., Le Quéré, C., Boden, T., Canadell, J. G., & Raupach, M. R. (2011). Rapid growth in CO2 emissions after the 2008-2009 global financial crisis. *Nature Climate Change*, *2*(1), 2–4. doi:10.1038/nclimate1332

Pongratz, J., Reick, C., Raddatz, T., & Claussen, M. (2008). A reconstruction of global agricultural areas and land cover for the last millennium. *Global Biogeochemical Cycles*, *22*(3), GB3018. Advance online publication. doi:10.1029/2007GB003153

Poudel, S., Funakawa, S., & Shinjo, H. (2017). Household Perceptions about the Impacts of Climate Change on Food Security in the Mountainous Region of Nepal. *Sustainability (Basel)*, *9*(4), 1–20. doi:10.3390/su9040641

Pounds, J. A. (2001). Climate and amphibian declines. *Nature*, *410*(6829), 639–640. doi:10.1038/35070683 PMID:11287931

Prakash, S. (2014). Global warming and its impact on ecosystem, agriculture and human health. *International Journal of Applied Research*, *1*(1), 292–296.

Ramakrishnan, T., Mohan Gift, M. D., Chitradevi, S., Jegan, R., & Subha Hency Jose, P. (2022). Study of Numerous Resins Used in Polymer Matrix Composite Materials. Advances in Materials Science and Engineering. Hindawi. doi:10.1155/2022/1088926

Selvakumar, P. (2024). Prediction of Earthquakes, Volcanic Eruptions, Tornadoes, Wildfires, and Droughts Through Artificial Intelligence. In D. Satishkumar, M. Sivaraja, & I. G. I. Global (Eds.), *Utilizing AI and Machine Learning for Natural Disaster Management* (pp. 133–145)., doi:10.4018/979-8-3693-3362-4.ch008

Chapter 5
Current Scenario of Biodiversity Loss Due to Developmental Activities

Prashantkumar Sathvara
https://orcid.org/0000-0002-3548-5584
Nims Institute of Allied Medical Science and Technology, India

J. Anuradha
Nims Institute of Allied Medical Science and Technology, India

Sandeep Tripathi
Nims Institute of Allied Medical Science and Technology, India

R. Sanjeevi
Nims Institute of Allied Medical Science and Technology, India

ABSTRACT

In the era of rapid global development, the delicate balance between progress and environmental preservation is strained. "Vanishing Harmony" addresses biodiversity loss, revealing the link between development and Earth's ecosystems decline. The book examines causes, consequences, and solutions against urbanization, industrialization, and infrastructure expansion. It unravels interconnected factors like deforestation, pollution, climate change, and resource exploitation. Biodiversity loss affects ecological stability and human well-being, impacting food security, medicine, and ecosystem resilience. Addressing root causes is vital for sustainability. Policy, international collaboration, and community engagement are key. The book advocates for a shift in societal attitudes and economic models to prioritize development and ecological preservation. It aims to inform policymakers, environmentalists, researchers, and the public, fostering understanding and inspiring collective commitment to preserve our planet's biological heritage.
DOI: 10.4018/979-8-3693-6950-0.ch005

INTRODUCTION

Biodiversity is declining rapidly due to land use change, climate change, invasive species, overexploitation, and pollution. Between 1970 and 2016, the population of vertebrate species fell by 68% on average worldwide. The biggest threats to Earth's biodiversity include climate change, changes in land and sea use, and pollution. Changes in land and sea use account for the largest portion of loss, making up 50% of recorded threats to biodiversity on average. The extent of biodiversity decline varies in different parts of the world. Latin America & Caribbean has seen the biggest drop in biodiversity at 94%, mainly driven by declining reptile, amphibian, and fish populations. Scenarios consistently indicate that biodiversity will continue to decline over the 21st century. However, there are major opportunities to intervene through better policies. The WWF's Living Planet Report 2022 calls for the world to adopt a goal of being "nature positive" - actively restoring the planet's living species rather than just halting their loss - by 2030 (Prashantkumar B. Sathvara *et al.*, 2023). Urgent action is needed to protect and restore our planet's biodiversity. This includes transformative shifts in how we produce, consume, govern, and finance.

BRIEF OVERVIEW OF BIODIVERSITY

Biodiversity, short for biological diversity, encompasses the variety of life on Earth, from the smallest microorganisms to the largest ecosystems. It includes the diversity of species, genes, and ecosystems, as well as the ecological and evolutionary processes that sustain life. Biodiversity is a fundamental characteristic of our planet, shaping the intricate web of relationships between organisms and their environments. At its core, biodiversity is a measure of the richness and complexity of life, encompassing the millions of species that inhabit the Earth's diverse habitats. It reflects the incredible diversity of forms, functions, and interactions that exist within and between species, contributing to the resilience, stability, and adaptability of ecosystems. Biodiversity is not only a source of wonder and beauty but also a vital component of the Earth's life support systems. It provides essential ecosystem services that sustain human well-being, including clean air and water, fertile soils, pollination of crops, and regulation of climate and disease. Biodiversity also underpins cultural and spiritual values, shaping identities, traditions, and connections to the natural world (Gaston & Spicer, 2013). However, biodiversity faces unprecedented threats from human activities, including habitat destruction, overexploitation of natural resources, pollution, invasive species, and climate change. These pressures are driving species to extinction at an alarming rate, with profound implications for ecosystems, economies, and societies. As we strive to address the challenges of the 21st century, understanding and conserving biodiversity is paramount. By recognizing the value and importance of biodiversity, and taking action to protect and restore it, we can safeguard the health of our planet and secure a sustainable future for generations to come (J. S. Singh, 2012).

Importance of Biodiversity for Ecosystem Balance

The importance of biodiversity for ecosystem balance cannot be overstated. Biodiversity plays a crucial role in maintaining the health, resilience, and functioning of ecosystems. Biodiversity ensures the interconnectedness of species within ecosystems. Each organism, from microorganisms to apex predators, plays a unique role in the ecosystem, contributing to nutrient cycling, energy flow, and ecosystem stability. Interactions between species, such as predation, competition, and mutualism, help regulate populations and maintain ecological balance (Vanneste *et al.*, 2024).

Ecosystems with high biodiversity are more resilient to environmental changes and disturbances. Diverse ecosystems can better withstand and recover from natural disasters, such as hurricanes, droughts, and wildfires, as they contain a variety of species with different traits and abilities to cope with changing conditions. Biodiversity enhances ecosystem stability and reduces the risk of ecosystem collapse in the face of environmental stressors. Biodiversity underpins the provision of ecosystem services that are essential for human well-being. These services include air and water purification, soil fertility, pollination of crops, pest control, climate regulation, and provision of food, medicine, and raw materials (Onoh *et al.*, n.d.-a). Maintaining biodiversity is essential for sustaining these ecosystem services, which support agriculture, fisheries, forestry, and other economic activities.

Biodiversity encompasses genetic diversity within and between species. Genetic variation is the raw material for evolution and adaptation, allowing populations to respond to changing environmental conditions, such as climate change and disease outbreaks. Genetic diversity enhances the resilience of species to environmental stressors, reduces the risk of inbreeding depression, and provides opportunities for species to evolve and thrive over time. Biodiversity contributes to ecosystem stability and productivity (Bridgewater & Schmeller, 2024). Studies have shown that diverse ecosystems are more productive, with higher rates of biomass production, nutrient cycling, and energy capture. Species diversity enhances ecosystem functioning by increasing resource use efficiency, resilience to disturbances, and the ability to exploit a wider range of resources. Biodiversity is essential for maintaining the balance, resilience, and functioning of ecosystems. Protecting and conserving biodiversity is not only crucial for the health of ecosystems but also for the well-being of humanity, as we depend on ecosystems for essential services and resources (Kevin J. Gaston & John I. Spicer, n.d.; Sanjeevi *et al.*, 2017). Recognizing the importance of biodiversity and taking action to conserve and restore it is essential for ensuring a sustainable future for all life on Earth.

CONCEPT OF DEVELOPMENTAL ACTIVITIES

Developmental activities encompass a wide range of human endeavors aimed at improving living standards, economic prosperity, and societal well-being. These activities involve the utilization of natural resources, construction of infrastructure, implementation of policies, and investment in various sectors to promote growth and development. At its core, developmental activities seek to address societal needs and aspirations by fostering progress and advancement across different domains, including economic, social, environmental, and cultural (Yanagi, 2024). These activities are driven by the desire to enhance quality of life, reduce poverty, create employment opportunities, and promote sustainable development.

Developmental activities can take various forms like Construction of roads, bridges, ports, airports, railways, and other transportation networks to facilitate the movement of goods, people, and services. Infrastructure development is essential for connecting communities, promoting trade and commerce, and supporting economic growth. Expansion of urban areas and development of cities to accommodate growing populations, provide housing, and offer access to essential services such as education, healthcare, and sanitation. Urbanization is a key driver of economic development but can also lead to challenges such as congestion, pollution, and loss of green spaces (Sathvara *et al.*, 2023; Sunita *et al.*, n.d.). Establishment of industries and manufacturing facilities to produce goods and services, generate employment, and stimulate economic growth. Industrialization contributes to technological innovation, wealth creation, and economic diversification but can also result in environmental pollution and resource depletion if not managed sustainably.

Promotion of agricultural activities to increase food production, improve food security, and enhance rural livelihoods. Agricultural development involves modernizing farming practices, improving irrigation and crop management techniques, and promoting sustainable agriculture to ensure long-term food security and environmental sustainability (Sokół & Łaska, 2024). Development of energy infrastructure and investment in renewable and non-renewable energy sources to meet growing energy demands, reduce reliance on fossil fuels, and mitigate climate change. Energy production is essential for powering industries, transportation, and households, but also poses environmental challenges such as air and water pollution and habitat destruction. Implementation of policies, regulations, and incentives to promote sustainable development, protect natural resources, and address social inequalities (Ramakrishnan & Jayaraman, 2019; Saleh *et al.*, 2024a). Policy interventions play a critical role in guiding and regulating developmental activities to ensure they are socially equitable, environmentally responsible, and economically viable. Developmental activities are essential for driving progress and improving quality of life but must be pursued in a manner that is sustainable, inclusive, and respectful of environmental and social considerations (Bhattacharyya & Univerity, n.d.-a). Balancing the pursuit of development with the need to conserve natural resources, protect ecosystems, and promote social equity is essential for achieving sustainable development and ensuring a better future for present and future generations.

Developmental Activities and Their Impact

Developmental activities encompass a wide range of human endeavors aimed at promoting economic growth, social progress, and infrastructure development. These activities can have significant implications for biodiversity, as they often involve the utilization of natural resources, habitat modification, and alteration of ecosystems. Here, we will define developmental activities, provide examples, and explore how they impact biodiversity.

Definition and Examples of Developmental Activities

Developmental activities refer to actions and initiatives undertaken to promote economic, social, and environmental progress within a society (figure 1).

Explanation of How These Activities Impact Biodiversity

Developmental activities can have significant impacts on biodiversity due to habitat destruction, fragmentation, pollution, and overexploitation of natural resources (Xiao *et al.*, 2024). These activities affect biodiversity:

1. Habitat Loss and Fragmentation: Developmental activities often involve clearing natural habitats, such as forests, wetlands, and grasslands, to make way for infrastructure, agriculture, urbanization, and industrial projects. This habitat loss can result in the displacement and extinction of wildlife species, as well as fragmentation of ecosystems, which reduces connectivity and disrupts ecological processes (Ankitkumar B Rathod *et al.*, 2022; Velado-Alonso *et al.*, 2024).
2. Pollution: Industrialization, urbanization, and agricultural activities can lead to pollution of air, water, and soil, which can have detrimental effects on biodiversity. Pollution from industries, mining operations, and urban runoff

Figure 1. Developmental activities

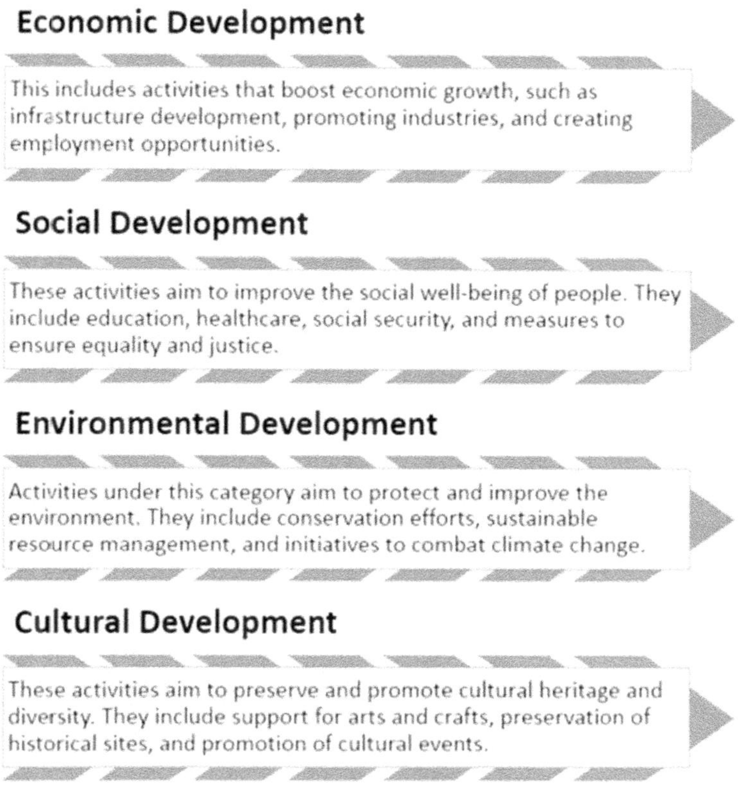

can contaminate water bodies, degrade habitats, and harm aquatic organisms and terrestrial wildlife (Mwale *et al.*, 2024).

3. Overexploitation of Natural Resources: Developmental activities often involve the unsustainable extraction of natural resources, such as timber, minerals, and fish, leading to overexploitation and depletion of biodiversity. Overfishing, deforestation, and illegal wildlife trade can result in declines in population sizes, loss of genetic diversity, and ecosystem degradation.
4. Introduction of Invasive Species: Developmental activities can facilitate the introduction and spread of invasive alien species, which can outcompete native species, disrupt ecosystems, and cause biodiversity loss. Infrastructure development, international trade, and tourism can serve as pathways for the introduction of invasive species, posing threats to native biodiversity.

Developmental activities can have profound impacts on biodiversity, threatening the survival of species, altering ecosystems, and undermining the provision of

Table 1. Development activity impact on biodiversity

Case	Impact on Biodiversity	Value	Reference
Global Biodiversity Decline	The Living Planet Index reports a 68% decline in global biodiversity between 1970 and 2016.	This decline has significant implications for ecosystem stability and services.	(*Living Planet Report 2020 | Publications | WWF*, n.d.)
Regional Differences	Latin America shows the greatest regional decline in average population abundance at 94%, while freshwater species populations have seen the greatest overall global decline (83%).	These regional and habitat differences highlight the need for targeted conservation efforts.	(*Living Planet Report 2020 | Publications | WWF*, n.d.)
Impact of Land Use Change	Land-use change, primarily driven by agriculture, has caused 70% of global biodiversity loss and half of all loss in tree cover.	This loss of biodiversity and tree cover has significant implications for climate regulation and other ecosystem services.	(secretariat, 2019)

ecosystem services. It is essential to consider biodiversity conservation in development planning and decision-making processes to minimize negative impacts and promote sustainable development (Jackson *et al.*, 2024). Effective mitigation measures, such as habitat restoration, conservation planning, and sustainable land use practices, are needed to safeguard biodiversity and ensure the long-term health and resilience of ecosystems.

CASE STUDIES OF BIODIVERSITY LOSS

Detailed Examples of Specific Regions or Species Affected by Developmental Activities

Biodiversity loss due to developmental activities is a global phenomenon with significant impacts on ecosystems, species, and habitats. Below are detailed examples of specific regions or species affected by developmental activities:

1. Amazon Rainforest, Brazil

According to data from Brazil's National Institute for Space Research (INPE), deforestation in the Brazilian Amazon reached a 12-year high in 2020, with 11,088 square kilometers (4,281 square miles) of forest lost. The International Union for Conservation of Nature (IUCN) Red List identifies numerous species in the Amazon rainforest as threatened or endangered, including the jaguar (Panthera onca), giant otter (Pteronura brasiliensis), and harpy eagle (Harpia harpyja). A study published in the journal Nature in 2015 found that roads constructed for infrastructure development in the Amazon increase deforestation by providing access to remote areas and facilitating illegal logging and land conversion (da Silva *et al.*, 2024; Furtado *et al.*, 2024).

2. Coral Reefs, Southeast Asia

The Global Coral Reef Monitoring Network (GCRMN) reports that approximately 75% of the world's coral reefs are currently threatened by a combination of local and global stressors, including coastal development, overfishing, pollution, and climate change. Data from the Food and Agriculture Organization (FAO) indicate that Southeast Asia has experienced significant declines in fish populations due to overfishing and habitat degradation, leading to reduced biodiversity and ecosystem collapse in coral reef ecosystems. The National Oceanic and Atmospheric Administration (NOAA) reports that coral bleaching events, caused by elevated sea temperatures, have become more frequent and severe in recent decades, resulting in widespread coral mortality and loss of biodiversity in Southeast Asia's coral reefs (Baciero *et al.*, 2024; Razak *et al.*, 2024).

3. Sundarbans Mangrove Forest, Bangladesh and India

A study published in the journal Science of the Total Environment in 2019 estimated that the Sundarbans mangrove forest has lost approximately 124,418 hectares (307,508 acres) of mangrove cover between 1986 and 2014 due to deforestation and conversion for shrimp farming and agriculture. Data from the World Bank indicate that industrial pollution from factories and oil spills is a major threat to water quality and biodiversity in the Sundarbans, with high levels of heavy metals and toxins detected in sediments and aquatic organisms. The Intergovernmental Panel on Climate Change (IPCC) reports that sea-level rise and increased frequency of cyclones associated with climate change pose significant threats to the Sundarbans mangrove ecosystem, leading to habitat loss, salinization, and biodiversity decline (Ahmad *et al.*, 2024).

These case studies highlight the complex interactions between developmental activities and biodiversity loss in different regions around the world. It underscores the urgent need for sustainable development practices that balance economic growth with environmental conservation to mitigate the impacts of developmental activities on biodiversity and ensure the long-term health and resilience of ecosystems.

Table 2. Different aspect impact on biodiversity

Aspect	Description
Declining Wildlife Populations	Wildlife populations have declined significantly in recent decades, with some species facing the risk of extinction due to habitat loss and other human-induced threats.
Habitat Destruction	Deforestation, urbanization, agriculture, and infrastructure development continue to degrade and fragment natural habitats, reducing available space for wildlife.
Overexploitation	Overexploitation of natural resources, including overfishing, poaching, and illegal wildlife trade, threatens the survival of many species.
Pollution	Pollution from industrial activities, agriculture, and waste disposal poses significant threats to biodiversity and ecosystem health.
Climate Change	Climate change is exacerbating biodiversity loss, impacting ecosystems and species worldwide through rising temperatures and extreme weather events.
Invasive Species	Invasive alien species threaten native biodiversity by outcompeting native species and altering ecosystem dynamics.
Loss of Ecosystem Services	Biodiversity loss undermines the provision of essential ecosystem services, compromising human well-being and livelihoods.

THE ROLE OF POLICIES AND REGULATIONS

Discussion on Existing Policies Related to Biodiversity Conservation

Policies and regulations play a crucial role in biodiversity conservation by providing frameworks, guidelines, and incentives to promote sustainable use of natural resources, protect ecosystems, and mitigate threats to biodiversity (Ankitkumar B Rathod *et al.*, 2023; R. Singh, n.d.). Existing policies related to biodiversity conservation:

- International Agreements: Convention on Biological Diversity (CBD): Adopted in 1992, the CBD is a global treaty aimed at promoting the conservation and sustainable use of biodiversity. It sets out key objectives, including the conservation of biological diversity, sustainable use of its components, and fair and equitable sharing of benefits derived from genetic resources. The CBD has been ratified by 196 countries and provides a framework for national biodiversity strategies and action plans.
- National Biodiversity Strategies and Action Plans (NBSAPs): Many countries have developed NBSAPs to implement the objectives of the CBD at the national level. These strategies outline priorities, targets, and actions for biodiversity conservation, including measures to protect ecosystems, conserve

species, promote sustainable land use practices, and engage stakeholders in conservation efforts.

- Protected Areas and Conservation Reserves: Governments establish protected areas, such as national parks, wildlife sanctuaries, and marine reserves, to conserve biodiversity and ecosystem integrity. These areas are legally designated and managed to safeguard habitats, protect species, and promote scientific research and ecotourism. Policies related to protected areas include regulations governing access, use, and management of natural resources within these areas.
- Environmental Impact Assessment (EIA) Regulations: Many countries have laws and regulations requiring developers to conduct EIAs before undertaking major projects that may have significant environmental impacts, including biodiversity loss. EIAs assess the potential environmental consequences of proposed developments, identify mitigation measures, and ensure compliance with environmental standards and regulations.
- Sustainable Development Goals (SDGs): The United Nations' Sustainable Development Goals (SDGs), adopted in 2015, include targets related to biodiversity conservation, sustainable land use, and ecosystem restoration. SDG 15 focuses specifically on protecting, restoring, and promoting the sustainable use of terrestrial ecosystems, including forests, wetlands, and mountains, as well as halting biodiversity loss by 2030.
- Payments for Ecosystem Services (PES) Programs: PES programs are mechanisms that incentivize landowners and communities to conserve ecosystems and biodiversity by providing financial rewards for maintaining ecosystem services, such as carbon sequestration, water purification, and habitat preservation. These programs help internalize the value of biodiversity and promote sustainable land management practices.

Existing policies and regulations related to biodiversity conservation provide important frameworks and mechanisms for addressing threats to biodiversity, promoting sustainable development, and achieving global conservation goals. However, effective implementation, enforcement, and monitoring of these policies are essential to ensure their success in safeguarding biodiversity and securing the well-being of current and future generations (Sengupta *et al.*, 2024). Continued collaboration and coordination among governments, stakeholders, and civil society are needed to address emerging challenges and promote biodiversity conservation at all levels (Abbasi *et al.*, 2012; Sailu & Unnisa, 2024).

Analysis of the Effectiveness of These Policies

The effectiveness of policies and regulations in biodiversity conservation can vary significantly, and it's often influenced by a range of factors (Dhyani, 2024).

- Effectiveness of Biodiversity Conservation Laws: While laws and policies have been introduced globally to prevent species extinction, they are often failing due to various reasons. Extinction rates continue to climb despite these laws. However, when implemented effectively, these laws can regulate human-environment interactions that drive extinction and biodiversity loss.
- Role of Evaluation and Review: Best practice biodiversity law and policy should include provisions that allow for effective evaluation and review of the law itself, as well as any function, activity, or program carried out under the law.
- Protected Areas: Designation and management of protected areas is a cornerstone of biodiversity conservation. However, despite an increase in the total number of protected areas in the world, biodiversity continues to decline. Many protected areas are not functioning as effectively as originally intended, due in part to limited resources to maintain these areas and/or enforce relevant legal frameworks.
- Need for Capacity Building: Lack of adequate technical and financial resources and capacity can limit the upscaling of innovative solutions, demonstrating further the need for regional and subregional cooperation.

While policies and regulations are crucial for biodiversity conservation, their effectiveness often depends on factors such as implementation, enforcement, regular review, and availability of resources (Bhattacharyya & Univerity, n.d.-b). More needs to be done to enhance the effectiveness of these policies and regulations in conserving biodiversity.

Suggestions for Policy Improvements

Improving policies and regulations for biodiversity conservation is crucial (Bhattacharyya & Univerity, n.d.-b).

- Strengthen enforcement mechanisms to ensure compliance with existing biodiversity conservation policies and regulations. Implement penalties for non-compliance and incentivize adherence to biodiversity-friendly practices through rewards or incentives.

- Integrate biodiversity considerations into sectorial policies and plans, such as agriculture, forestry, energy, infrastructure, and urban planning. Promote cross-sectoral coordination and collaboration to ensure that development activities are aligned with biodiversity conservation objectives.
- Involve local communities, indigenous peoples, and other stakeholders in decision-making processes related to biodiversity conservation. Ensure meaningful participation, consultation, and consent of affected communities in the design, implementation, and monitoring of biodiversity policies and projects.
- Invest in capacity-building programs to enhance technical expertise, knowledge, and skills related to biodiversity conservation and sustainable development. Raise awareness among policymakers, businesses, civil society organizations, and the general public about the importance of biodiversity and the need for conservation action.
- Allocate adequate funding and resources for biodiversity conservation efforts, including protected area management, species recovery programs, and sustainable land use practices. Develop financial mechanisms, such as grants, loans, tax incentives, and subsidies, to support biodiversity-friendly initiatives and investments.
- Adopt adaptive management approaches that allow for flexibility, learning, and adjustment based on monitoring, evaluation, and feedback. Foster a culture of learning and knowledge sharing among policymakers, practitioners, researchers, and communities to improve the effectiveness of biodiversity conservation policies and practices.
- Recognize and respect indigenous knowledge systems, traditional practices, and cultural values related to biodiversity conservation. Indigenous perspectives and practices into biodiversity policies and management strategies to enhance their relevance, effectiveness, and sustainability.
- Strengthen international cooperation and collaboration on biodiversity conservation through partnerships, knowledge exchange, and joint initiatives. Align national biodiversity strategies and actions with global biodiversity goals and targets, such as those outlined in the Convention on Biological Diversity (CBD) and the Sustainable Development Goals (SDGs).

By implementing these policy improvements, governments, stakeholders, and civil society organizations can enhance the effectiveness of biodiversity conservation efforts, address emerging challenges, and promote sustainable development for present and future generations (Anjanappa, 2024; Sanjeevi, 2011).

MITIGATION STRATEGIES

Strategies to Reduce the Impact of Developmental Activities on Biodiversity

Mitigation strategies are crucial for minimizing the impact of developmental activities on biodiversity. Avoidance involves changing site locations or implementing alternative development practices to completely avoid impacts on biodiversity. In the Minimization step, project impacts are reduced in intensity through the use of new technologies or alterations in timing. Restoration, if impacts still occur, restoration aims to repair the damage done to the environment (Vasiliev *et al.*, 2024). Offsetting is considered a last resort, balancing the overall negative impacts through actions such as funding for national parks or restoration projects. Eliminating Harmful Subsidies that support local and national activities harmful to biodiversity – such as deforestation, over-fertilization, and over-fishing, can also support climate change mitigation and adaptation. Changing Consumption Patterns individual consumption patterns, reducing loss and waste, and shifting diets, especially in rich countries, toward more plant-based options can also help mitigate impacts on biodiversity (Mishra *et al.*, 2016; Rey *et al.*, 2024).

These strategies, when implemented effectively, can help achieve no net loss of biodiversity or even a net gain. They guide environmental professionals, offering practical examples and innovative approaches. By prioritizing biodiversity conservation, development projects contribute to the overall goal of sustainable development (Onoh *et al.*, n.d.-b).

Role of Technology in Biodiversity Conservation

Remote sensing technology, combined with Geographic Information Systems (GIS), allows researchers to monitor changes in land cover, habitat fragmentation, and biodiversity hotspots over large spatial scales (Qing *et al.*, 2024; Abbasi *et al.*, 2016). This enables better-informed conservation planning and management decisions. DNA barcoding techniques help identify and classify species based on genetic markers, aiding in species identification, tracking, and monitoring. Molecular tools also facilitate the detection of endangered species and illegal wildlife trade, supporting conservation efforts. Unmanned aerial vehicles (UAVs) or drones equipped with cameras and sensors enable researchers to survey and monitor remote or inaccessible habitats, track wildlife populations, and detect illegal activities such as poaching and deforestation (Millner *et al.*, 2024). Technology platforms and mobile applications allow citizen scientists to contribute valuable biodiversity data, such as species observations and habitat information. This crowd-sourced data enhances biodiversity monitoring and research efforts. Advances in genetic sequencing technologies enable researchers to study the genetic diversity, population structure, and evolutionary history of species (de Mori *et al.*, 2024). This information is crucial for understanding species resilience, adaptation to environmental changes, and informing conservation strategies.

Importance of Sustainable Development

Sustainable development aims to balance economic growth with environmental conservation, ensuring the preservation of natural resources, ecosystems, and biodiversity for future generations (Yasin *et al.*, 2024). Sustainable development prioritizes social equity and inclusivity, aiming to address poverty, inequality, and social injustice while promoting access to education, healthcare, and basic amenities for all. Sustainable development promotes long-term economic prosperity by fostering innovation, resource efficiency, and green technologies. It encourages sustainable business practices that minimize environmental impacts and maximize social benefits. Sustainable development strategies contribute to mitigating climate change by promoting renewable energy sources, reducing greenhouse gas emissions, and enhancing resilience to climate-related risks and disasters. Sustainable development recognizes the value of ecosystem services, such as clean water, air purification, pollination, and soil fertility, and seeks to conserve and sustainably manage natural ecosystems to ensure their continued provision (Yang *et al.*, 2024). Sustainable development emphasizes inter-generational equity, ensuring that present actions do not compromise the ability of future generations to meet their needs. It encourages responsible stewardship of natural resources and the environment. Sustainable development requires international cooperation and collaboration to address global challenges such as climate change, biodiversity loss, and sustainable resource management (Saleh *et al.*, 2024b). It promotes partnerships between governments, businesses, civil society, and academia to achieve common goals.

CONCLUSION

In conclusion, it is imperative to recap the significant impacts and the pressing urgency of addressing the current biodiversity crisis. The rapid loss of biodiversity due to human activities such as habitat destruction, pollution, climate change, and overexploitation of natural resources poses severe threats to ecosystems, wildlife populations, and the services they provide to humanity. The urgency of the situation cannot be overstated, as biodiversity loss not only undermines the resilience of ecosystems but also jeopardizes human well-being and future socio-economic development.

Restating the importance of biodiversity is crucial. Biodiversity is the foundation of healthy ecosystems, playing essential roles in regulating climate, purifying air and water, pollinating crops, and providing food, medicine, and other ecosystem services essential for human survival. Protecting and conserving biodiversity is not only a moral imperative but also a strategic investment in securing our planet's long-term sustainability and resilience.

As we navigate the complex challenges of the 21st century, it is essential to strike a delicate balance between development and conservation. While economic growth and human development are important goals, they must be pursued in harmony with nature and with full consideration of the impacts on biodiversity and ecosystem health. Sustainable development approaches that integrate environmental conservation, social equity, and economic prosperity offer a pathway towards achieving this balance.

Addressing the biodiversity crisis requires urgent and concerted efforts from all sectors of society, including governments, businesses, civil society organizations, and individuals. By embracing conservation practices, adopting sustainable development strategies, and fostering a deeper appreciation for the intrinsic value of biodiversity, we can pave the way toward a more resilient, equitable, and sustainable future for both people and the planet.

REFERENCES

Abbasi, T., Anuradha, J., & Abbasi, S. (2016). Utilization of the pernicious aquatic weed salvinia (Salvinia molesta DS Mitchell) in generating gold nanoparticles. *Indian Journal of Biotechnology*, *15*, 382–391.

Abbasi, T., Sanjeevi, R., Makhija, M., & Abbasi, S. A. (2012). Role of Vitamins B-3 and C in the Fashioning of Granules in UASB Reactor Sludge. *Applied Biochemistry and Biotechnology*, *167*(2), 348–357. doi:10.1007/s12010-012-9691-y PMID:22549583

Ahmad, H., Jose, F., & Shoemaker, D. J. (2024). Mapping, Dynamics, and Future Change Analysis of Sundarbans delta using Cellular Automata and Artificial Neural Network Modeling. *IEEE Journal of Selected Topics in Applied Earth Observations and Remote Sensing*, *17*, 5594–5603. doi:10.1109/JSTARS.2024.3367116

Anjanappa, J. (2024). Unveiling the Green Paradox-Biodiversity Risk in Renewable Energy Investments in India. *SSRN*, *4722554*. https://papers.ssrn.com/sol3/papers.cfm?abstract_id=4722554. doi:10.2139/ssrn.4722554

Ankitkumar, B. Rathod, Prashantkumar B Sathvara, Aviral Tripathi, J. Anuradha, Sandeep Tripathi, & R. Sanjeevi. (2022). Vegetational Cartography Analysis Utilizing Multi-temporal Ndvi Data Series: A Case Study from Rajkot District (Gujarat), India. *Tianjin Daxue Xuebao, 55*(04), 490–497. https://doi.org/ doi:10.17605/OSF.IO/UGJYM

Ankitkumar,, B. (2023). Vegetational Change Detection Using Machine Learning in GIS Technique: A Case Study from Jamnagar (Gujarat). *Journal of Data Acquisition and Processing*, *38*(1), 1046–1061. doi:10.5281/zenodo.7700655

Baciero, A. G., Herrero, Á. G., Anderson, M., Berná, E. H., Gómez, G. G., & Zamorano, J. G. (2024). *Reef fish abundance enhancement through low-cost coral reef restoration practices: A case study in Tioman island, Malaysia*. Research Square. https://www.researchsquare.com/article/rs-3852467/latest

Bhattacharyya, P., & Univerity, B. (n.d.-a). 7. A Sustainable Approach to Harmonize Urbanization and Biodiversity Conservation in India. *The Symphony Of Knowledge*, *105*.

Bridgewater, P., & Schmeller, D. S. (2024). Nature abhors a vacuum: Invasive themes and other outputs from the tenth plenary of the Intergovernmental Platform on Biodiversity and Ecosystem Services (IPBES-10). *Biodiversity and Conservation*, *33*(3), 867–873. doi:10.1007/s10531-024-02797-x

da Silva, Q. P., Moreira, D. S., de Freitas, H. C., & Domingues, L. M. (2024). Discrepancies between observation and ERA5 reanalysis in the Amazon deforestation context: A case study. *Dynamics of Atmospheres and Oceans*, *106*, 101442. doi:10.1016/j.dynatmoce.2024.101442

de Mori, B., Mercugliano, E., Biasetti, P., Pollastri, I., Spiriti, M. M., Florio, D., Andreucci, F., Göritz, F., Holtze, S., Galli, C., Stejskal, J., Colleoni, S., Lazzari, G., Seet, S., Zwilling, J., Ndeereh, D., Lekolool, I., Ngulu, S., Mijele, D., & Hildebrandt, T. B. (2024). The ethical assessment of Assisted Reproductive Technologies (ART) in wildlife conservation. *Biological Conservation*, *290*, 110423. doi:10.1016/j.biocon.2023.110423

Dhyani, S. (2024). *Strengthening Science Diplomacy for Reducing Biodiversity Loss and Restoring Degraded Ecosystems*. NOPR. https://nopr.niscpr.res.in/bitstream/123456789/63145/1/SD%207%282%29%2019-22.pdf

Furtado, M. D. S. C., Queiroz, J. C. B., Bentes, B., Gouveia, N. D. A., De Lima, M. J. A., Ruffino, M. L., & Isaac, V. (2024). How does climate change affect small scale fisheries? A case study of the Lower Amazon in Brazil. *Fisheries Management and Ecology*, *31*(1), e12654. doi:10.1111/fme.12654

Gaston, K. J., & Spicer, J. I. (2013). *Biodiversity: An introduction*. John Wiley & Sons. https://books.google.com/books?hl=en&lr=&id=0Bjp2o5CVnQC&oi=fnd&pg=PT6&dq=Biodiversity:+An+overview&ots=Y2X6YSmcEx&sig=KHEre9jRGafSgjzx7GfRWb1tK-w

Jackson, M. C., Friberg, N., Moliner Cachazo, L., Clark, D. R., Mutinova, P. T., O'Gorman, E. J., Kordas, R. L., Gallo, B., Pichler, D. E., Bespalaya, Y., Aksenova, O. V., Milner, A., Brooks, S. J., Dunn, N., Lee, K. W. K., Ólafsson, J. S., Gíslason, G. M., Millan, L., Bell, T., & Woodward, G. (2024). Regional impacts of warming on biodiversity and biomass in high latitude stream ecosystems across the Northern Hemisphere. *Communications Biology*, *7*(1), 1–8. doi:10.1038/s42003-024-05936-w PMID:38480906

Kevin, J. Gaston & Spicer, J. (n.d.). *Biodiversity: An Introduction*. Wiley. https://www.wiley.com/en-in/Biodiversity:+An+Introduction,+2nd+Edition-p-9781405118576

Living Planet Report 2020. (n.d.). World Wildlife Fund. https://www.worldwildlife.org/publications/living-planet-report-2020

Millner, N., Newport, B., Sandbrook, C., & Simlai, T. (2024). Between monitoring and surveillance: Geographies of emerging drone technologies in contemporary conservation. *Progress in Environmental Geography*, *3*(1), 17–39. doi:10.1177/27539687241229739

Mishra, S., Anuradha, J., Tripathi, S., & Kumar, S. (2016). In vitro antioxidant and antimicrobial efficacy of Triphala constituents: Emblica officinalis, Terminalia belerica and Terminalia chebula. *Journal of Pharmacognosy and Phytochemistry*, *5*(6), 273–277.

Mwale, V. D., To, L. S., Gogoda, C., Ngonda, T., & Nkhoma, R. (2024). Community resilience in Bondo community, Southern Malawi: Balancing energy, water and biodiversity. *International Journal of Energy Sector Management*. https://www.emerald.com/insight/content/doi/10.1108/IJESM-10-2023-0007/full/html

Onoh, U. C., Ogunade, J., Owoeye, E., Awakessien, S., & Asomah, J. K. (n.d.-a). *Impact of Climate Change on Biodiversity and Ecosystems Services*. IIARD. https://www.iiardjournals.org/get/IJGEM/VOL.%2010%20NO.%201%202024/Impact%20of%20Climate%20Change%2077-93.pdf

Onoh, U. C., Ogunade, J., Owoeye, E., Awakessien, S., & Asomah, J. K. (n.d.-b). *Impact of Climate Change on Biodiversity and Ecosystems Services*. IIARD. https://www.iiardjournals.org/get/IJGEM/VOL.%2010%20NO.%201%202024/Impact%20of%20Climate%20Change%2077-93.pdf

Prashantkumar, B. (2023). Impact of climate change and its importance on human performance. In Insights on Impact of Climate Change and Adaptation of Biodiversity (1st ed., pp. 1–9). KD Publication.

Qing, L., Usman, M., Radulescu, M., & Haseeb, M. (2024). Towards the vision of going green in South Asian region: The role of technological innovations, renewable energy and natural resources in ecological footprint during globalization mode. *Resources Policy*, *88*, 104506. doi:10.1016/j.resourpol.2023.104506

Ramakrishnan, S., & Jayaraman, A. (2019). Global Warming and Pesticides in Water Bodies. In *Handbook of Research on the Adverse Effects of Pesticide Pollution in Aquatic Ecosystems* (pp. 421–436). IGI Global. https://www.igi-global.com/chapter/global-warming-and-pesticides-in-water-bodies/213519 doi:10.4018/978-1-5225-6111-8.ch023

Razak, F., Lua, W. Y., Abd Rasid, N. H., Aziz, N., Repin, I. M., Xue, X.-Z., Ashraf, A. R. M., Bachok, Z., Afiq-Firdaus, A., & Talaat, W. I. A. W. (2024). Adopting Marine Spatial Planning (MSP) using coral health assessment as indicator: A case study in Pulau Redang Marine Park, Malaysia. *Ocean and Coastal Management*, *248*, 106943. doi:10.1016/j.ocecoaman.2023.106943

Rey, F., Dupire, S., & Berger, F. (2024). Forest-based solutions for reconciling natural hazard reduction with biodiversity benefits. *Nature-Based Solutions*, 100114.

Sailu, G., & Unnisa, S. A. (2024). *Conserving precious urban freshwater bodies: A case study of Telangana, India*. Academic Journals. https://academicjournals.org/journal/IJBC/article-full-text-pdf/701C19771854.pdf

Saleh, M., Ashqar, H. I., Alary, R., Bouchareb, E. M., Bouchareb, R., Dizge, N., & Balakrishnan, D. (2024a). Biodiversity for ecosystem services and sustainable development goals. In *Biodiversity and Bioeconomy* (pp. 81–110). Elsevier. https://www.sciencedirect.com/science/article/pii/B9780323954822000055 doi:10.1016/B978-0-323-95482-2.00005-5

Saleh, M., Ashqar, H. I., Alary, R., Bouchareb, E. M., Bouchareb, R., Dizge, N., & Balakrishnan, D. (2024b). Biodiversity for ecosystem services and sustainable development goals. In *Biodiversity and Bioeconomy* (pp. 81–110). Elsevier. https://www.sciencedirect.com/science/article/pii/B9780323954822000055 doi:10.1016/B978-0-323-95482-2.00005-5

Sanjeevi, R. (2011). *Studies on the treatment of low-strength wastewaters with upflow anaerobic sludge blanket (UASB) reactor: With emphasis on granulation studies*. Centre for Pollution Control and Environmental Engineering.

Sanjeevi, R., Haruna, M., Tripathi, S., Singh, B., & Jayaraman, A. (2017). Impacts of Global Carbon Foot Print on the Marine Environment. *International Journal of Engineering Research & Technology (Ahmedabad), 5*, 51–54.

Sathvara, P. B., Anuradha, J., Sanjeevi, R., Tripathi, S., & Rathod, A. B. (2023). Spatial Analysis of Carbon Sequestration Mapping Using Remote Sensing and Satellite Image Processing. In *Multimodal Biometric and Machine Learning Technologies* (pp. 71–83). John Wiley & Sons, Ltd. doi:10.1002/9781119785491.ch4

Secretariat. (2019, May 17). *Global Assessment Report on Biodiversity and Ecosystem Services*. IPBES secretariat. https://www.ipbes.net/node/35274

Sengupta, A., Bhan, M., Bhatia, S., Joshi, A., Kuriakose, S., & Seshadri, K. S. (2024). Realizing "30 × 30" in India: The potential, the challenges, and the way forward. *Conservation Letters, 13004*(2), e13004. doi:10.1111/conl.13004

Singh, J. S. (2012). Biodiversity: An Overview. *Proceedings of the National Academy of Sciences. India. Section B, Biological Sciences, 82*(2), 239–250. doi:10.1007/s40011-012-0112-3

Singh, R. (n.d.). *Environmental policies in India*. Research Gate. https://www.researchgate.net/profile/Ravikant-Singh-2/publication/378590048_ENVIRONMENTAL_POLICIES_IN_INDIA/links/65e0ad05adf2362b635e43c3/ENVIRONMENTAL-POLICIES-IN-INDIA.pdf

Sokół, M., & Łaska, G. (2024). Biodiversity and ecosystem services: Complementary approach for catchment protection and land use in northeastern Poland. *Ecological Indicators, 159*, 111649. doi:10.1016/j.ecolind.2024.111649

Sunita, A., Daisy, R., & Singh, R. (n.d.). Sarus crane, biodiversity and pesticides: A review. *Fauna Journal*. https://www.faunajournal.com/archives/2024/vol11issue1/PartA/11-1-5-363.pdf

Vanneste, T., Depauw, L., De Lombaerde, E., Meeussen, C., Govaert, S., De Pauw, K., Sanczuk, P., Bollmann, K., Brunet, J., Calders, K., Cousins, S. A. O., Diekmann, M., Gasperini, C., Graae, B. J., Hedwall, P.-O., Iacopetti, G., Lenoir, J., Lindmo, S., Orczewska, A., & De Frenne, P. (2024). Trade-offs in biodiversity and ecosystem services between edges and interiors in European forests. *Nature Ecology & Evolution, 8*(5), 1–8. doi:10.1038/s41559-024-02335-6 PMID:38424266

Vasiliev, D., Hazlett, R., Hutchinson, K. L., & Bornmalm, L. (2024). Light at the end of the tunnel: Innovative opportunities for saving tropical biodiversity. *Ambio, 53*(5), 702–717. doi:10.1007/s13280-023-01970-w PMID:38353913

Velado-Alonso, E., Kleijn, D., & Bartomeus, I. (2024). Reassessing science communication for effective farmland biodiversity conservation. *Trends in Ecology & Evolution*. https://www.cell.com/trends/ecology-evolution/abstract/S0169-5347(24)00032-6

Xiao, Y., Zhou, G., Qiu, X., Liu, F., Chen, L., & Zhang, J. (2024). Biodiversity of network modules drives ecosystem functioning in biochar-amended paddy soil. *Frontiers in Microbiology*, *15*, 1341251. doi:10.3389/fmicb.2024.1341251 PMID:38328424

Yanagi, M. (2024). Understanding the ecological significance of wheat: Roles, climate change impacts, and biodiversity conservation strategies. *Geographical Research Bulletin*, *3*, 6–9.

Yang, H., Suntornvongsagul, K., Wang, J., Wang, Q., & Zhang, W. (2024). Indigenous knowledge in the sustainable management of a biodiversity-based millennium-old tea forest in the highlands of Yunnan Province, southwestern China. *Biodiversity (Nepean)*, *25*(2), 1–11. doi:10.1080/14888386.2023.2290176

Yasin, E. H., Siddig, A. A., & Kornel, C. (2024). *Forests at the Crossroads: Biodiversity Conservation in the Era of Climate Change*. InTech Open. https://www.intechopen.com/online-first/1180035

Chapter 6
Impact of Road Widening and New Highway Construction on Biodiversity and Its Solution

Rajiv Dutta
Shobhit University, Gangoh, India

Vinay Kumar
Shobhit University, Gangoh, India

ABSTRACT

Today, the global decline of biological diversity, or biodiversity, is one of the key environmental problems. Keeping in the view of day by day reduction of biodiversity, Convention on Biological Diversity advised a comprehensive analysis that how developmental activity specially, road widening and new highway affect biodiversity must be included in the environmental impact assessment (EIA) process: "Each Contracting Party shall introduce appropriate procedures requiring environmental impact assessment of its proposed projects that are likely to have significant adverse impact on biological diversity." In response to this suggestion, a number of governmental organizations have released guidelines on biodiversity and impact assessments and several non-governmental organizations, like The World Conservation Union and the International Association for Impact Assessment, are also working in this field. As a result, a new field of study called biodiversity impact assessment (BIA) was created.

DOI: 10.4018/979-8-3693-6950-0.ch006

BACKGROUND

Today globally decline of biological diversity, or biodiversity is one of the key environmental problems. Keeping in the view of day by day reduction of biodiversity, Convention on Biological Diversity advised a comprehensive analysis that how developmental activity specially, road widening and new highway affect biodiversity must be included in the Environmental Impact Assessment (EIA) process: "Each Contracting Party shall introduce appropriate procedures requiring environmental impact assessment of its proposed projects that are likely to have significant adverse impact on biological diversity." In response to this suggestion, a number of governmental organizations have released guidelines on biodiversity and impact assessments and several non-governmental organizations, like The World Conservation Union and the International Association for Impact Assessment, are also working in this field. As a result, a new field of study called Biodiversity Impact Assessment (BIA) was created. Its goal is to create and implement methods for analyzing the effects on biodiversity in Environmental Impact Assessments (EIAs). Road widening and new highway construction can have significant impacts on biodiversity, ecosystems, and wildlife habitats. Some of the key impacts include habitat fragmentation, loss of natural vegetation, disruption of wildlife migration routes, increased roadkill incidents, and changes in hydrology and water quality due to altered drainage patterns. These activities can also contribute to air and noise pollution, further affecting the surrounding environment and wildlife.

INTRODUCTION

Worldwide agreement has been reached that the primary threat to biodiversity protection at the moment is habitat reduction. When a natural ecosystem is transformed into an artificial system, as is the case with road building, the most severe decline in habitat occurs. One of the most common ways that the environment has changed in the last century, especially since World War II, is via the construction of new highways through the biodiversity rich area. The habitat conditions are damaged and modified by road constructions, and this has an impact on the biodiversity the quantity and distribution of plant and animal species in the affected regions. According to Byron et al. (2000) and Thompson et al. (1997), road designs have a significantly higher likelihood of affecting natural areas than other types of development because of their linear layout. The extent of the phenomena may be visualized by keeping in mind that roads and roadsides occupy between 1% and 2% of the land in many industrialized nations, making road networks a ubiquitous element of almost every landscape. Both a direct and indirect loss of habitat is brought about by roads. When a road and its verges reduce the overall size of an ecosystem, it means the natural land cover such as woods, grassland, wetland, etc. has been replaced with an artificial surface. This is referred to as a direct loss. The term "indirect loss" describes consequences like ecosystem fragmentation, which is the division of an ecosystem into smaller, more isolated areas, and ecosystem degradation, which is the biophysical modification of an ecosystem brought on by factors like artificial light, noise, air and water pollution, etc. These consequences minimize an ecosystem's capacity to maintain its initial biodiversity, which results in an indirect loss of habitat. In order to map the various ecosystem types and determine their significance for biodiversity conservation, basic criteria must be established. This will allow for a reliable assessment of the impact of ecosystem loss. With the current framework of international treaties and regulations for the promotion of sustainable development, biodiversity has emerged as one of the major environmental challenges. The primary challenge to the protection of biodiversity is now thought to be the global decrease of habitat. The decline in the amount and quality of natural habitat is mostly attributed to transportation facilities, particularly road networks. As a result, while planning and developing new roads, a thorough Biodiversity Impact Assessment (BIA) must be combined with other widely accepted factors. The fact that the preservation of endangered ecosystems is frequently cited as the primary goal of biodiversity conservation serves as evidence of its applicability. Consequently, by spatially estimating the predicted losses of each ecosystem type, the impact of a road project may be measured.

COMPLICATIONS

The present, unparalleled pace of biodiversity loss worldwide is mostly due to human activity, including habitat degradation, fragmentation, pollution, and climate change (Vitousek et al., 1997). Convention on Biological Diversity (CBD 2019) defined Biodiversity as "the variety of life on Earth, including plants, animals, and microorganisms, as well as the ecosystem of which they are part". In other word, biodiversity is described as biological diversity in a range of contexts, such as genetic differences with and within the species and ecological variety. The economic growth of any country is depends on the amount and quality of the road infrastructure. Numerous significant economic variables, including commerce, power, communication, and health, have been found to positively correlate with road infrastructure. For this reason, every government believes that investing in road infrastructure is crucial to enhancing the nation's potential. The effects of road building, however, cannot be disregarded. Building roads for human movement and accessibility invariably converts natural habitat into a sealed, highly disturbed environment, which always results in a net loss of habitat for wildlife. Physical invasion results in disturbance and barrier effects, which in turn fuel the loss, fragmentation, and deterioration of habitats overall. Because of the architecture of highways, humans are altering the natural structure of ecological communities, increasing the number of species that are vanishing. The alteration of terrestrial biological communities raises serious concerns about the loss of ecosystem diversity.

The majority of developing nations, including India, are working to build roads for accessibility and transportation. Governmental and non-governmental road development initiatives decrease the permeability of the terrain. Thus, the conservation of species and biodiversity is impacted both directly and indirectly by the basic changes in landscape structure. Here the term "direct loss" describes how the existence of highways and their verges reduces the overall size of an ecosystem, converting its natural land cover into man-made (artificial) surfaces. The term "indirect loss" describes consequences like ecosystem fragmentation, which is the division of an ecosystem into smaller, more isolated sections, and ecological deterioration brought on by artificial light, noise, air, and water pollution, among other things. These consequences result in both direct and indirect habitat loss, which minimizes the ecosystem's ability to maintain its initial biodiversity. It is clearly stated that, the majority of species are threatened by road construction. The use of bulldozers and other machinery to clear trees, level the land, and flatten the ground are also destroy and displace a great number of living things from their natural habitat, altering their entire way of life, and make it impossible for the various species to adapt to significant changes in their natural habitat (Zoker et al., 2022).

One of the most common ways that natural landscape modification occurs is through the building of roads (Gao & Liu, 2012; Noss & Cooperrider, 1994; Patarasuk & Binford, 2012; Van der Ree, Jaeger, van der Grift, & Clevenger, 2011). Various threats to biological diversity, such as habitat destruction and fragmentation, edge effect, invasion by exotic species, pollution, overhunting, and genetic barriers, have been shown by studies conducted in a range of terrestrial and aquatic ecosystems over the past few decades. The biological features of edges connected to traffic corridors are thus of greater importance (Reed et al., 1996; Holderegger & Di Giulio, 2010; Jackson & Fahrig, 2011; Lapaix, Harper, & Freedman, 2012).

Natural habitats are divided into "islands" by roads and utility corridors, which reduces the amount of interior habitats across the terrain (Andrews, 1990). Furthermore, living things near roadways are subjected to the environmental conditions of an alternative surrounding habitat. The edges may alter the biotic and abiotic conditions, such as species composition, temperature, moisture, light availability, and wind speed, on the organisms in a fragment due to what have been referred to as "edge effects" (Fahrig, 1997; Hill & Curran, 2001; Murcia, 1995, Delgado, Arroyo, Arévalo, & Fernández-Palacios, 2007; Flory & Clay, 2009; Parendes & Jones, 2000; Watkins et al., 2003).

Specifically, road margins are often dominated by foreign and disturbance-tolerant species, have a relatively high species richness, and have few regionally unique species (Forman & Alexander, 1998; Tyser & Worley, 1992). Additionally, highways have the potential to modify surface ecosystem processes and functions (such as production), and these changes are often closely correlated with the width of the roadways (Brothers & Spingarn, 1992; Flory & Clay, 2009).

Natural road margins are regarded as important repositories of biological diversity in landscapes where nearly all native vegetation has been removed for farming, as they may preserve a variety of native plant communities (Reed et al., 1996). Rather, the disruption associated with roadways is particularly noticeable in forest environments, presenting a unique combination of significant ecological consequences (Forman & Alexander, 1998). The persistence of a core forest habitat is actually hampered by road-induced discontinuities in the forest cover (McGarigal, Romme, Crist, & Roworth, 2001). This is because rare species depend on old-growth interior and environmental conditions, which are essential for the functioning of the entire ecosystem (Reed et al., 1996; Wei & Hoganson, 2005).

The management of forest resources served as the primary justification for road extension into forest ecosystems, but with the introduction of mechanisation in recent decades, this has drastically changed (Kivinen, Moen, Berg, & Eriksson, 2010). Forest roads are recognised to have an impact on soil erosion, the sedimentation of aquatic ecosystems, plants, and animals, despite often being narrow and unpaved (Donaldson & Bennett, 2004). Altered habitat, altered soil chemistry and water balance, increased fragmentation of formerly continuous habitats, dispersal of exotic species, and altered species behaviour are all direct ecological effects of road construction and maintenance on forest ecosystems (Benítez-López, Alkemade, & Verweij, 2010).

Roads can therefore alter the spatial patterns of forests by reducing the critical variability in natural processes and disturbances, interrupting horizontal natural processes, and fragmenting the original habitats (Hawbaker, Radeloff, Clayton, Hammer, & Gonzalez-Abraham, 2006). Roads also create linear gaps in the forest environment, resulting in the creation of a new ecosystem that differs from the inner forest in both positive and negative ways.

In rural systems, roads play a critical role in the economy and society. They connect rural and urban areas, manage forests, prevent forest fires, promote tourism, and provide country residents with access to forest products (Badia, Serra, & Modugno, 2011; Patarasuk & Binford, 2012). Nonetheless, they frequently have a significant impact on the processes and functions of ecosystems as well as the makeup of plant communities. However, the bulk of research has been done on how they interact with invasive and exotic species (Flory & Clay, 2009).

There are currently few studies in the Mediterranean region, but the majority of studies examining the impact of roads on plant diversity focus on tropical or temperate regions of North America and Europe (Patarsuk &Binford, 2012).The Mediterranean basin is a hot spot for biodiversity that is especially vulnerable to changes in land use and cover, high levels of human activity and pressure, total and extreme precipitation, and seasonal temperature extremes. One of the biggest threats to the biodiversity of Mediterranean forests is habitat fragmentation, which is mostly encouraged by road openings, usage, and maintenance (Jordán-López, Martínez-Zavala, & Bellinfante, 2009; Riitters, Coulston, & Wickham, 2012). Thus, it is essential to evaluate how highways affect the variety of complex forest systems and disturb the biodiversity. Thus this study focuses the impact of highways on plant variety in Mediterranean woodlands. The main theory that has to be investigated is that highways reduce the integrity of the forest ecosystem and, consequently, the variety of plants and finally impacted on biodiversity in all these ways.

SOLUTIONS

In addition, in India, mitigation measures specific to the site are being implemented in collaboration with the Forest Authorities. These measures include the construction of culverts, underpasses, overpasses (ecoducts), viaducts, tunnels, guard walls, fencing, vegetative barriers, anti-light glare, sound barriers, and other options based on the specific requirements of the site. Additionally, funds are given to the relevant Forest Authorities so they can implement the measures outlined in their approved WildLife Management Plan, such as building waterholes, planting and landscaping specifically for the site, animal conservation units, rescue operations, anti-poaching units, watch towers, monitoring, raising awareness, involving the local community, building post guards, lighting and fencing around the perimeter of Protected Areas (PAs) or their Eco Sensitive Zones (ESZ). In collaboration with the Forest Authorities, cautionary notice boards and rumble strips are also erected to warn the public and other drivers as well as to save wildlife. Said the Honourable Union Minister of the Government of India for Road Transport and Highways.

One of the finest strategies used by the Government of India's NHAI is the construction of eco-bridges and elevated corridors.

Here are some potential solutions to mitigate these impacts:

1. Environmental Impact Assessment (EIA): Conduct comprehensive EIAs before any road widening or highway construction project begins. This involves identifying potential environmental impacts and proposing mitigation measures.

2. Habitat Restoration and Conservation: Implement habitat restoration projects to compensate for habitat loss. This may include reforestation, wetland restoration, and creation of wildlife corridors to facilitate the movement of animals across fragmented landscapes.
3. Wildlife Crossings: Design and construct wildlife crossings such as overpasses, underpasses, and culverts to allow animals to safely cross roads without risking collisions with vehicles. These structures should be strategically located based on wildlife movement patterns.
4. Speed Limit Reductions and Traffic Calming Measures: Implement speed limit reductions and traffic calming measures in areas where roads intersect with important wildlife habitats or migration corridors. This can reduce the likelihood of wildlife-vehicle collisions.
5. Fencing and Barrier Installation: Install wildlife-proof fencing and barriers along roads to prevent animals from accessing high-traffic areas and reduce roadkill incidents.
6. Land-Use Planning and Zoning: Incorporate biodiversity conservation considerations into land-use planning and zoning decisions. Designate protected areas and green corridors to maintain connectivity between natural habitats.
7. Public Awareness and Education: Raise public awareness about the importance of biodiversity conservation and the impacts of road infrastructure on wildlife. Educate drivers about the need to watch out for wildlife on roads and to drive cautiously in areas with known wildlife crossings.
8. Green Infrastructure: Integrate green infrastructure elements such as vegetated road shoulders, bioswales, and roadside vegetation buffers into road design to help mitigate the effects of habitat loss and fragmentation.
9. Collaboration and Stakeholder Engagement: Foster collaboration among government agencies, transportation authorities, conservation organizations, and local communities to develop and implement effective biodiversity conservation measures.
10. Monitoring and Adaptive Management: Continuously monitor the effectiveness of mitigation measures and adjust them as needed based on monitoring data and feedback from stakeholders and experts.

By implementing a combination of these solutions and adopting a holistic approach to infrastructure development, it is possible to minimize the negative impacts of road widening and highway construction on biodiversity while still meeting transportation needs.

REFERENCES

Andrews, A. (1990). Fragmentation of habitat by roads and utility corridors: A review. *Australian Zoologist*, *26*(3-4), 130–141. doi:10.7882/AZ.1990.005

Badia, A., Serra, P., & Modugno, S. (2011). Identifying dynamics of fire ignition probabilities in two representative Mediterranean wildland-urban interface areas. *Applied Geography (Sevenoaks, England)*, *31*(3), 930–940. doi:10.1016/j.apgeog.2011.01.016

Benítez-López, A., Alkemade, R., & Verweij, P. A. (2010). The impacts of roads and other infrastructure on mammal and bird populations: A meta-analysis. *Biological Conservation*, *143*(6), 1307–1316. doi:10.1016/j.biocon.2010.02.009

Brothers, T. S., & Spingarn, A. (1992). Forest fragmentation and alien plant invasion of central Indiana old-growth forests. *Conservation Biology*, *6*(1), 91–100. doi:10.1046/j.1523-1739.1992.610091.x

Byron, H. J., Treweek, J. R., Sheate, W. R., & Thompson, S. (2000). Road developments in the UK: An analysis of ecological assessment in environmental impact statements produced between 1993 and 1997. *Journal of Environmental Planning and Management*, *43*(1), 71–97. doi:10.1080/09640560010775

Delgado, J. D., Arroyo, N. L., Arévalo, J. R., & Fernández-Palacios, J. M. (2007). Edge effects of roads on temperature, light, canopy cover, and canopy height in laurel and pine forests (Tenerife, Canary Islands). *Landscape and Urban Planning*, *81*(4), 328–340. doi:10.1016/j.landurbplan.2007.01.005

Donaldson, A., & Bennett, A. (2004). *Ecological effects of roads: Implications for the internal fragmentation of Australian parks and reserves.*

Fahrig, L. (1997). Relative effects of habitat loss and fragmentation on population extinction. *The Journal of Wildlife Management*, *61*(3), 603–610. doi:10.2307/3802168

Flory, S. L., & Clay, K. (2009). Invasive plant removal method determines native plant community responses. *Journal of Applied Ecology*, *46*(2), 434–442. doi:10.1111/j.1365-2664.2009.01610.x

Flory, S. L., & Clay, K. (2009). Effects of roads and forest successional age on experimental plant invasions. *Biological Conservation*, *142*(11), 2531–2537. doi:10.1016/j.biocon.2009.05.024

Forman, R. T., & Alexander, L. E. (1998). Roads and their major ecological effects. *Annual Review of Ecology and Systematics*, *29*(1), 207–231. doi:10.1146/annurev.ecolsys.29.1.207

Gao, W., Liu, S., & Huang, L. (2012). A global best artificial bee colony algorithm for global optimization. *Journal of Computational and Applied Mathematics*, *236*(11), 2741–2753. doi:10.1016/j.cam.2012.01.013

Hawbaker, T. J., Radeloff, V. C., Clayton, M. K., Hammer, R. B., & Gonzalez-Abraham, C. E. (2006). Road development, housing growth, and landscape fragmentation in northern Wisconsin: 1937–1999. *Ecological Applications*, *16*(3), 1222–1237. doi:10.1890/1051-0761(2006)016[1222:RDHGAL]2.0.CO;2 PMID:16827014

Hill, J. L., & Curran, P. J. (2001). Species composition in fragmented forests: Conservation implications of changing forest area. *Applied Geography (Sevenoaks, England)*, *21*(2), 157–174. doi:10.1016/S0143-6228(01)00002-9

Holderegger, R., & Di Giulio, M. (2010). The genetic effects of roads: A review of empirical evidence. *Basic and Applied Ecology*, *11*(6), 522–531. doi:10.1016/j.baae.2010.06.006

Jackson, N. D., & Fahrig, L. (2011). Relative effects of road mortality and decreased connectivity on population genetic diversity. *Biological Conservation*, *144*(12), 3143–3148. doi:10.1016/j.biocon.2011.09.010

Jordán-López, A., Martínez-Zavala, L., & Bellinfante, N. (2009). Impact of different parts of unpaved forest roads on runoff and sediment yield in a Mediterranean area. *The Science of the Total Environment*, *407*(2), 937–944. doi:10.1016/j.scitotenv.2008.09.047 PMID:18992920

Kivinen, S., Moen, J., Berg, A., & Eriksson, Å. (2010). Effects of modern forest management on winter grazing resources for reindeer in Sweden. *Ambio*, *39*(4), 269–278. doi:10.1007/s13280-010-0044-1 PMID:20799676

LaPaix, R., Harper, K., & Freedman, B. (2012). Patterns of exotic plants in relation to anthropogenic edges within urban forest remnants. *Applied Vegetation Science*, *15*(4), 525–535. doi:10.1111/j.1654-109X.2012.01195.x

Marcantonio, M., Rocchini, D., Geri, F., Bacaro, G., & Amici, V. (2013). Biodiversity, roads, & landscape fragmentation: Two Mediterranean cases. *Applied Geography (Sevenoaks, England)*, *42*, 63–72. doi:10.1016/j.apgeog.2013.05.001

McGarigal, K., Romme, W. H., Crist, M., & Roworth, E. (2001). Cumulative effects of roads and logging on landscape structure in the San Juan Mountains, Colorado (USA). *Landscape Ecology*, *16*(4), 327–349. doi:10.1023/A:1011185409347

Murcia, C. (1995). Edge effects in fragmented forests: Implications for conservation. *Trends in Ecology & Evolution*, *10*(2), 58–62. doi:10.1016/S0169-5347(00)88977-6 PMID:21236953

Noss, R. F., & Cooperrider, A. Y. (1994). *Saving nature's legacy: protecting and restoring biodiversity*. Island Press.

Parendes, L. A., & Jones, J. A. (2000). Role of light availability and dispersal in exotic plant invasion along roads and streams in the HJ Andrews Experimental Forest, Oregon. *Conservation Biology*, *14*(1), 64–75. doi:10.1046/j.1523-1739.2000.99089.x

Patarasuk, R., & Binford, M. W. (2012). Longitudinal analysis of the road network development and land-cover change in Lop Buri province, Thailand, 1989–2006. *Applied Geography (Sevenoaks, England)*, *32*(2), 228–239. doi:10.1016/j.apgeog.2011.05.009

Patarasuk, R., & Binford, M. W. (2012). Longitudinal analysis of the road network development and land-cover change in Lop Buri province, Thailand, 1989–2006. *Applied Geography (Sevenoaks, England)*, *32*(2), 228–239. doi:10.1016/j.apgeog.2011.05.009

Reed, E. S. (1996). *Encountering the world: Toward an ecological psychology*. Oxford University Press.

Riitters, K. H., Coulston, J. W., & Wickham, J. D. (2012). Fragmentation of forest communities in the eastern United States. *Forest Ecology and Management*, *263*, 85–93. doi:10.1016/j.foreco.2011.09.022

Thompson, K., Bakker, J. P., & Bekker, R. M. (1997). *The soil seed banks of North West Europe: methodology, density and longevity*. Cambridge university press.

Tyser, R. W., & Worley, C. A. (1992). Alien flora in grasslands adjacent to road and trail corridors in Glacier National Park, Montana (USA). *Conservation Biology*, *6*(2), 253–262. doi:10.1046/j.1523-1739.1992.620253.x

Van Der Ree, R., Jaeger, J. A., van der Grift, E. A., & Clevenger, A. P. (2011). Effects of roads and traffic on wildlife populations and landscape function: Road ecology is moving toward larger scales. *Ecology and Society*, *16*(1), art48. doi:10.5751/ES-03982-160148

Vitousek, P. M., Mooney, H. A., Lubchenco, J., & Melillo, J. M. (1997). Human domination of Earth's ecosystems. *Science*, *277*(5325), 494–499. doi:10.1126/science.277.5325.494

Watkins, P. C., Woodward, K., Stone, T., & Kolts, R. L. (2003). Gratitude and happiness: Development of a measure of gratitude, and relationships with subjective well-being. *Social Behavior and Personality*, *31*(5), 431–451. doi:10.2224/sbp.2003.31.5.431

Wei, Y., & Hoganson, H. M. (2005). Landscape impacts from valuing core area in national forest planning. *Forest Ecology and Management*, *218*(1-3), 89–106. doi:10.1016/j.foreco.2005.07.002

Zoker, M. E., Karim, S., Kargbo, B., Kemoh, R., & Yorpoi, L. D. (2022). Impacts of road constructions on ecological biodiversity and livelihood in Sierra Leone. *Resaerch Gate*, *3*, 280–296. doi:10.54660/anfo.2022.3.3.18

Chapter 7
Biodiversity Loss in Streams Due to Human Diversion Activities

Bharti Sadhiyan
https://orcid.org/0009-0001-8983-708X
Gurukul Kangri University, Haridwar, India

Monika Kumari
Kanya Gurukul Campus, Haridwar, India

Sangeeta Madan
Kanya Gurukul Campus, Haridwar, India

Ritika Maurya
Kanya Gurukul Campus, Haridwar, India

ABSTRACT

Stream biodiversity loss is a major environmental issue that directly affects the delicate ecological balance of these aquatic habitats. The stability and health of the surrounding environment are supported by the wide variety of plants, fish, and invertebrates that may be found in streams. These habitats are becoming less healthy largely because of human activity, which may lead to a decline in biodiversity and the extinction of many species. Understanding the origins, impacts, and challenges involved, as well as increasing awareness of the issue, are crucial goals for the preservation of stream ecosystems.

DOI: 10.4018/979-8-3693-6950-0.ch007

INTRODUCTION TO BIODIVERSITY LOSS IN STREAMS

The term "**biodiversity**" was first used in 1988 by American entomologist ***Edward O. Wilson,*** and it was later adopted by ISPRA to refer to **the variety of life forms on Earth,** including all species of fungi, animals, plants and microorganisms, as well as the ecological links that exist between them. An ecosystem's size may be estimated in a few different ways, such as by calculating the genetic variety within each species' population or by looking at the distribution of the species across the ecosystem's various settings. Biodiversity can be defined as *the richness of life on Earth.* The concept of "***biodiversity loss***" describes the decrease in the variety and richness of aquatic life, as well as the health of the ecosystem and species diversity. Because they provide as habitat for a diverse range of plants, animals and microbes, water bodies—such as lakes, rivers, seas, oceans and wetlands-are essential parts of the Earth's ecosystems. Water bodies, such as streams, serve as reservoirs for biodiversity. They are home to a remarkable variety of organisms, ranging from famous aquatic species to microbial communities, all of which have a substantial impact on the general resilience and health of our planet.

A stream is a body of water on Earth's surface. Although larger streams are often referred to as rivers, the words "stream" and "river" are sometimes used synonymously.

A stream is a brief, usually uninterrupted stream of water. When describing the water flow, the terms "river," "stream" and "creek" are used depending on the size of the river basin. There would be three types of waterways: rivers, the biggest, streams in the centre and the smallest. Streams cannot exist without both water and gravity. Much of the precipitation runs off the surface and collects in streams, but some of it seeps into groundwater. A watershed, commonly referred to as a drainage basin, is the area that holds water for a stream. Smaller streams are referred as tributaries which often combine to for bigger ones as they run downward. The type of stream varies with the terrain it flows over. Streams carry rock and other material downstream, eroding the rock and creating channels this process is called erosion.

Streams provide an essential habitat for a diverse range of flora and fauna, such as fish, birds, salamanders, snails, animals, and more. Different kinds of plants and animals have evolved to live in streams. Common characteristics of plants include long, flexible branches that may flow with the stream and strong root systems that serve as anchors. Fly larvae feed on leaves when they fall into the water, and the eaten leaves end up as fish food. Mussels burrow into the sediment on the bottom of streams to stay there as they filter the water for food. The streamlined bodies of many fish help to stabilise them and prevent strong currents from sweeping them away. Streams are beneficial to people in numerous ways. Streams can produce electricity via hydropower, eliminate waste, irrigate farmland, and provide drinking water. Many people utilize streams for swimming, boating, and fishing as leisure activities. Moreover, streams provide vital habitat for species. Quality of stream water is significantly compromised by various factors. With over 140 000 species (plants, animals, invertebrates, and vertebrates; roughly 12% of all described species) crammed into just * 2% of the planet's surface area and a minute * 0.007% of its total water supply, freshwater ecosystems are also among the most diverse per unit habitat volume on Earth (Reid *et al.*, 2019).

Many human activities, such as habitat modification, water pollution, overfishing, introduction of exotic species, river diversions, fragmentation and flow regulation, growth of agricultural and urban landscapes, climate change, rising sea levels, and altered precipitation regimes, are putting freshwater species and ecosystems in danger (Dudgeon 2019; Grill *et al.* 2019; IPBES 2019).

This prehistoric human custom creates manmade canals to link two river basins. As beneficial as this practice is for the local populace, there are both short- and long-term risks to the environment, the economy, and society (Moreira Filho and Buckup 2005; Zhang *et al.*, 2018). Water diversions can significantly reduce the amount of water in the donor basin and alter the dynamics of the river flow, in addition to causing habitat destruction, fishery collapse, parasite spread, disease transmission, loss of genetic variability between populations, biological invasions, extinction of species, and water pollution (Merciai *et al.*, 2017; Qin *et al.*, 2018).

The ***diversion of streams*** and its impact on biodiversity loss in water bodies is a concerning issue that requires attention. Biodiversity loss is the term used to describe the decline in the number and variety of species found in a particular ecosystem. This occurrence, known as stream diversion, happens when streams are changed or rerouted, significantly altering aquatic ecosystems and habitats. The decline in biodiversity linked to stream diversion has several reasons. The disturbance of natural flow patterns is one of the main causes. Changes in a stream's path may affect sediment movement, flow velocity, and water temperature. Many aquatic species may be directly impacted by these changes in terms of their ability to survive and procreate.

The fragmentation of habitat is another common outcome of stream diversion. In isolated pockets that restrict communication between various portions of a network of rivers or streams, the interruption of constant water flow may occur. Movement, gene flow, and availability of vital resources like food and breeding grounds are all impeded by this fragmentation of the species. In addition, increased sedimentation or pollutant inputs from nearby land use practices might cause changes in the chemistry of the water as a consequence of stream diversions. By detrimentally influencing the survival and health of delicate aquatic creatures, these changes in water quality further aggravate the loss of biodiversity.

EXPLANATION OF HUMAN DIVERSION ACTIVITIES AND THEIR IMPACT ON STREAMS

Water Diversion

Water diversion often results in the introduction of new stresses that have an impact on water body ecosystems concurrently Sabater et al. (2018). During project construction or diversion, clean surface water runoff is collected upstream of a project site, redirected around the work area, and released downstream to reduce the degree of water quality degradation. This technique, known as water diversion, is accomplished by a network of structures and procedures. A water diversion is often built when the required permits are secured and work has to be done in a live stream or other body of water. Additional stressors are often added when water is redirected, impacting the ecosystems of water bodies concurrently (Sabater et al., 2018).

Construction of dams, irrigation and water extraction for urban use are examples of human diversion activities that significantly affect the natural flow and volume of streams. The total level of biodiversity is decreased since there is less water available for plants and animals as a result of the disruption of many species' habitats.

For stream habitats to remain ecologically intact, water flow is a crucial need. Due to water diversion, man-induced drought is endangering the populations of numerous freshwater fish in the Mediterranean area. These constructions' water diversion causes significant habitat constriction or even the complete loss of surface water flow (von Schiller *et al.,* 2017). This has an impact on community composition as well since, in freshwater ecosystems, ecosystem size has a significant role in determining the structure of the food web (McHugh *et al.,* 2015; McIntosh *et al.,* 2018).

The complexity of the food chain will change depending on where water is diverted. The quantity of coarse debris and its contribution to the diets of primary consumers are expected to decrease in diverted reaches. Feeding more on the uncommon biofilm leads to increased trophic diversity and reduced trophic redundancy. However, a reduced storage of baseline resources (FCL) will lead to a shorter food chain, (de Guzman et al., 2021).

Impact on Stream Ecosystems and Stream Flow

Water flow is a crucial element required to preserve the biological integrity of stream habitats. These actions disrupt the normal flow of streams, harming the habitats of several aquatic species and upsetting the balance of the environment. A significant effect of changed stream flow on stream ecosystems is the availability and quality of habitat. The substrate composition, channel architecture, and water depth of stream habitats may all be directly impacted by changes in flow regimes. For instance, by reducing water depth and changing substrate stability, decreased flow brought on by drought conditions or water diversion may result in the loss of habitat quality. On the other hand, higher flow brought on by storms or dam releases may damage and erode habitat. In addition, changes in stream flow patterns have the potential to upset aquatic creatures' life-history and reproductive plans. For spawning, migrating, and growing, many fish, invertebrate and amphibian species need certain flow conditions. Reductions in population variety and abundance may result from disturbances to these crucial life cycle events caused by variations in flow size and timing (Richter *et al.*, 2003). Reduced flow, for instance, might impede fish movement and successful reproduction during the spawning season (Arthington *et al.*, 2010). Stream flow may affect energy dynamics and nutrient cycling in stream ecosystems. The movement and deposit of sediment, organic matter and nutrients in streams may be affected by changes in flow, and this can have an effect on the rates of primary production and decomposition (Webster, 2014). Changes in flow may also have an impact on the quantity of dissolved oxygen, which is essential for the survival of aquatic life and the well-functioning of aerobic microbial communities.

HUMAN DIVERSION ACTIVITIES AND THEIR EFFECTS ON STREAM BIODIVERSITY

Urban Development

Urbanization usually results in a stream receiving more water after a storm and the water entering the stream from changed land surfaces in less time. When water is swiftly drained from parking lots and roadways in an attempt to lessen floods, more water may rapidly enter a stream, changing its course and making it flashier and more pronounced. Rapid runoff also reduces the quantity of water that may replenish aquifers and enter the soil, which often leads to lower sustained stream flows, particularly in the summer. Increasing stream flows may erode stream banks and streambeds, which damages fish breeding and feeding grounds and reduces invertebrate dwelling space.

Agricultural Irrigation

Agriculture is the main industry in developing countries. 68% of the annual decrease in stream flow may be attributed to direct human activity; the impact is amplified during the irrigation season due to water withdrawal for irrigation. Situated in northern India, the State of Assam is the most populous state in the northeast and ranks second in terms of land area. Extensive irrigation techniques may harm stream ecosystems and the species they sustain by increasing sedimentation, depleting water supplies, and contaminating water supplies with chemical.

Dam Construction

The natural flow of streams may be hampered by the building of dams. Big dams disrupt the flow of nutrients across ecosystems, flood land, split up habitat, isolate species, and obstruct migratory paths, all of which contribute to biological diversity loss. They modify the characteristics of a river's estuary, where many fish species throughout the globe breed, and decrease water and sediment flows to habitat downstream. The effects of dams make whole ecosystems more susceptible to external threats like climate change. According to the International Commission on major Dams (ICOLD 2019), there were 57,985 major dams in the globe as of 2019, with China accounting for 23,841, the United States for 9263, India for 4408 and Japan for 3130, making up around 53% of the total. Large dams are known to disrupt fish populations, release greenhouse gases, cause deforestation and the fragmentation of clean forests, destroy communities, interfere with rivers' natural flow, change the flow of sediments and nutrients, alter the physical-chemical parameters of the water, jeopardise the ecological integrity of aquatic systems, and have other detrimental effects (Arias *et al.* 2020; Benchimol and Peres 2015; Deemer *et al.* 2016; Kirchherr and Charles 2016; Aung *et al.* 2020; Xu *et al.* 2020).

The belief that rivers that empty into the sea are a waste of valuable freshwater resources has historically guided river development in India (Krishnaswamy et al., 2017). Shorter west-flowing rivers in the Western Ghats are coming under growing pressure from water diversion and hydroelectric projects intended to supply the needs of the arid eastern (rain shadow) areas.

The Narmada River is the biggest river in India to flow westward, covering an area of 97,410 square kilometres. The Narmada Valley Development Authority (NVDA) was founded in 1979 after the Narmada Water Tribunal's resolution of disputes about the allocation of water among the riparian states. A master plan for the building of 30 large, 135 medium, and 3000 small dams for irrigation and energy production on the Narmada and its tributaries was later created by the NVDA. One million people were expected to be displaced by these dams, according to WII (1994). As of 2019, 12 major and 21 minor dams have been constructed.

ECOLOGICAL CONSEQUENCES OF BIODIVERSITY LOSS IN STREAMS

Loss of Keystone Species

The decline of key species in stream ecosystems can have cascading effects, disrupting food webs and vital ecological processes, leading to imbalances and potential collapses within the ecosystem. Recently, there have been a few global projects addressing this problem. The Red data book propsed a category named "Collapse" (CO), which is equivalent to the IUCN's extinct (EX) category species. There is a 27% risk of extinction for each of the 29,500 freshwater-dependent species evaluated for the IUCN Red List. As a result, the IUCN 2019 has classified as endangered: about 62% of turtle species, 47% of gastropods, 42% of mammals, 33% of amphibians, 30% of decapod crustaceans (crabs, crayfish, and prawns), 28% of fish, and 20% of birds. Apart from having a distinct freshwater ecosystem with a high degree of endemicity for fish and amphibians specifically, the Western Ghats mountain range in India is acknowledged as a global hotspot for biodiversity. Approximately thirty west-flowing rivers and many east-flowing rivers originate and are nourished by these heavily wooded highlands. Over 440 SHPs are under different phases of commissioning or implementation in the Ghats area of Karnataka State alone (Ministry of New and Renewable Energy (MNRE), 2015).

Major significant species under the *schizothoracine* (*Schizothorax* and *Schizothoraichthys* spp.) and cyprinid (*Tor* spp. and *Labeo* spp.) taxa are exploited in the Himalayan nations' subsistence and commercial fisheries. More precise and current data on the variety and biological habitat of fish species in a few Indian Himalayan streams are needed to manage the environmental and socioeconomic needs (Sehgal, 1999; Singh and Akhtar, 2015; Singh and Sarma, 2017; Zaidi et al., 2018). The Central, Eastern, and Western Himalayan native and endemic fish species will be given priority for restoration and particular care when a database is created about them (Singh & Agrawal, 2017). Policy makers should also be required to get assistance from strategic environmental assessments and actions aimed at restoring environmental flow.

Altered Nutrient Cycling

Eutrophication, hypoxia, fish deaths, and a rise in the frequency and severity of toxic algal blooms are just a few of the detrimental effects that higher nutrient concentrations in water bodies have on the ecosystem and human health. Point sources—like wastewater treatment facilities and nonpoint sources—like the use of fertilisers in agriculture—can provide nutrients to the stream ecosystem. Planktonic, benthic and filamentous algae may develop excessively when there are high nitrogen levels in stream water. Other detrimental effects include a decrease in species diversity, frequent low dissolved oxygen events, and notable pH fluctuations. But it's crucial to understand that a water body cannot become eutrophic due to excessive nutrient concentrations alone.

Erosion and Sedimentation

Sedimentation is another significant issue. Sediments would typically flow into the ocean downstream. These sediments settle in reservoirs when a dam restricts their flow, which lessens the amount of space available for water and affects the plants, animals, and human settlements downstream that rely on the nutrients the sediments provide. When the regular flow of material from upstream is disrupted, rivers will attempt to rejuvenate themselves by eroding their banks and riverbeds. As such, a decline in the water table has an adverse effect on ecosystems. Furthermore, the stagnant reservoir may alter the river's temperature, dissolved oxygen content, and chemical composition, all of which might have catastrophic consequences for the nearby plant and animal life.

Variations in the supply and flow patterns of sediment may affect certain plant and animal species. Increases in sediment content may cause turbid streams with a smaller euphotic zone. Fish and bird species suffer as a result of this decreased plant output (Morris *et al.,* 2008). Additionally, it abrades the fish's gills, increasing the chance of disease or death.

After being pushed downslope, the "lost" material finally finds its way into stream channels. After that, the sediment may be taken to the ocean, placed in floodplains, along the canal's banks, or, as the water in the channel flows further, incorporated into the channel bed itself.

OVERVIEW OF CURRENT CONSERVATION EFFECTS AND INITIATIVES TO MITIGATES BIODIVERSITY LOSS

Globally, efforts to enhance freshwater ecosystems and establish environmental flows have made great progress thanks to a range of strategies and initiatives. Planning river basins, creating water distribution schedules, and integrating environmental flows into infrastructure management and design have all proven to be significant approaches. Improving farming practices, regulating polluting companies, and purifying water have all contributed to the preservation of aquatic life and water quality. Two examples of nature-based solutions that have been essential in recovering significant ecosystems are the creation of protected areas and habitat restoration.

Table 1. The emergency recovery plan for freshwater biodiversity includes a range of key initiatives aimed at addressing the critical decline in freshwater biodiversity

Implement environmental flows more quickly.	
River basin planning	Environmental flows are regulated by legally required watershed management agencies, such as those on the Crocodile River, and are integrated into South African water policy.
Water allocation.	189 rivers around the nation have sustainable water allocation restrictions defined by Mexico's Water Reserves programme, which takes environmental flows into consideration.
Design and operation of infrastructure	Environmental flows are now part of the Chinese Three Gorges Dam's operational regime, supporting downstream fisheries.
Enrich the water to keep aquatic life alive.	
Treatment of water	Widespread sewage contamination has decreased as a result of the EU Urban Waste Water Treatment Directive.
Regulation of polluting industries.	A major initiative to clean up the Singapore River and restore aquatic life was started in Singapore in the 1970s. It included relocating and removing pollutants from industry, pig and duck farms, and residential and commercial development along the shoreline.
Improved agricultural practices	Pesticides and fertilisers are now less likely to end up in watercourses thanks to the improved Cotton Initiative and Bonsucro, two market-based initiatives that encourage improved management practices on cotton and sugarcane fields in Pakistan and India.
Nature-based solutions	Improvements in lake water quality have led to an increase in fisheries and floodplain biodiversity in China as a consequence of restoration work for floodplain lakes along the middle Yangtze River.
Preserve and replenish vital environments	
Protected areas	The whole 825,000-acre Bita River basin, which is a subbasin of the Orinoco, has been recognised as a Ramsar site by the Colombian government. This activity is a noteworthy illustration of how to designate and administer protected areas in an efficient manner. One of the few free-flowing rivers in the world with complete protection for the whole basin is the Bita River basin..
Planning for land use or ecosystem service markets.	In order to preserve and restore natural processes in the Delaware and Castkill watersheds and to provide an economically secure metropolitan water supply, the New York City Watershed Agreement has sparked better land use planning and management.
Habitat restoration	A global agreement signed by ministers from Bulgaria, Romania, Moldova, and Ukraine has resulted in the restoration of some sixty thousand hectares of floodplain wetlands along the lower Danube River.
Control the use of freshwater resources and riverine aggregations.	
Science-driven management of fisheries	Since the 1990s, Malawi has institutionalised legally an ecosystem-based approach to fisheries management. The measures for implementation include comanagement with fishing communities, sustainable harvesting of high-value chambo *(Oreochromis lidole)* and nursery sanctuaries or breeding grounds for commercial species.
Community fisheries management	Thirty times more Arapaima, or Arapaima gigas, were found in the Western Brazilian Amazon as a consequence of a community protection and resource management programme implemented in oxbow lakes along the Juruá River.
By catch reduction.	The commercial eel and carp fisheries in New South Wales, Australia, have shown that platypus bycatch may be reduced by using a mix of closures and modified traps.
Reducing aggregates demand.	In India, harmless municipal trash is used as a replacement for aggregates in road construction, whereas Germany recycles 87% of its waste aggregates.
Improved regulation of riverine aggregate extraction.	The aggregates levy, a tax imposed on the sale of primary aggregates in the UK (sand, gravel, and crushed rock), has been implemented in conjunction with an effective regulatory framework to determine whether riverine aggregate extraction is acceptable or not. The levy has funded research aimed at improving extraction methods and developing a better understanding of how to minimise environmental effects.
Prevent and manage the influx of non-native organisms into freshwater environments.	
Identification and control of introduction pathways	It has been successful for nonnative carp species to invade the Great Lakes (the United States and Canada). This has been achieved by scientific risk assessments, live fish transit limitations, and the installation of an electrical barrier.

Implement environmental flows more quickly.	
Preserve and bring back freshwater connection.	
System-scale infrastructure planning.	The major stems of the Salween and Irrawaddy rivers should remain open for flow, according to a strategic environmental study that was conducted for hydropower planning in Myanmar.
Dam reoperation and removal	Reopening and removing dams that affected 1500 km of the Penobscot River in the United States led to an increase in migratory fish populations while preserving the river's ability to generate energy.
Levee repositioning	The Netherlands' Room for the Rivers initiative has promoted widespread levee reduction and the restoration of lateral connectivity along the Rhine. This has increased and enhanced freshwater biodiversity's habitat while also improving flood storage and transport.

Science-driven fisheries management, community-based conservation programmes, and methods to reduce bycatch and aggregate demand all contribute to the cause of freshwater conservation. Supporting initiatives to discourage the introduction of non-native species and promote system-scale infrastructure design has also been crucial. Together, we can demonstrate our commitment to protecting freshwater ecosystems and ensuring their sustainability for future generations (Tickner *et al.*, 2020)

CHALLENGES AND BARRIERS TO ADDRESSING BIODIVERSITY LOSS IN STREAMS

Pollution and Contamination

Pollution is the most detrimental effect of human activity. Anthropogenic pollutants—such as those from homes, cities, and industries—have a significant impact on the quality of water. These pollutants find their way into bodies of water via human activity. Washing, cooking, and taking a bath are examples of household activities that alter the quality of water by adding chemical residue (Kalal *et al.,* 2021). According to Doldec *et al.*, (2021), pollution is a prevalent stressor in freshwater ecosystems that may worsen the ecosystem's overall health and water quality when combined with other stressors (Lemm *et al.,* 2021). The nature of the pollutants, the target species, and the degree of dilution in the receiving water bodies are among the factors that may affect how chemical pollution affects ecosystems (Flores *et al.*, 2014; Artigas *et al.*, 2014). (Carey & Migliaccio, 2009). Moreover, certain substances are harmful mainly because of their unique effects on the biota, according to Patel *et al.,* 2020; Vasilachi *et al.,* 2021. Nutrients, for example, display dose-dependent rises and falls in biological activity. Given that algae provide higher-quality food than detritus, their increased abundance at moderate nutrient levels makes sense (Keck & Lepori, 2012).

OPPORTUNITIES FOR INDIVIDUALS AND COMMUNITIES TO CONTRIBUTE TO STREAM CONSERVATION

Education and Awareness

The main factor for biodiversity loss is a lack of public understanding about biodiversity protection. Stakeholders in the conservation effort should emphasise this, especially in certain significant conservation areas (such as the UNESCO World Heritage sites). The kind of respondents (e.g., members of the local community or tourists), their educational attainment, and the number of their dependents all have an impact on the degree of knowledge about biodiversity protection. Regression analysis, on the other hand, makes it evident which sociodemographic variables affect the awareness variable, as determined by the knowledge, attitude, and experience indicators. This research supports the association between community well-being and knowledge of biodiversity protection. Raising well-being via its metrics—the social and economic pillars—has a significant effect on raising awareness of conservation.

Volunteer Programs

Global changes in freshwater biodiversity are monitored by the volunteer community of practice known as the Freshwater Biodiversity Observation Network (FWBON). It was founded in 2016 as a GEOBON-connected network for observing biodiversity (https://geobon.org/). By assisting with the following: (1) enhancing the harmonised data collection; (2) developing data management and transmission standards and methodologies; (3) sharing information on biodiversity internationally without compromising national interests; (4) Integrating physical and chemical data with biodiversity data, (5) producing goods that are beneficial to the prudent management of rivers, lakes, wetlands, and their catchments, as well as subsurface aquatic ecosystems, and (6) Combine the goals of land and coastal conservation with freshwater research and practice.

Advocacy and Policy Support

In order to promote long-term beneficial benefits on biodiversity, it is essential that policies and regulations that prioritise stream protection and sustainable water management be supported. The CBD suggested using enhanced impact assessment methods for biodiversity protection, acknowledging the need of including biodiversity issues into development decision-making. In order to guarantee that biodiversity is taken into account in projects, plans, and policy choices, Article 14 of the CBD expressly mandates impact assessment procedures (CBD 2005). Since then, the most widely used method for planning and managing the effect of biodiversity on a project is an Environmental effect Assessment (EIA), which is conducted at the project level. UN Environment (2018) states that enhancing EIA is essential to promoting development that is ecologically conscious. Therefore, the adoption of EIA greatly benefits the 2030 Sustainable Development Goal and associated policy frameworks, such as the Strategic Plan for Biodiversity (Bawa *et al.,* 2021).

CONCLUSION

Stream diversion contributes to the loss of biodiversity in water bodies by posing a serious danger to the ecosystem's health and species variety. Natural stream flow patterns are disrupted by human actions including constructing dams, installing irrigation systems, and taking out water, which causes ecosystem fragmentation and modification. These modifications weaken stream ecosystems and endanger aquatic species. Furthermore, by resulting in erosion, sedimentation, and nutrient imbalances, stream diversion exacerbates the loss of biodiversity. Enacting policies like environmental flow controls and promoting sustainable water management techniques are essential to addressing these issues. Preserving biodiversity and guaranteeing the long-term health of freshwater ecosystems need an understanding of the complex relationships between human diversion activities and their effects on stream ecosystems.

REFERENCES

Arthington, Á. H., Naiman, R. J., Mcclain, M. E., & Nilsson, C. (2010). Preserving the biodiversity and ecological services of rivers: New challenges and research opportunities. *Freshwater Biology, 55*(1), 1–16. doi:10.1111/j.1365-2427.2009.02340.x

Artigas, J., Pascault, N., Bouchez, A., Chastain, J., Debroas, D., Humbert, J. F., Leloup, J., Tadonleke, R. D., ter Halle, A., & Pesce, S. (2014). Comparative sensitivity to the fungicide tebuconazole of biofilm and plankton microbial communities in freshwater ecosystems. *The Science of the Total Environment, 468*, 326–336. doi:10.1016/j.scitotenv.2013.08.074 PMID:24048021

Aung, M. T., Song, Y., Ferguson, K. K., Cantonwine, D. E., Zeng, L., McElrath, T. F., Pennathur, S., Meeker, J. D., & Mukherjee, B. (2020). Application of an analytical framework for multivariate mediation analysis of environmental data. *Nature Communications, 11*(1), 5624. doi:10.1038/s41467-020-19335-2 PMID:33159049

Benchimol, M., & Peres, C. A. (2015). Widespread forest vertebrate extinctions induced by a mega hydroelectric dam in lowland Amazonia. *PLoS One*, *10*(7), e0129818. doi:10.1371/journal.pone.0129818 PMID:26132139

Carey, R. O., & Migliaccio, K. W. (2009). Contribution of wastewater treatment plant effluents to nutrient dynamics in aquatic systems: A review. *Environmental Management*, *44*(2), 205–217. doi:10.1007/s00267-009-9309-5 PMID:19458999

de Guzman, I., Altieri, P., Elosegi, A., Pérez-Calpe, A. V., von Schiller, D., González, J. M., Brauns, M., Montoya, J. M., & Larrañaga, A. (2021). *Dataset: Water diversion and pollution interactively shape freshwater food webs through bottom-up mechanisms.*

Deemer, B. R., Harrison, J. A., Li, S., Beaulieu, J. J., DelSontro, T., Barros, N., Bezerra-Neto, J. F., Powers, S. M., dos Santos, M. A., & Vonk, J. A. (2016). Greenhouse gas emissions from reservoir water surfaces: A new global synthesis. *Bioscience*, *66*(11), 949–964. doi:10.1093/biosci/biw117 PMID:32801383

Dudgeon, D. (2019). Multiple threats imperil freshwater biodiversity in the Anthropocene. *Current Biology*, *29*(19), R960–R967. doi:10.1016/j.cub.2019.08.002 PMID:31593677

Flores, F., Garreaud, R., & Muñoz, R. C. (2014). OpenFOAM applied to the CFD simulation of turbulent buoyant atmospheric flows and pollutant dispersion inside large open pit mines under intense insolation. *Computers & Fluids*, *90*, 72–87. doi:10.1016/j.compfluid.2013.11.012

Grill, G., Lehner, B., Thieme, M., Geenen, B., Tickner, D., Antonelli, F., Babu, S., Borrelli, P., Cheng, L., Crochetiere, H., Ehalt Macedo, H., Filgueiras, R., Goichot, M., Higgins, J., Hogan, Z., Lip, B., McClain, M. E., Meng, J., Mulligan, M., & Zarfl, C. (2019). Mapping the world's free-flowing rivers. *Nature*, *569*(7755), 215–221. doi:10.1038/s41586-019-1111-9 PMID:31068722

Kalal, V., Giri, M. N., Baskar, S., & Kuba, R. (2021). Detection of Pollutants present in the surface water of Ayad River, Udaipur, Rajasthan. *IJBI, 3*(1).

Keck, F., & Lepori, F. (2012). Can we predict nutrient limitation in streams and rivers? *Freshwater Biology*, *57*(7), 1410–1421. doi:10.1111/j.1365-2427.2012.02802.x

Kirchherr, J., & Charles, K. J. (2016). The social impacts of dams: A new framework for scholarly analysis. *Environmental Impact Assessment Review*, *60*, 99–114. doi:10.1016/j.eiar.2016.02.005

Krishnaswamy, J., Kumar, M., Kelkar, N., Nair, T., Atkore, V., Hiremath, A. J., & Siddhartha, A. (2017). Moving from requiem to revival: India's rivers and riverine ecosystems. *Transcending boundaries: Reflecting on twenty years of action and research at ATREE*, 94-103.

Lemm, J. U., Venohr, M., Globevnik, L., Stefanidis, K., Panagopoulos, Y., van Gils, J., Posthuma, L., Kristensen, P., Feld, C. K., Mahnkopf, J., Hering, D., & Birk, S. (2021). Multiple stressors determine river ecological status at the European scale: Towards an integrated understanding of river status deterioration. *Global Change Biology*, *27*(9), 1962–1975. doi:10.1111/gcb.15504 PMID:33372367

McHugh, P. A., Thompson, R. M., Greig, H. S., Warburton, H. J., & McIntosh, A. R. (2015). Habitat size influences food web structure in drying streams. *Ecography*, *38*(7), 700–712. doi:10.1111/ecog.01193

McIntosh, A. R., McHugh, P. A., Plank, M. J., Jellyman, P. G., Warburton, H. J., & Greig, H. S. (2018). Capacity to support predators scales with habitat size. *Science Advances*, *4*(7), eaap7523. doi:10.1126/sciadv.aap7523 PMID:29978038

Merciai, R., Molons-Sierra, C., Sabater, S., & García-Berthou, E. (2017). Water abstraction affects abundance, size-structure and growth of two threatened cyprinid fishes. *PLoS One*, *12*(4), e0175932. doi:10.1371/journal.pone.0175932 PMID:28414787

Moreira Filho, O., & Buckup, P. A. (2005). A poorly known case of watershed transposition between the São Francisco and upper Paraná river basins. *Neotropical Ichthyology*, *3*(3), 449–452. doi:10.1590/S1679-62252005000300015

Qin, Y., Wen, Q., Ma, Y., Yang, C., & Liu, Z. (2018). Antibiotics pollution in Gonghu Bay in the period of water diversion from Yangtze River to Taihu Lake. *Environmental Earth Sciences*, *77*(11), 1–11. doi:10.1007/s12665-018-7558-4

Reid, A. J., Carlson, A. K., Creed, I. F., Eliason, E. J., Gell, P. A., Johnson, P. T., Kidd, K. A., MacCormack, T. J., Olden, J. D., Ormerod, S. J., Smol, J. P., Taylor, W. W., Tockner, K., Vermaire, J. C., Dudgeon, D., & Cooke, S. J. (2019). Emerging threats and persistent conservation challenges for freshwater biodiversity. *Biological Reviews of the Cambridge Philosophical Society*, *94*(3), 849–873. doi:10.1111/brv.12480 PMID:30467930

Richter, B. D., Mathews, R., Harrison, D. L., & Wigington, R. (2003). Ecologically sustainable water management: Managing river flows for ecological integrity. *Ecological Applications*, *13*(1), 206–224. doi:10.1890/1051-0761(2003)013[0206:ESWMMR]2.0.CO;2

Sabater, S., Elosegi, A., & Ludwig, R. (Eds.). (2018). *Multiple stressors in river ecosystems: Status, impacts and prospects for the future*. Elsevier.

Sehgal, K. L. (1999). Coldwater fish and fisheries in the Indian Himalayas: rivers and streams. Fish and fisheries at higher altitudes: Asia. *Food and Agriculture Organization of the United Nations Technical Paper*, *385*, 41–63.

Singh, A. K., & Akhtar, M. S. (2015). *Coldwater fish diversity of India and its sustainable development. Biodiversity for Sustainable development.* UP Biodiversity Board, Lucknow.

Singh, G., & Agarwal, N. K. (2017). Impact of hydropower project (RoR) on the ichthyofaunal diversity of river Birahiganga in Central Himalaya (India). *Journal of Fisheries*, *5*(2), 507–512. doi:10.17017/j.fish.38

Tickner, D., Opperman, J. J., Abell, R., Acreman, M., Arthington, A. H., Bunn, S. E., Cooke, S. J., Dalton, J., Darwall, W., Edwards, G., Harrison, I., Hughes, K., Jones, T., Leclère, D., Lynch, A. J., Leonard, P., McClain, M. E., Muruven, D., Olden, J. D., & Young, L. (2020). Bending the curve of global freshwater biodiversity loss: An emergency recovery plan. *Bioscience*, *70*(4), 330–342. doi:10.1093/biosci/biaa002 PMID:32284631

Vasilachi, I. C., Asiminicesei, D. M., Fertu, D. I., & Gavrilescu, M. (2021). Occurrence and fate of emerging pollutants in water environment and options for their removal. *Water (Basel)*, *13*(2), 181. doi:10.3390/w13020181

von Schiller, D., Bernal, S., Dahm, C. N., & Martí, E. (2017). Nutrient and organic matter dynamics in intermittent rivers and ephemeral streams. In *Intermittent rivers and ephemeral streams* (pp. 135–160). Academic Press. doi:10.1016/B978-0-12-803835-2.00006-1

Webster, K. L. (2014). *Sediment dispersal and accumulation in an insular sea: deltas of Puget Sound* [Doctoral dissertation, UW].

Zaidi, S. G. S., Singh, A. K., & Sarma, D. (2018). Breeding, seed production and rearing of coldwater ornamental fishes in Aquarium. *Published by ICAR-DCFR Bulletin*, (27), 43.

ADDITIONAL READING

Daga, V. S., Azevedo-Santos, V. M., Pelicice, F. M., Fearnside, P. M., Perbiche-Neves, G., Paschoal, L. R., Cavallari, D. C., Erickson, J., Ruocco, A. M. C., Oliveira, I., Padial, A. A., & Vitule, J. R. (2020). Water diversion in Brazil threatens biodiversity. *Ambio*, *49*(1), 165–172. doi:10.1007/s13280-019-01189-8 PMID:31030418

de Guzman, I., Altieri, P., Elosegi, A., Pérez-Calpe, A. V., von Schiller, D., González, J. M., Brauns, M., Montoya, J. M., & Larrañaga, A. (2022). Water diversion and pollution interactively shape freshwater food webs through bottom-up mechanisms. *Global Change Biology*, *28*(3), 859–876. doi:10.1111/gcb.16026 PMID:34862833

de Guzman, I., Montoya, J. M., Elosegi, A., Pérez-Calpe, A. V., von Schiller, D., González, J. M., & Larrañaga, A. (2024). Food-web energy fluxes, energy transfer efficiency, and diversity respond distinctively to pollution and water diversion in rivers. *Freshwater Biology*, *69*(3), 351–364. doi:10.1111/fwb.14215

Greig, H. S., McHugh, P. A., Thompson, R. M., Warburton, H. J., & McIntosh, A. R. (2022). Habitat size influences community stability. *Ecology*, *103*(1), e03545. doi:10.1002/ecy.3545 PMID:34614210

Ibrahim, M. S. N., Assim, M. I. S. A., Johari, S., Mohammad, S. K. W., Afandi, S. H. M., & Hassan, S. (2023). Public awareness on biodiversity conservation and well-being: Case of Gunung Mulu National Park, Sarawak. *GeoJournal*, *88*(3), 3471–3496. doi:10.1007/s10708-022-10818-x PMID:36591568

Meter, K., Thompson, S., & Basu, N. (2016). Human Impacts on Stream Hydrology and Water Quality., 441-490. https://doi.org/. doi:10.1016/B978-0-12-405890-3.00011-7

Sharma, T., Singh, A., Kumar, N., Chauhan, G., Singh, D. P., Singh, A., & Rana, B. B. (2023). Emerging Pollutants in the Environment and Ecological Risks. In *Management and Mitigation of Emerging Pollutants* (pp. 1–20). Springer International Publishing. doi:10.1007/978-3-031-41005-5_1

Shivaprakash, K. N., Sen, S., Paul, S., Kiesecker, J. M., & Bawa, K. S. (2021). Mammals, wildlife trade, and the next global pandemic. *Current Biology*, *31*(16), 3671–3677. doi:10.1016/j.cub.2021.06.006 PMID:34237267

Singh, A. K. (2018). Major challenges through the transforming coldwater fisheries in India. *Coldwater Fisheries Society of India*, *1*(1), 1–3.

Tampo, L., Kaboré, I., Alhassan, E. H., Ouéda, A., Bawa, L. M., & Djaneye-Boundjou, G. (2021). Benthic macroinvertebrates as ecological indicators: Their sensitivity to the water quality and human disturbances in a tropical river. *Frontiers in Water*, *3*, 662765. doi:10.3389/frwa.2021.662765

Xin, Z., Li, Y., Zhang, L., Ding, W., Ye, L., Wu, J., & Zhang, C. (2019). Quantifying the relative contribution of climate and human impacts on seasonal streamflow. *Journal of Hydrology (Amsterdam)*, *574*, 936–945. doi:10.1016/j.jhydrol.2019.04.095

Chapter 8
Biodiversity Loss Due to Construction of Dams and Hydropower Projects and Its Impact on Aquatic Ecosystems

Sharda Dhadse
https://orcid.org/0000-0001-9818-786X
CSIR-National Environmental Engineering Research Institute, Nagpur, India

Ashwini Kardbhajne
CSIR-National Environmental Engineering Research Institute, Nagpur, India

ABSTRACT

Due to the growing need for freshwater resources, water diversion projects are becoming an essential part of international water management plans. While hydropower is considered renewable, its environmental impact includes altered river flows, disrupted fish migration, and sediment retention affecting ecosystems. Water diversion projects impact aquatic flora and fauna and their habitats. Sustainable water management is crucial, considering the potential effects such as variations in temperature, sediment movement, and flow patterns caused by dam construction. In aquatic habitats building hydropower projects reduced connectivity, altered nutrient dynamics, and habitat fragmentation can decrease organism diversity and numbers, posing a significant threat to biodiversity in affected areas. A comprehensive approach that includes adaptive management and environmental impact evaluations is crucial for water diversion projects, as this abstract highlights. Addressing water diversion impacts on biodiversity is essential for long-term environmental sustainability.

DOI: 10.4018/979-8-3693-6950-0.ch008

INTRODUCTION

Utilizing the energy of falling or flowing water, hydropower is a renewable energy source that produces electricity. The infrastructure needed to capture and regulate water flow is provided by dams, which are frequently essential parts of hydropower systems. Redirecting a river's flow through channels that are specifically designed for that purpose is known as water diversion. By giving extra water a different path, peak flows are reduced and the likelihood of floods is partially controlled. Generally, established cities have diversion canals built around them to minimize major flood damage. This leads to change in the natural flow of river water into the sea, which can lead to sedimentation which is a serious problem that affects entire ecosystems, fish migration, plant death, and animal displacement from their native habitat. Normally, sediments would flow downstream and into the ocean. Excessive water diversion projects may result in a lack of ecological water in diversion zones, resulting in poor water quality in the lower river course and warm, stagnant circumstances that promote cyanobacteria growth (Paerl, 2014). Dams, levees, pumping stations, irrigation canals, and other man-made structures that alter the natural flow of a waterway are all examples of water diversion projects. Diversion projects can be designed for hydroelectric power generation, agriculture irrigation, consumer and industrial water supply, and flood management. Throughout human history, societies have altered river systems to their own benefit. Freshwaters support a higher proportion of species (animals, plants, and microbes) per unit area than any other environment (10% more than land and 15% more than seas). However, other areas, such as the Amazon, Congo, Nile, and Mekong basins, are heavily inhabited by species and referred to as "hotspots." The long-term environmental impact of several of these major river projects became apparent when western wetlands disappeared and dozens of migrating salmon and other species were declared as endangered or vulnerable under the Endangered Species Act (ESA).

Construction of Hydropower Dams

A lot of major rivers in the globe have dams that help with navigation, generate hydroelectricity, regulate floods, and redirect water (Walker, 1985; Dynesius & Nilsson 1994; Power et al., 1995). Since less water is now reaching floodplain wetlands, these changes have also had an impact on the ecosystem of estuaries and coastal areas (Milliman, 1997). The most well-known example in terms of ecology is the Aral Sea, which is the floodplain terminus of the Amu-Darya and Syr-Darya Rivers in Kazakhstan and Uzbekistan. Water levels in this massive inland sea (68 000 km) dropped by 13 m over 27 years from 1960 to 1987 due to upstream diversion of water for agriculture. This resulted in a 40% decrease in wetland area and had a significant impact on biodiversity (Mickling, 1998). One common strategy for addressing the increasing scarcity of freshwater is to construct water-diversion projects (Liu et al. 2015; Zhang et al. 2015). This long-standing human activity using artificial canals to connect two river basins often benefits nearby populations, but it also entails dangers that are typically disregarded in terms of the immediate and long-term repercussions on the environment, the economy, and society (Moreira-Filho and Buckup 2005; Zhang et al., 2018). Water diversions may result in significant hydrological deficits and significant changes in the patterns of river flow in the donor basin, in addition to habitat destruction, fishery collapse, parasite spread, disease transmission, loss of genetic diversity between populations, biological invasions, extinction of species, and water pollution (Liang et al., 2012; Vitule et al., 2015; Merciai et al., 2017; Qin et al., 2018). Over 10% of all known species are found in freshwater habitats while making up less than 0.1 percent of the Earth's surface (Balian et al., 2008). In addition to providing energy generation, industrial uses, and drinking water, freshwater habitats are hotspots for biodiversity. There are very few free-flowing rivers remaining in industrialized countries, with humans having altered about 75% of the world's major rivers (Grill et al., 2019, Nilssonet al., 2005, Dynesius and Nilsson, 1994). According to Poff et al. (1997), freshwater species have adapted to a natural streamflow regime that is changed by diversion. Fish populations may decline and their composition may alter as a result of water diversion (Poff & Zimmerman, 2010). Water diversion and climate change are two pressures that deteriorate salmonids' freshwater habitat by changing temperature and streamflow. Although making up less than 0.1 percent of the Earth's surface, freshwater ecosystems are home to over 10% of all known species (Balian et al., 2008). Survival and carrying capacity are critical components for the recovery of vulnerable salmon populations, in addition to being biodiversity hotspots (Kareiva et al., 2000; Zabel et al., 2006) as well as size structural reductions (Walters & Post, 2008; Freeman & Marcinek, 2006). However, the creation of hydropower breaks up channels and modifies river flows, drastically changing rivers and their ecosystems. On the other hand, hydropower generation has changed natural rivers. While strongly supporting hydropower in this new era, it is important to weigh the trade-off between freshwater species protection and hydropower generation (Hastik et al. 2015; Song et al. 2020).

These modifications lessen flow velocity, decrease the frequency of rapids, and modify the ecosystems of marshes, floodplains, and deltas. River ecosystems are drastically changed by hydroelectric dams, which has serious implications for aquatic biodiversity. This can lead to a decline in fish populations by upsetting important life cycles. Water flow is regulated by dams, which can cause abnormal variations in sediment movement and water temperature. The habitat of fish, insects, and other species that require specific conditions for survival is badly impacted by these changes. The construction of reservoirs by dams results in widespread flooding that submerges neighboring vegetation and habitats, which also contributes to habitat loss. Fish aren't the only animals affected by this habitat loss; amphibians, reptiles, and many bird species also depend on these areas. Many fish species have seen a decrease in population as a result of these effects. For example, the Grand Coulee Dam's construction in the Pacific Northwest resulted in the extinction of multiple salmon species. Such losses can hurt other wildlife that depends on fish as a food source and have cascade consequences across the food chain. Hydroelectric dams have effects on land environments in addition to aquatic ones. Dams split habitats and separate river systems, separating various species' populations from one another. This fragmentation raises the possibility of local extinctions and limits gene flow. Redevelopment of vegetation and the natural dynamics of riparian habitats can be affected by altered flood cycles caused by changes in water flow patterns downstream of dams. The loss of riparian zones, which are essential habitats for numerous species, including birds, animals, and insects, is frequently brought on by the floods that dams create. These places offer refuge from predators, food supplies, and breeding grounds. Terrestrial habitats around hydropower dams may see a major decline in biodiversity as a result of these changes taken together. Pollinators and herbivores may be impacted by decreased plant diversity, and different animal species may have fewer options for breeding and grazing due to changes in the vegetation's structure. According to Halls and Moyle (2018), diadromous species are especially susceptible to the effects of hydropower development due to their life cycles, which frequently entail lengthy migrations, natal homing, and the existence of genetically different stocks. In several parts of the world, including North America, Europe, Asia, and New Zealand, the populations of diadromous species have drastically decreased (Zhang, H.I. et al., 2020; Halls A.S. and Moyle P.B. 2018). This pattern of decrease indicates that hydropower and other types of river fragmentation probably provide comparable dangers to the diadromous species of the Mekong.

Additionally, small hydropower projects (SHPs) modify natural flow regimes in two ways: (i) by diverting impounded water for power generation, which frequently results in flow reduction or dewatering for several kilometres immediately below the dam; and (ii) by causing frequent and severe fluctuations in the downstream flow caused by sudden releases of impounded water. The changes made have an impact on both hydrology and ecology. (Anderson et al., 2014) state that a decrease in flows beneath the dam may have an impact on water chemistry and the availability of habitat for aquatic biota. According to (Carlisle et al., 2010), dewatering is also known to result in changes in the communities it affects by preferring "eurytopic" fishes those with life histories that are well suited to lentic conditions over "rheophilic" fishes, which are naturally adapted to flowing river waters. It is well established that rapidly fluctuating flow affects fish swimming behaviour, age and size-class composition, spawning, fish survival, and fisheries yields (Cushman, 1985; Liao, 2007). Owing to their sensitivity to flow conditions, freshwater fishes are good indicators of altered hydrological regimes (Poff & Zimmerman, 2010).

Variation in Composition of Fishes in Dammed and Undammed Streams

Streams with and without dams would have different fish communities. In particular, Rheophilic and migratory fish species would be less common in dammed streams, while Eurytopic along with imnophilic species would not alter (Poulet, 2007). The composition of fish populations would be negatively impacted in dammed streams, especially in the dammed site group, and the streams would have lower species richness and variety than the undammed stream (Freeman, Anderson, et al., 2006). In addition to providing some data on the abstraction effects on the phenology of sensitive species of macroinvertebrates, which are the primary consumers, detritivores, and prey in aquatic ecosystems, the researcher highlighted some of the effects of water diversion on the headwaters of the Western Himalaya, including a decrease in the number of rheophiles (Trichoptera taxa) and increased frequency and abundance of non-rheophilic taxa of Coleoptera, Odonata, and Lepidoptera in disturbed sites.

Dammed Streams

1. ***Modified Flow Regime:*** Dams usually control water flow, which modifies the hydrological patterns that naturally occur. Fish species that are acclimated to certain flow conditions may be affected by this.
2. ***Formation of Reservoirs:*** Dams create reservoirs, which may differ from natural stream habitats in terms of their environmental features. Reservoirs frequently contain varied substrate types, slower currents, and various nutrition and temperature levels.
3. ***Alterations to the Habitat:*** When a stream is dammed, upstream habitats may become flooded, which may result in adjustments to the accessibility of hiding places, feeding locations, and eggs. While certain fish species may find it difficult to adapt, others may flourish in reservoirs.
4. ***Variation in Temperature***: might be caused by the retention of water behind a dam. Stratification in reservoirs, where various water layers have varying temperatures, can have an impact on the distribution and habits of fish.
5. ***Introduction of Non-native Species:*** Whether done knowingly or unknowingly, dams can provide a way for the introduction of non-native fish species. Competition, predation, and other ecological disturbances may result from this.

Undammed Streams

1. ***Natural Flow Regime:*** Undammed streams tend to maintain a more natural flow regime, which is crucial for the survival and reproduction of certain fish species. These streams may exhibit seasonal variations in flow, creating diverse habitats.
2. ***Uninterrupted Migration:*** Dams can impede the migration of fish, especially those that rely on upstream or downstream movement for spawning or feeding. Undammed streams provide a more continuous pathway for fish migration.
3. ***Varied Habitat:*** Undammed streams often have a variety of habitats, including riffles, pools, and runs, which are essential for different life stages of various fish species. These diverse habitats support a higher overall biodiversity.
4. ***Water Quality:*** The absence of a dam can contribute to more natural water quality conditions. Water temperature, oxygen levels, and nutrient cycling may follow more natural patterns in undammed streams.
5. ***Native Species Dominance:*** undammed streams are more likely to support native fish species that have evolved in those environments. These species may be better adapted to the natural conditions of the streams.

Many stressors are having a growing effect on stream ecosystems, which is causing a loss of sensitive species as well as a general decline in variety and the integrity of floodplains (Palmer et al., 2010). As stated by (Gaston et al., 2016) anthropogenic modifications like channelization have resulted in changes in habitat flow regimes, which have also been linked to changes in fish community structure. The process of channelization has led to the loss of place-based fluvial ecosystems, resulting in severely concretized active floodplains. The filling of wetlands and the channelization and concretization of floodplains have resulted in significant environmental deterioration, which has generated a physical, emotional, and spiritual alienation from the Gomti riverfront's cultural landscape. According to (Saini et al., 2010), substantial changes in habitat, sewage contamination, and water abstraction pose a serious threat to the fish variety in the river. The loss of habitat variety and degradation in channelized streams (Pedersen, 2009) poses a serious threat to the biodiversity of moving water (Kendle, 1970; Johansson, 2013; Hartson et al., 2014). Natural channels sustain more species, higher densities, and larger biomass; recovery from these conditions is typically gradual (Brooker, 1985). Researchers studied channelized and unchannelized Gomti River species from Lucknow given in (Table No. 1).

Table 1. Habit matrix and species preference and their status in channelized and unchannelized sections of the Gomti River

Sr. No	Habitat Types	Characteristics	Species Preference	Unchannelized Section	Channelized Section
1	Open river	River without channel confluences and pool	A diverse group of fish	Present	Present
2	Deep pools	Deep pools with a high turbidity level and coarse substrate	This region is home to species including Wallago, *Ompok*, *Channa*, and *Chitala*	Present	Present
3	Backwater and shallow pool	Medium size pool with low water flow	*Rita*, *Labeo*, *Cirrhinus*, and *Notopterus* favor this environment	Present	Absent
4	Scour pools	Channel spanning pools	Fish susceptible to prolonged low flow phases prefer the pool, which contains juvenile fish utilized for spawning	Present	Absent
5	Wetland connected ditches	Medium-sized to small ditches connected to riverside wetlands	Numerous specialized fish species find homes in small natural ditches.	Present	Absent
6	Grassed banks	Stabilize stream banks	Grass-covered banks support a greater number of fish species and individuals.	Present	Absent
7	Channel confluence	Junction of two rivers	The ideal habitat for fish	Present	Absent
8	An effective vegetative bank	Temperature of the stream is maintained by lush banks.	Fish go to the areas with a lot of vegetation because their assemblage composition improves in vegetated areas.	Present	Absent

Source: (Dutta et al., 2018)

Impact on Benthic Macroinvertebrates in River Ecosystems

Because of their ability to respond either positively or negatively to various stressors, including water pollution, climate change, and flow disturbance, benthic macroinvertebrates are thought to be among the most reliable bio-indicators in river ecosystems (Dewson et al., 2007). Worldwide, benthic macroinvertebrate-based biotic indicators are widely used (DePauw and Hawkes, 1993; Campbell et al., 2020). A decrease in taxonomic diversity, a shift in the relative abundance of particular animals, and a fall in biomass output are the most common alterations to the ecosystem linked with an observed negative reaction of these indicator species, according to (Holt et al., 2014). Most of these aquatic insect species are less common in low flows due to hydrological changes. This is because river flows apply physical forces that affect the availability of habitat, the nutrient cycle, and the chemistry of the water. Orders Ephemeroptera, Plecoptera, and Trichoptera (EPT) are useful for ecological assessments of rivers to discover the effects of development of water resources since they are reliable bio-indicators (Holt et al., 2014). In this study, researchers show the ecological consequences of water diversion projects (especially irrigation projects, water mills, and micro-hydropower) on the condition of rivers by looking at how indicator species respond to changes in flow regimes. The health of river ecosystems and instream biotic communities may be impacted by changes in stream flow characteristics brought about by water diversion projects. These changes may result in the loss of sensitive biota and the increase of tolerant biota, sediment transport, and alterations to riverbeds and nutrient availability (Bunn and Arthington, 2002; Ward and Stanford, 1983).

Studies on Hippo Trail

In floodplain wetlands, sediment deposition produces a natural disturbance regime where a reduction in river gradient is typically associated with a loss in stream competence. The distribution of species is impacted by this regime in addition to the distribution of nutrients and other solutes (Kottelat, M. 2019; Diederichs and Ellery, 2001). The most noticeable effect was caused by the excavation of canals during the 1970s and 1980s, which significantly changed the water flow inside the marsh (Goodman, 1987). In African marshes, diversionary routes have been observed beside hippos' tracks (McCarthy et al., 1998). These aquatic creatures travel back and forth on routes that are aligned parallel to local hydraulic gradients, which connect their nocturnal grazing grounds in terrestrial environments to their daytime resting spots in aquatic environments. Frequent hippos' movement between these places has a significant impact on the local hydraulic conditions; by deepening channels (raising the hydraulic radius) and removing aquatic vegetation, roughness is reduced and water flow efficiency is raised. Hippos' activity in the Okavango Delta has been identified as the primary determinant of a new channel's path during an avulsion (McCarthy et al., 1998). Thus, the net transfer of clay from the floodplain onto the terrestrial boundary will result from hippos' frequent migration between sites of rest and feeding. There won't be much sediment transfer from terrestrial to aquatic ecosystems since sand does not stick to these animals' bodies in the same way that clay does. Thus, even in peaceful backwaters, hippos are likely to carve out routes in the wetland habitats that offer little resistance to water flow over extended periods of time. Within the semicontinuous levees, hippos trailing off the main channel into backswamp areas cause depressions.

Environmental Impact on River Diversion

River floodplains are dynamic habitats that support a high level of biodiversity in areas disturbed by human activity. Dams and Construction of Hydropower Project generally pose more problems in low gradient channels such as.,

Alteration to Riparian Vegetation: Through terrestrial "insect drop," riparian vegetation directly supplies vital fish food and nutrients to food webs. It shields riverbanks from erosion and offers shade. Reduced instream flows can lead to lower moisture levels, which can impact riparian vegetation and ecosystems. When it is eliminated to create room for powerhouses, highways, dams, and power lines, it raises concerns.

Variation in Temperature: Fish and other aquatic species' habitat quality and quantity are directly impacted by the temperature of the water. In one study, the advantages of warmer seas caused people to migrate to the sea earlier, which decreased the chances of survival for marine life. Fish may experience stress or may perish if summertime waters heat up excessively due to decreased flow rates.

Permanent Vegetation Clearance causing Habitat Loss: The project's thick deciduous forest region has been cleared of vegetation, which damages the local species and disrupts ecosystems. Changes to even minor locations may have disastrous effects, as the western ghats, for instance, are home to uncommon ecosystems that support vulnerable species. Predators exist in this area as well. Because they have a significant impact on a wide range of other species, predators are an essential component of healthy ecosystems. The project's introduction of new land uses makes it crucial to take big predator risk into account. Genetic diversity may be impacted if any of the remaining predator populations disappear, which would reduce the species' ability to withstand effects like climate change.

Effects of Transmission lines and Roads: Transmission lines and roads link population centres to dam and diversion sites. These "linear disturbances" have the potential to make up the majority of the project's footprint for this river diversion project. Common consequences include increased hunting pressure, heightened confrontations between humans and wildlife, drainage and erosion issues, and landslides. Because vegetation is removed and managed differently along transmission lines and other linear corridors, a new type of habitat is created, and the effects are mostly similar to those of roadways. Electrocution bird and bat collisions with power wires are the effects specific to transmission lines. Certain species are more vulnerable to harm and even death when they collide with electricity wires.

During the Construction: There may be a lot of human disturbance and presence during the several-year-long construction phase. Unintentional spills, landslides, and soil erosion are all possible. Large equipment will be utilized to construct or repair roads, rock blasting will be done for roads or penstocks, leaving behind enormous "spoil sites," and chainsaws will be used to remove vegetation.

Direct Impacts of Dams and Barrages on Marine Fish and Invertebrates: The main method for regulating freshwater is through the construction of dams. The obstruction of anadromous and catadromous fish species' migration paths is one of the direct results of such barriers. Members of the Salmonidae and Clupeidae families, such as shad, are examples of anadromous fish, which spend most of their adult lives in saline water but spawn in freshwater. Anadromous fish populations that are prevented from reaching their spawning grounds by obstacles will eventually become extinct (Watt, 1989).

Cause of Fish and Invertebrate Migration: It is thought that variations in river run-off cause many anadromous fish stocks to undergo upstream spawning migrations (Fraser, 1972; Kominoski J. S. et al., 2018). When run-off is at its highest or increasing, salmon migrate most frequently. Furthermore, many juvenile salmon stocks prefer to migrate downstream in response to excessive freshwater discharge (Northcote, 1982; Vu A. V. et al., 2022). Migrational cues can alter because of a general drop brought on by diversions or a temporal shift brought on by regulation in the usual discharge cycle.

Lower the Success of Spawning: A large number of marine fish species spawn in flood plains or estuaries, usually at the time of high run-off. Asynchrony between the peak run-off and spawning period may arise from significant changes in the timing of flood waters. The spawning grounds may get coated in sediment, and suffocate the eggs, as a result of changes in the amount of freshwater discharge rates that create an increase in sediment concentrations or settling rates of the river-borne materials. On the other hand, inadequate sediment transport can result in less food being available for developing larvae. Reduced oxygenation of eggs stuck to the bottom due to low flow rates may increase stress and may increase mortality. According to (Zimpfer et al., 1987), dams of certain species may lower current speeds below the necessary thresholds to maintain their eggs suspended, causing the eggs to sink to the bottom. Certain species' eggs may also suffer from the presence of high salinity water from decreased river flow (Peters, 1982). It is possible for spawning locations to dry up when freshwater flow is severely reduced.

Distribution and Competition among Species: Many marine fish species have made freshwater-dominated estuaries and deltas, like mangrove swamps and saltwater marshes, their natal grounds (McHugh, 1967; Kominoski J. S. et.al., 2018; Naughton et.al., 2021). The main advantages of these nearshore nurseries for young fishes are reduced predation pressure and increased food availability (Miller et al.,

1985). A decrease in the amount of silt and an increase in salinity from less fresh water may allow adult fish to have easier access to nearshore areas, which could also lead to a decline in the number of juvenile fish. Severe conditions, including river diversions, can result in salinity changes large enough to drive out native species and introduce new ones.

In addition to their direct effects on food, predators, and parasites, changes in salinity can also have an indirect influence on fish and shellfish (Copeland, 1966). According to Davies et al. (1992), water transfers between basins have the potential to introduce illnesses, parasites, and new plant and animal species into the receiving river basin.

Impacts on Society and Economy: Human-animal conflict will increase as habitat continues to decline and wildlife corridors become more fragmented. Elephants and tigers, for example, require expansive habitats to meet their needs and have a suitable place to live. However, the growing number of haphazard development projects, such as building highways, railroads, and hydroelectric dams, upset the natural ecology and result in the fragmentation of vast forest regions into smaller ones, a phenomenon known as "habitat fragmentation." there are often confrontations between humans and animals since wild animals cannot find adequate supplies in their fragmented habitats and frequently come into human communities.

Food Web Modifications: River diversions are anticipated to result in decreased habitat quantity and quality, which will have an impact on aquatic food webs. Because they represent the main food supply for fish in the area and are crucial to the proper operation of aquatic ecosystems, benthic invertebrates are the most well-known component of the food web.

Restoration and Mitigation Strategies

Diversion of water can lead to biodiversity loss, which can have major ecological effects. To solve these problems, mitigation and restoration techniques must be put into practice. Here are a few successful methods for reviving biodiversity.

Habitat Restoration: Restoring natural habitats requires planting native vegetation in places impacted by water diversion. This aids with soil stabilization, inhibits erosion, and provides habitat for a variety of organisms.

Wetland Restoration: If wetlands have been affected, think about establishing new wetlands or repairing existing ones. A variety of plants and animals are supported by wetlands.

***Fish Migration and Passage*:** Fish Steps: To assist fish in moving around obstacles like dams, build fish ladders or other fish passageways. Maintaining fish populations and sustaining the aquatic ecosystem depend on this.

Management of Flow Regime: Imitating Natural Flow Patterns Establish water release schedules that take seasonal fluctuations into account and imitate natural flow patterns. This preserves the ecosystems down stream's ecological integrity.

Enhancing Water Quality: Put pollution control and water quality improvement initiatives into action. Aquatic biodiversity may be negatively impacted by pollution and elevated nutrient levels.

Allocation of Environmental Flow: Set aside some diverted water just to keep the ecosystem flowing in the right direction. By doing this, it is made sure that ecosystems downstream would get enough water to sustain biodiversity.

Monitoring of Hydrology: Provide a reliable mechanism for tracking changes in the hydrological cycle, such as sediment movement, water levels, and flow rates. The application of adaptive management requires this knowledge.

***Community Engagement*:** Involve local communities, conservation organizations, and other stakeholders in the design and execution of restoration projects as part of community engagement and stakeholder involvement. This may result in more long-lasting and efficient solutions.

Habitat Connectivity: To connect disparate habitats, establish corridors for habitats. This promotes genetic variety and permits the migration of species.

Management of Invasive Species: Changed environments can support the growth of invasive species. Take action to manage and control invasive species that might have spread because of altered water supply.

Law and Policy Interventions: To guarantee that environmental factors are sufficiently considered in water management practices, strengthen and implement regulations pertaining to water diversion.

CONCLUSION

The loss of various habitats and the species that depend on them is a consequence of water diversion, which is a danger to biodiversity. Affected aquatic and terrestrial habitats may suffer significant effects from the modification of natural water flow patterns, whether as a result of the building of dams, irrigation schemes, or other water management initiatives. The effects of water diversion on biodiversity loss are extensive. The ability of different species to survive and procreate can be negatively impacted by habitat deterioration, fragmentation, and loss caused by disrupted water flows. Particular difficulties for aquatic ecosystems include decreased oxygen levels, fluctuating temperatures, and altered sediment transport, all of which put the survival of many different species at risk. The extinction of one species can harm other species, disrupting complex ecological balances and making ecosystems less resilient overall. Furthermore, when biodiversity is lost, it compromises the services that different ecosystems offer, such as pollination, temperature regulation, and water purification. An all-encompassing and sustainable approach to water management is needed to address the problem of biodiversity loss brought on by water diversion. It is crucial to strike a balance between protecting biodiversity and satisfying human demands for water resources. Implementing adaptive management measures, considering the biological needs of various species, and encouraging conservation and restoration activities are critical steps towards reducing the negative impacts of water diversion on biodiversity. The proposal suggests that to optimize the conservation and restoration potential of riparian ecosystems, riverbank development initiatives should pay greater attention to river corridors, the channel, and the surrounding local environment. A complex freshwater environment cannot remain stable over the long term without species diversity and habitat heterogeneity. Therefore, conservation policies ought to support management plans that use ecohydrological techniques to preserve the integrity of aquatic ecosystems. Restoring local fish variety may be aided by appropriate environmental impact assessments for any riverfront development projects in the basin, habitat restoration plans, and time-bound species recovery plans for endangered species in the study region. A comprehensive strategy to prevent biodiversity loss due to water diversion and ensure the long-term health and resilience of our planet's ecosystems must include conservation initiatives, international cooperation, and sustainable water management practices.

REFERENCES

Acuña, V., Muñoz, I., Giorgi, A., Omella, M., Sabater, F., & Sabater, S. (2005). Drought and post drought recovery cycles in an intermittent Mediterranean stream: Structural and functional aspects. *Journal of the North American Benthological Society*, *24*(4), 919–933. doi:10.1899/04-078.1

Allan, J. D., & Flecker, A. S. (1993). Biodiversity conservation in running waters. *Bioscience*, *43*(1), 32–43. doi:10.2307/1312104

Almodóvar, A., & Nicola, G. G. (1999). Effects of a small hydropower station on brown trout *Salmo trutta* L. in the River Hoz Seca (Tagus basin, Spain) one year after regulation. *Regulated Rivers*, *154*(5), 477–484. doi:10.1002/(SICI)1099-1646(199909/10)15:5<477::AID-RRR560>3.0.CO;2-B

Anderson, D., Moggridge, H., Shucksmith, J. D., & Warren, P. H. (2017). Quantifying the impact of water abstraction for low head 'run of the river hydropower on localized river channel hydraulics and benthic macroinvertebrates. *River Research and Applications*, *33*(2), 202–213. doi:10.1002/rra.2992

Anderson, D., Moggridge, H., Warren, P., & Shucksmith, J. (2014). The impacts of "run-of-river" hydropower on the physical and ecological condition of rivers. *Water and Environment Journal : the Journal / the Chartered Institution of Water and Environmental Management, 29*(2), 268–276. doi:10.1111/wej.12101

Barnes, K., Ellery, W., & Kindness, A. (2002). A preliminary analysis of water chemistry of the Mkuze Wetland System, KwaZulu-Natal: A mass balance approach. *Water S.A., 28*(1), 1–12. doi:10.4314/wsa.v28i1.4861

Bunn, S. E., & Arthington, A. H. (2002). Basic principles and ecological consequences of altered flow regimes for aquatic biodiversity. *Environmental Management, 30*(4), 492–507. doi:10.1007/s00267-002-2737-0 PMID:12481916

Campbell, T., Pin, K., Ngor, P. B., & Hogan, Z. (2020). Conserving Mekong megafishes: Current status and critical threats in Cambodia. *Water (Basel), 12*(6), 1820. doi:10.3390/w12061820

Carlisle, D. M., Wolock, D. M., & Meador, M. R. (2010). Alteration of streamflow magnitudes and potential ecological consequences: A multiregional assessment. *Frontiers in Ecology and the Environment, 9*(5), 264–270. doi:10.1890/100053

Dewson, Z. S., James, A. B., & Death, R. G. (2007). Invertebrate responses to short-term water abstraction in small New Zealand streams. *Freshwater Biology, 52*(2), 357–369. doi:10.1111/j.1365-2427.2006.01682.x

Diederichs, N. J., & Ellery, W. N. (2001). An analysis of plant species distributions on the floodplains of the Okavango River, Namibia, with respect to impacts of possible water abstraction. African Journal of Aquatic Sciences, 26, 139–147.

Drinkwater, K. F., & Frank, K. T. (1994). Effects of river regulation and diversion on marine fish and invertebrates. *Aquatic Conservation, 4*(2), 135–151. doi:10.1002/aqc.3270040205

Dutta, V., Sharma, U., Iqbal, K., Adeeba, Kumar, R., & Pathak, A. K. (2018). Impact of river channelization and riverfront development on fluvial habitat: Evidence from Gomti River, a tributary of Ganges, India. *Environmental Sustainability, 1*(2), 167–184. doi:10.1007/s42398-018-0016-0

Dynesius, M., & Nilsson, C. (1994). Fragmentation and flow regulation of river systems in the northern third of the world. *Science, 266*(5186), 753–762. doi:10.1126/science.266.5186.753 PMID:17730396

Ellery, W. N., Dahlberg, A. C., Strydom, R., Neal, M. J., & Jackson, J. (2003). Diversion of water flow from a floodplain wetland stream: An analysis of geomorphological setting and hydrological and ecological consequences. *Journal of Environmental Management, 68*(1), 51–71. doi:10.1016/S0301-4797(03)00002-1 PMID:12767862

Ellery, W. N., Ellery, K., Rogers, K. H., McCarthy, T. S., & Walker, B. H. (1993). Vegetation, hydrology, and sedimentation processes as determinants of channel form and dynamics in the northeastern Okavango Delta, Botswana. *African Journal of Ecology, 31*, 10–25. doi:10.1111/j.1365-2028.1993.tb00514.x

Ellery, W. N., & McCarthy, T. S. (1998). Environmental change over two decades since dredging and excavation of the lower Boro River, Okavango Delta, Botswana. *Journal of Biogeography, 25*, 361–378. doi:10.1046/j.1365-2699.1998.252168.x

Fontaine, T. D., & Bartell, S. M. (Eds.). (1983). *Dynamics of lotic ecosystems* (p. 494). Ann Arbor Science.

Gaston, K. A., Eft, J. A., & Lauer, T. E. (2012). Morphology and its effect on habitat selection of stream fishes. *Proceedings of the Indiana Academy of Sciences, 121*(1), 71–78.

Goodman, P. S. (1987). Mkuze floodplain and swamps. Wetland management case study. *African Wildlife, 41*, 269–270.

. Halls, A. S., & Moyle, P. B. (2018). Comment on "Designing River flows to improve food security futures in the Lower Mekong Basin. *Science, 361*(6398).

Jumani, S., Rao, S., Kelkar, N., Machado, S., Krishnaswamy, J., & Vaidyanathan, S. (2018). Fish community responses to stream flow alterations and habitat modifications by small hydropower projects in the Western Ghats biodiversity hotspot, India. *Aquatic Conservation, 28*(4), 979–993. doi:10.1002/aqc.2904

Kominoski, J. S., Ruhí, A., Hagler, M. M., Petersen, K., Sabo, J. L., Sinha, T., Sankarasubramanian, A., & Olden, J. D. (2018). Patterns and drivers of fish extirpations in rivers of the American Southwest and Southeast. *Global Change Biology, 24*(3), 1175–1185. doi:10.1111/gcb.13940 PMID:29139216

. Kottelat, M. (2019). Rhyacoschistura larreci, a new genus and species of loach from Laos and redescription of R. suber (Teleostei: Nemacheilidae). *Zootaxa, 4612*(2).

McCarthy, T. S., Cooper, G. R. J., Tyson, P. D., & Ellery, W. N. (2000). Seasonal flooding in the Okavango Delta, Botswana—Recent history and future prospects. *South African Journal of Science, 96*, 25–33.

McKay, S. F., & King, A. J. (2006). Potential ecological effects of water extraction in small, unregulated streams. *River Research and Applications, 22*(9), 1023–1037. doi:10.1002/rra.958

Micklin, P. P. (1988). Desiccation of the Aral Sea: A water management disaster in the Soviet Union. *Science, 241*(4870), 1170–1176. doi:10.1126/science.241.4870.1170 PMID:17740781

Milliman, J. D. (1997). Blessed dams or damned dams? *Nature, 386*(6623), 325–327. doi:10.1038/386325a0

Naughton, G. P., Hogan, Z. S., Campbell, T., Graf, P. J., Farwell, C., & Sukumasavin, N. (2021). Acoustic telemetry monitors movements of wild adult catfishes in the Mekong River, Thailand and Laos. *Water (Basel), 13*(5), 641. doi:10.3390/w13050641

Neal, M. (1998). *An investigation of altered wetland inflow partitioning due to landuse change in the catchment of Alfred Park, New Germany. BSc (Hons.) Research Report*. School of Life and Environmental Sciences, University of Natal.

NMFS (National Marine Fisheries Service). (2011). *Endangered species act status of west coast salmon and steelhead*. NMFS, Seattle, Washington. http://www.nwr.noaa.gov/ ESA-Salmon-Listings/upload/1-pgr-8--11.pdf

NPPC (Northwest Power Planning Council). (1989). *Smolt density model*. StreamNet, Portland. ftp://ftp.streamnet. org/pub/streamnet/projman_files/sdmdoc.pdf

Paerl, H. W., Hall, N. S., Peierls, B. L., Rossignol, K. L., & Joyner, A. R. (2014). Hydrologic variability and its control of phytoplankton community structure and function in two shallow, coastal, lagoonal ecosystems: The Neuse and New River Estuaries, North Carolina, USA. *Estuaries and Coasts, 37*(S1), 31–45. doi:10.1007/s12237-013-9686-0

Palmer, M. A., Menninger, H. L., & Bernhardt, E. (2010). River restoration, habitat heterogeneity and biodiversity: A failure of theory or practice? *Freshwater Biology, 55*(s1), 205–222. doi:10.1111/j.1365-2427.2009.02372.x

Poff, N. L., Allan, J. D., Bain, M. B., Karr, J. R., Prestegaard, K. L., Richter, B. D., Sparks, R. E., & Stromberg, J. C. (1997). The natural flow regime. *Bioscience, 47*(11), 769–784. doi:10.2307/1313099

Poff, N. L., Allan, J. D., Bain, M. B., Karr, J. R., Prestegaard, K. L., Richter, B. D., & Stromberg, J. C. (1997). The natural flow regime. *Bioscience*, *47*(11), 769–784. doi:10.2307/1313099

Poff, N. L., & Zimmerman, J. H. (2010). Ecological responses to altered flow regimes: A literature review to inform the science and management of environmental flows. *Freshwater Biology*, *55*(1), 194–205. doi:10.1111/j.1365-2427.2009.02272.x

Poff, N. L. R., Allan, J. D., Bain, M. B., Karr, J. R., Prestegaard, K. L., Richter, B. D., Sparks, R. E., & Stromberg, J. C. (1997). The Natural Flow Regime. *Bioscience*, *47*(11), 769–784. doi:10.2307/1313099

Poole, G. C., & Berman, C. H. (2001). An ecological perspective on in-stream temperature: Natural heat dynamics and mechanisms of human-caused thermal degradation. *Environmental Management*, *27*(6), 787–802. doi:10.1007/s002670010188 PMID:11393314

Postel, S. L., Daily, G. C., & Ehrlich, P. R. (1996). Human appropriation of renewable fresh water. *Science*, *271*(5250), 785–788. doi:10.1126/science.271.5250.785

Power, M. E., Sun, A., Parker, G., Dietrich, W. E., & Wootton, J. T. (1995). Hydraulic food-chain models. *Bioscience*, *45*(3), 159–167. doi:10.2307/1312555

Richardson, C. J., & McCarthy, E. J. (1994). Effect of land development and forest management on hydrologic response in south-eastern coastal wetlands: A review. *Wetlands*, *14*(1), 56–71. doi:10.1007/BF03160622

Rogers, K. H. (1995). Riparian wetlands. In G. I. Cowan (Ed.), *Wetlands of South Africa* (pp. 41–52). Department of Environmental Affairs and Tourism.

Rogers, K. H. (1997). Freshwater wetlands. In R. M. Cowling, D. M. Richardson, & S. M. Pierce (Eds.), *Vegetation of Southern Africa* (pp. 322–347). Cambridge University Press.

Shah, R. D. T., Sharma, S., & Bharati, L. (2020). Water diversion induced changes in aquatic biodiversity in monsoon-dominated rivers of Western Himalayas in Nepal: Implications for environmental flows. *Ecological Indicators*, *108*, 105735. doi:10.1016/j.ecolind.2019.105735

Stromberg, J. C., Lite, S. J., Marler, R., Paradzick, C., Shafroth, P. B., Shorrock, D., White, J. M., & White, M. S. (2007). Altered stream-flow regimes and invasive plant species: The Tamarix case. *Global Ecology and Biogeography*, *16*(3), 381–393. doi:10.1111/j.1466-8238.2007.00297.x

Sutton, R., & Morris, C. (2006). *Instream flow assessment upper Lemhi River and Canyon Creek, Idaho. U. S.* Department of the Interior, Bureau of Reclamation.

Tipping, J. M. (2011). Effect of juvenile length on Chinook Salmon survivals at four hatcheries in Washington state. *North American Journal of Aquaculture, 73*(2), 164–167. doi:10.1080/15222055.2011.568862

Tonkin, J. D., & Death, R. G. (2013). Macroinvertebrate drift-benthos trends in a regulated river. *Fundamental and Applied Limnology, 182*(3), 231–245. doi:10.1127/1863-9135/2013/0404

Tonkin, J. D., & Death, R. G. (2013). Macroinvertebrate drift-benthos trends in a regulated river. *Fundamental and Applied Limnology, 182*(3), 231–245. doi:10.1127/1863-9135/2013/0404

. Torgersen, C. E. (1999). Multiscale thermal refugia and stream associations of chinook salmon in northeastern Oregon. *Ecological Applications, 9*, 301–319.

Vu, A. V., Baumgartner, L. J., Mallen-Cooper, M., Doran, G. S., Limburg, K. E., Gillanders, B. M., & Cowx, I. G. (2022). Diverse migration tactics of fishes within the large tropical Mekong River system. *Fisheries Management and Ecology, 29*(5), 708–723. doi:10.1111/fme.12566

Walker, K. F. (1985). A review of the ecological effects of river regulation in Australia. In *Perspectives in Southern Hemisphere Limnology: Proceedings of a Symposium, held in Wilderness,* (pp. 111-129). Springer Netherlands. 10.1007/978-94-009-5522-6_8

Watts, R. J., Richter, B. D., Opperman, J. J., & Bowmer, K. H. (2011). Dam reoperation in an era of climate change. *Marine and Freshwater Research, 62*(3), 321–327. doi:10.1071/MF10047

Zhang, H., Jarić, I., Roberts, D. L., He, Y., Du, H., Wu, J., Wang, C., & Wei, Q. (2020). Extinction of one of the world's largest freshwater fishes: Lessons for conserving the endangered Yangtze fauna. *The Science of the Total Environment, 710*, 136242. doi:10.1016/j.scitotenv.2019.136242 PMID:31911255

KEY TERMS AND DEFINITIONS

Biodiversity Hotspot: A biodiversity hotspot refers to a biogeographic region that is characterized by a high level of species richness and a significant degree of threat to that biodiversity.

Estuaries: Freshwater from rivers and streams combines with saltwater from the ocean to form estuaries, which are partially enclosed coastal bodies of water.

Eurytopic Fish: Fish are considered to be "eurytopic" organisms because of their broad tolerance and capacity to flourish in a variety of environmental circumstances.

Floodplain: A flat or gently sloping section of land next to a river, stream, or other body of water that occasionally floods is known as a floodplain.

Fluvial Ecosystems: The ecological community and interactions that take place inside a river or stream and its surrounding environment are referred to as a fluvial ecosystem.

Hydroelectric Dams: Buildings known as hydroelectric dams are placed across rivers to capture the energy of the flowing water and use it to produce power. Hydroelectric power plants depend heavily on these dams.

Rheophiles: Rheophiles are organisms that have adapted to live in turbulent or fast-moving water, including some types of fish. The Greek word "rheos," which means "flow" or "stream," is where the word "rheophile" originates.

Riparian Habitat: A riparian habitat refers to an ecosystem or area of land that is situated along the banks of rivers, streams, lakes, or other water bodies. Riparian zones are transitional areas between aquatic environments and upland terrestrial areas.

Water Diversion: Redirecting the natural flow of water from one place, like a river or lake, to another for a variety of reasons is known as water diversion. This could involve building infrastructure like pipelines, canals, dams, etc.

Wetland: A wetland is a region of land where seasonal or permanent water saturation of the soil results in the emergence of distinct ecosystems with vegetation that thrives in wet conditions.

Chapter 9
Biodiversity Loss Due to Climate Change and Pollution

Ritika Maurya
Gurukul Kangri University, Haridwar, India

Sangeeta Madan
Gurukul Kangri University, Haridwar, India

Monika Kumari
Gurukul Kangri University, Haridwar, India

Bharti Sadhiyan
https://orcid.org/0009-0001-8983-708X
Gurukul Kangri University, Haridwar, India

ABSTRACT

The variety among living things from various environments, such as desert, marine, and terrestrial ecosystems, as well as the ecological complexes they are a part of, is known as biodiversity. The wellbeing of the earth and all of its people depends on the preservation of biodiversity. Human-induced long-term changes in temperatures and weather patterns disrupt the delicate life-supporting equilibrium, causing climate change. Pollution of land, water, and air jeopardizes ecosystems and species globally. According to the World Health Organization, air pollution affects 92% of the population and contributes to 11.6% of global fatalities. Climate change impacts biodiversity through altered breeding and migration behaviors, shifting distribution patterns, invasive species incursions, increased insect attacks, and heightened forest fire risks. India can enhance coordination by enacting eco-friendly laws, integrating these issues into national plans, and adopting biodiversity-based climate change mitigation strategies.

DOI: 10.4018/979-8-3693-6950-0.ch009

INTRODUCTION

Definition of Biodiversity

"The diversity of living forms found in all settings, including terrestrial, marine, and other aquatic ecosystems as well as the ecological complexes that comprise them. Biodiversity is defined by the Convention on Biological Diversity as "the diversity of ecosystems, between species, and within species." A specific population of organisms interacting with the soil, rocks, water, and air to form an ecosystem is what gives rise to an area's biodiversity. The relationships between top predators in the wild, phytoplankton in the ocean, and microbes in soils control the state of the planet (Abrahms *et al.*, 2017).

Importance of Biodiversity

Numerous benefits to humanity stem from biodiversity, including cultural, economic, and environmental ones. From an economic perspective, biodiversity benefits businesses that depend on natural resources; nations that primarily depend on oil and gas exploration to drive economic growth stand to gain from biodiversity as a supply of clean air, food, shelter, fresh water, wood, and fiber for their people.

- Fuel, fodder, and food supplies are only a few of the environmental services that biodiversity supports.
- Health-Related Sources
- Pollution control;
- Soil erosion reduction
- Improvement of soil conditions
- Creating a Stable Environment.
- Preservation of Resources
- Nutrient recycling
- Leisure Source
- Essential to the progress of culture
- Reusing water
- Economic growth
- Being aware of the significance of the species.

It contributes significantly to our mental and physical health, upholds our sense of cultural identity, and provides comfort and spiritual inspiration. Additionally, it offers recreational, ethical, artistic, and medical advantages as well as educational values for the growth of ecotourism and conservation. (Palita *et al.*, 2023).

ENVIRONMENTAL CHALLENGES AND BIODIVERSITY

All life on Earth is significantly impacted by its surroundings.

Everywhere you look, you may find environmental problems, many of which are detrimental to both current and future generations. The seven key environmental issues are deforestation, soil degradation, depletion of natural resources, air and water pollution, biodiversity loss and species extinction, and climate change.

Pollution

Numerous pollutants have contaminated the land, water, and air. Air pollution affects 92% of the world's population and is thought to be the reason behind 11.6% of deaths worldwide. The quality of the air in cities is quite terrible, and this issue will only get worse as more people move there. Chemical pollution from business and agricultural operations is a worry since it can cause toxins to kill or harm plants and animals. In addition, nutrient pollution from fertilizers, residences, and other sources that finds its way into ponds, lakes, and the ocean causes eutrophication. There are 500 dead zones, or regions where oxygen is not detectable, in the ocean as a result of global warming and fertilizer pollution.

Global Warming and Climate Change

The term "climate change" refers to long-term, human-caused variations in local, global, or regional temperature and weather patterns. The delicate balance that has existed between living forms and the weather for thousands of years is what has allowed all life forms on this planet to survive. In the past two millennia, there has never been climate change as extreme as it is now, and it is getting worse everywhere in the world. Recent increases in human activity have produced a significant amount of greenhouse gases into the atmosphere, including carbon dioxide and methane. This has raised the possibility of global warming and exacerbated the effects of greenhouse gases. According to the Intergovernmental Panel on Climate Change (IPCC), temperatures will rise by 1.4 to 5.8 degrees Celsius this century. By 2030, carbon dioxide concentrations will have doubled from pre-industrial levels, predicts the (IPCC., 2013). It is anticipated that biodiversity would be impacted by the various facets of climate change on all scales, from species to biomes. Climate change is a development challenge that requires a strategic, integrated response because of the harm it has caused to biodiversity. Including biodiversity-based mitigation strategies in climate change mitigation plans is one such strategy. Targeted initiatives like REDD (reducing emissions from deforestation and forest degradation) have been implemented by the United Nations Framework Convention on Climate Change

(UNFCCC). "Conservation, sustainable management of forests, and enhancement of forest carbon" are now included in these programs"(IPCC ., 2013)

POLLUTION AND BIODIVERSITY

Pollution significantly alters all ecosystems, but it has a particularly negative effect on freshwater and marine settings. Pollution is detrimental for the environment and is increasing in all forms, including soil, water, and air. The land, freshwater, oceans, and atmosphere that sustain life have been negatively damaged by pollution in many forms, including oil spills, plastic waste, greenhouse gas emissions, and agricultural fertilizers. The health and welfare of humans as well as many other animals are directly or indirectly endangered by pollutants, bright lights, and loud noises. Pollution was a key contributing cause when over 430 species were designated as "threatened" or "endangered" under the U.S. Endangered Species Act.

Water Pollution's Silent Assault on Aquatic Biota

The aquatic environment is contaminated by a wide range of toxins and pollutants, which disrupt the life cycle of aquatic creatures and lower water quality. Certain pollutants are very capable of causing harm to the morphology and metabolism of aquatic organisms. Water pollution lowers DO, which can harm eggs and embryos. A multitude of physiological processes are interfered with by pollutants, however they do not always result in mortality. Every mucous membrane of a fish's gills is encrusted with toxic materials and suspended debris, which makes breathing difficult. Essentially, lead and mercury impair the functions of digestive enzymes. POPs, or particulates. Among the worst POPs are polychlorinated biphenyls (PCBs), which are used in military and electrical equipment like submarines.

In the past, dioxins and furans—by-products of burning polymers, wood, and fossil fuels—were produced by burning DDT, a pesticide used in forestry and agriculture. Polybrominated diphenyl ethers (PBDEs), on the other hand, are widely employed as flame retardants. The fatty layer of blubber on whales accumulates these pollutants. Certain chemicals have the potential to affect the reproductive system, increasing the chance of unwanted pregnancies. The Pacific Northwest's orcas are one massive and enthralling example of a species whose recovery is threatened by pollution.

But pollution also endangers the lives of countless other animals by threatening their health. From fish and seabirds with tummies full of plastic trash to frogs consuming dangerous compounds in pond water, marine turtles, and right whales becoming entangled in abandoned fishing gear .Many studies have periodically reported on the various impacts of pollutants on the populations of distinct fish species, and they anticipate a chronic level that results in a range of effects on aquatic life, including modifications to the histopathology, physiological harm, migration, and changes to the processes of embryonic and developmental development, especially in fish species.

Air Borne Perils on Biodiversity

The three primary pollutants that affect biodiversity are ground-level ozone, nitrogen fallout, and acidification. The specific pollutants are nitrogen compounds, sulphur dioxide, and volatile organic compounds. Air pollution directly affects higher animals by producing gasses or acidity. Reproductive process disturbance and food supply depletion are typically the main indirect impacts. Certain bird species, such the pied flycatcher and great tit, have thinned their eggshells as a result of acidification's high aluminium content. Lichens have the most responses to pollution, accumulating sulphur dioxide and acids in both wet and dry forms. Numerous delicate species have vanished from the local and, in certain cases, national landscapes. Air pollution affects the respiratory systems of animals, particularly turtles, where it can lead to internal damage and eventually death. It has been noted that seabirds who consume plastic, like the *Layson albatross*, die before they even leave their nests. The leaves of plants are the most severely harmed by air pollution. Chlorosis and leaf necrosis are two of these harms.

Light Pollution Impact on Nocturnal Fauna

ALAN, or artificial light at night, denotes the lighting of nighttime landscapes using man-made sources such as street lamps, pathway lights, and vehicle headlights. This illumination is prone to disturbing the habitats of nocturnal and crepuscular animal species that reside in these regions (Davies & Smyth, 2018; Gaston & Holt, 2018). Over 30% of vertebrates and nearly 60% of invertebrates in the natural world are nocturnal. These species are arguably the most susceptible to artificial lighting, and the scientific literature contains fewer examples of them. ALAN can desynchronize organisms from their usual biorhythms and induce temporal disorientation. The overwhelming majority of land-dwelling creatures follow circadian, circamensual, and/or circannual rhythms of behavior, including activities such as foraging, reproduction, and migration. These rhythms are synchronized with the natural cycles of light that occur on a daily, monthly, and yearly basis.Using monochromatic yellow light from low-pressure sodium (LPS) lamps with UV irradiation to attract the diurnal *Bombus terrestris* bum-blebees is one method of artificially attracting the pink bollworm *Pectinophora gossypiella.* Fruit flies (*Drosophila melanogaster*) have a lower reproductive fitness when they desynchronize. During dark nights without moonlight, *Noctua pronuba* yellow under-wing moths align themselves with the North Star, while *Scarabaeus satyrus* dung beetles utilize the Milky Way as a navigational aid to move away from dung piles in the shortest and most efficient route available for migration. There are two mechanisms through which artificial light at night (ALAN) could disrupt these nocturnal navigation methods: Ecological light pollution distinguishes itself from atmospheric light pollution by the abundance of artificial light sources in the night sky, which might be confused with the moon or stars. Meanwhile, evidence of prevalent light pollution is obscured by the former type of pollution.

In cities, skyglow significantly reduces the sight of stars (Falchi *et al.*, 2016), and artificial lights are known to disrupt polarization signals (Owens *et al.*, 2018). Bats come into direct touch with ALAN because of their nocturnal behaviour. Various bat species' responses to ALAN vary depending on how they fly and forage. *Plecotus spp.*, *Myotis spp*, and *Rhinolophus spp.* are examples of slow-moving animals that are specialized in grazing along or among dense foliage; they avoid lit regions because they perceive an inherent increase in the risk of predation. However, regardless of feeding technique, light pollution has a deleterious impact on bat species occurrence at the landscape scale. (Azam ., *et al* 2018).

Silent Victims: Impact of Noise Pollution on Small Animals

Not only can noise harm people, but it also has an adverse effect on a wide variety of nonhuman creatures, such as fish, amphibians, birds, mammals, and arthropods. Animals are generally affected by noise at various levels of the biological system, including physiology, behaviour and ecological processes. For example, noise pollution from industry and traffic interferes with birds' ability to perceive warning cries, increasing the danger of predation in locations affected by noise pollution. The recognition of acoustic cues by greater mouse-eared bats (*Myotis myotis*) to locate their insect prey is similarly interfered with by traffic noise, resulting in a decreased hunting effectiveness near motorways (Framond *et al.,* 2022). The masking effect increases with increasing noise level. Animal signals, including calls used for social communication, sounds used by mothers to recognize their young, echolocation signals, noises from the surroundings, and sounds made by predators and prey, can all be affected by masking (Dooling and Leek 2018). Urban noise also affects birds' choices of habitats, often having negative effects on their ability to live and nest. For example, compared to individuals living in quieter areas, Eastern bluebirds (*Sialia sialis*) reported to have lower reproductive production and smaller broods. (Erbe *et al.*, 2022). Bird reproduction and mating patterns have been found to be impacted by urban noise in a number of ways. Urban noise can interfere with the elaborate courtship displays of male greater sage grouse during lekking. Additionally, it disrupts the preference of female sage grouse for low-frequency music produced by male great tits and canaries. Since males of higher quality typically sing songs with lower frequencies, females across various species, including sage grouse, exhibit a preference for males emitting lower-frequency songs over those producing higher-frequency ones. (Erbe *et al.*, 2022).

CLIMATE CHANGE: A GLOBAL CHALLENGE TO BIODIVERSITY

Climate change and biodiversity are significantly correlated. Even though the temperature has fluctuated throughout Earth's history in response to the evolution of ecological groups and the extinction of species, the disruption of ecological systems and species' ability to adapt due to accelerated climate change contributes to the loss of biodiversity (Singh *et al.*, 2021). Many species' habitats are altered and their survival is threatened, increasing the likelihood of their extinction. As a result, even a small change in the climatic pattern can have a big impact on biodiversity. Biogeography and ecosystem distribution are strongly influenced by climate, as are the distribution of plant vegetation zones and ecosystems, or biomes. The loss of entire species and ecosystems could occur from certain plants and animals' inability to adapt to the changing environment. Climate change might be the reason for these distributions to change. This results in a loss of some biodiversity. When one species goes extinct, so do species that are closely related to it.

Shifting Landscapes and Changing Canopies

Changes in the frequency of disturbances, interactions with invasive species, and habitat degradation are some of the factors that can lead to the extinction of rare plant populations. Globally, species ranges are being impacted by climate change in particular, with creatures moving toward higher latitudes and elevations. For instance, over the past century, plant ranges have shifted upslope at a pace of 29 m/decade in Western Europe and at a similar rate in California. The consequences of climate change on moisture and temperature may endanger Europe's plant diversity, particularly in the highlands.

As environmental conditions undergo alterations, species found in low-elevation habitats become increasingly at risk of extinction, primarily due to the tendency for their habitats to shift upwards in elevation. This heightened vulnerability is particularly acute for rare species, which already occupy limited geographic ranges. Any diminishment in the extent of their distribution could have far-reaching implications for their long-term survival and complicate conservation efforts aimed at safeguarding these species (Kaye *et al.*, 2019). Due to the escalation in temperatures, a considerable number of plant species within the Nainital region have undertaken a relocation towards elevated altitudes. Notable among these are *Berberis asiatica, Taraxacum officinale, and Jasminum officinale*, alongside other varieties. Projections indicate an imminent shift in the dominant flora of central India, wherein teak woods are anticipated to gradually replace the incumbent Sal trees. Furthermore, there exists a conjecture that deciduous species might progressively encroach upon habitats traditionally occupied by conifers, thus signalling a potential alteration in the ecological dynamics of the area. Tree species may have a restricted ability to regenerate because tree seedlings are extremely vulnerable to climatic changes, such as drastic changes in temperature and water availability.

Research in Australia's eucalyptus forest has shown that in the next decades, changes in temperature of 1.0 to 4.5 C and reductions in precipitation of 3 to 25% will affect the regeneration of tree species at both the landscape and occurrence levels. It is predicted that alpine and boreal forests will extend further northward and upward along the peripheries of alpine ecosystems and lower tundra regions as elevations and latitudes rise. This anticipated movement, implies a substantial geographical shift in forest habitats (Srivastava et al. 2019). Furthermore, according to FAO estimates from 2000, around 9% of all plant species are believed to be at risk of extinction as a consequence of these environmental alterations.

Rising Tides, Vanishing Roots: Impact of Climate Change on Mangrove Forest

In addition to serving as vital breeding grounds and nurseries for a variety of semi-terrestrial and estuarine organisms, mangroves also mitigate the effects of extreme weather events like cyclones and tsunamis, provide a renewable supply of wood for housing and fuel, as well as food and traditional medicines, and are significant sites for the accumulation of sediment, carbon, nutrients, and pollutants. The melting of land and polar ice combined with the thermal expansion of ocean water brought on by rising temperatures is currently causing mean sea levels to rise.

In numerous low-lying islands in Micronesia and Melanesia, rising sea levels will cause mangroves to go extinct locally. A prime example is the Sundarbans, an area of coastal India and Bangladesh where numerous deltaic islands are continuously subsiding and disappearing. Mangroves in other river deltas have shifted landward, but their net contraction has resulted from increased subsidence and decreased sediment supply due to sea level rise, storms, and cyclones. 28% of the mangrove ecosystem has disappeared as a result of the sea level increase that has been observed during the last 40 years. Based on calculations, it is possible that up to 96% of the viable tiger habitat in the Sundarbans could vanish within the next 50–90 years.

Mangroves respond to changes in their environment by exhibiting variations in their morphology and anatomy. Many species' leaf structure, vascular vessel density, diameter, grouping and length, and fiber wall thickness are affected by variations in salinity and flooding. (Alongi *et al.*, 2015).

Oceanic Crisis

Corals stand out as one of the marine species profoundly affected by the increasing temperatures and acidification of the oceans. Within their symbiotic relationship, corals and algae share a crucial bond wherein the algae provide coloration to the corals and supply them with photosynthates. This mutualistic association plays a pivotal role in the health and vitality of coral reef ecosystems. Due to their great sensitivity to heat and acidity, corals can bleach when ocean temperatures rise by as little as 2-3°F above usual. This is because the symbiotic algae are forced out of the tissues of the corals (Hoegh-Guldberg *et al.*, 2017). Up to 95% of corals in the Indian Ocean bleach over this period of time, and corals eventually die from it; ultimately, 16% of corals are declared extinct globally. Experts anticipate that if greenhouse gas emissions continue on their current course, the reef will radically change in fifty years. Sharks, turtles, and whales are just a few examples of the diverse array of species that rely on coral reefs as their habitat. The demise of corals would have far-reaching consequences for the entire ecosystem. The North Atlantic whale's primary food source, planktons, are declining, which could lead to its extinction (Brat *et al.,* 2020). The high temperature causes more female turtles to be born, which disturbs the sea turtle population's sex ratio. In certain regions, the location and size of fish populations are shifting due to climate change, which presents a significant danger to marine fisheries used for both commercial and recreational purposes. The geographic distribution of several fish species has already shifted in response to climate change. Pollock and cod, for example, migrate northward to colder waters when local ocean temperatures rise. Fisheries may be impacted by the timing of seasonal events, which is also being affected. For example, certain species—like striped bass—are procreating earlier in the year (Brat *et al.,* 2020).

Terrestrial Biodiversity in Peril

Climate change has a variety of impacts on biodiversity, with even minor shifts in climate capable of driving certain animal species to extinction. Examples include the disappearance of species like the golden toad and the *Monteverde harlequin* frog. Additionally, the reduction in Arctic ice coverage endangers animals such as seals, penguins, polar bears, and Arctic foxes. Modern research indicates that there is a greater chance of extinction for native species such as the *Nilgiri tahr*. It is observed that several species, like the Black and Rufous Flycatcher (*Ficedula nigrorufa*), are shifting to higher elevations from their lower range boundaries. Moreover, climate change alters the behaviour of sick animals.

The terrible sickness that affects amphibians which warmer weather has probably made the *chytrid* fungus worse, which has caused many amphibian populations to decline or go extinct. The phenological shift that plants have experienced is another effect of climate change. Numerous reproductive processes in flowering plants, including the length, intensity, and timing of fruiting and flowering, are significant phonological events. The *Rhododendron arboretum*, a significant Himalayan tree species, typically flowers from early February to mid-March. Over the past century, temperature fluctuations have caused this plant to bloom 88–97 days earlier, as determined by a generalized additive model incorporating real-time field data (2009–2011) and historical herbarium records (1893–2003). Climate change is reshaping the distribution patterns of plants, animals, and insects. Global studies indicate that a mere 2 °C rise in temperature could lead to a significant loss of habitat for approximately 18% of plant species, 16% of insect species, and 8% of vertebrate species, with even more dire consequences anticipated with a 1.5°C increase, albeit to a somewhat lesser extent, affecting 6% of insects, 8% of plants, and 4% of vertebrates. (Warren *et al.*, 2018).

Extinction of Various Species Globally

Given the reported rapid loss of climatically suitable area, it should come as no surprise that mountain-restricted species have experienced the first total extinctions connected to climate change. It is conceivable that 20% of species may go extinct in the coming decades given the growing number of causes contributing to species extinction. A startling 67% of harlequin frogs in the tropical regions of Central and South America have vanished in the last 20 to 30 years. Throughout the past 40 years, all populations of the cool-adapted Apollo butterfly (*Parnassius apollo*) in southern France have become extinct. Seven of the twenty-five pika (*Ochotona princeps, Lagomorpha*) populations that had been surveyed in the Great Basin of the western United States had gone extinct since the 1930s. The primary food source for North Atlantic right whales, plankton numbers are decreased by climate variations in North America. There are currently just about 300 individuals left, and a growing contributing factor to death is the decreased food supply brought on by climate change. Approximately 27,000 species are thought to go extinct annually. By 2050, thirty percent of the world's species might become extinct if this trend keeps up. The current pace of extinction is 100–1000 times higher than the rate of extinction that occurs naturally. (Prakash and Srivastava., 2019). The rising maximum air temperature approaching the physiological limit has led to the extinction of 48 lizard species (genus *Sceloporus*), presumably as a result of decreased surface activity during Mexico's breeding season. The decline in the temporal overlap between the larvae and host plants in San Francisco, USA, coupled with an increase in precipitation variability has resulted in the extinction of the checker spot butterfly (*Euphydryas edithabayensis*). Due to a decrease in precipitation, the plant community in California has changed, resulting in the extinction of desert bighorn sheep (*Ovis canadensis*). In New Britain, Papua, and New Guinea, fish (*Gobiodon* sp. A) hav e become extinct as a result of coral bleaching brought on by rising sea temperatures, which has destroyed their obligatory coral home. In the Baltic Sea, Mundey Eelpout (*Zoarces viviparus*) are limited in oxygen because to high temperatures (Prakash and Srivastava., 2019).

CASE STUDIES

Climate Change is Turning Antarctica Green

The planet's thermostat, the polar ice caps, are impacted by climate change due to their high albedo. The percentage of sunlight that a surface reflects back is called its albedo. Greening will change the ice caps from "net reflective" to "net absorptive," disrupting the world's energy equilibrium and hastening the melting of the polar ice caps. The Arctic has been changing dramatically and warming twice as fast as the rest of the planet during the past few decades. Over the past 38 years, there has been a roughly 39% decrease in Arctic sea ice.

Antarctica's ice also decreased by 6.2% during same time. According to a recent study, moss has been steadily growing in Antarctica over the past 50 years as a result of rising temperatures brought on by climate change. Phytoplankton, free-floating microscopic marine algae, are flourishing as warming seas reduce ice cover. Because of the migration of species from the south, it is predicted that the overall number of moss species in the Arctic will grow under a warmer environment. The Antarctic Peninsula, which has been warming more quickly than the rest of the continent, will become a considerably greener region in the future due to the glaciers' ongoing retreat.

Receding Ice Creates Risks for Arctic and Antarctic Species

Sea ice cover in the Arctic is continuing to thin and age, with a 13% annual decrease in summertime sea ice. Over the previous 30 years, there has been a more than 33% reduction in the amount of snow on sea ice in the western Arctic. If the rise in global temperatures is constrained to 1.5°C, it is projected that the crucial summer sea ice in the Arctic's marine ecosystems will persist. On the other hand, if the temperature rises above 1.5°C, we will lose the Arctic summer ice within a few decades. Animals that depend on sea ice, such as walruses, polar bears, and narwhals, are especially vulnerable because of variations in the rate and timing of sea ice melting. By 2100, polar bears might go hungry.

The distribution of fish populations is undergoing shifts, with species traditionally found in the southern Arctic, such as orcas, expanding their range further north (Barrett et al., 2023). During 1970, the Adelie and emperor penguins (*Pygoscelisadeliae* and *Aptenodytes forsteri*, respectively), reliants on sea ice, have almost disappeared from their northernmost habitats in the Antarctic. According to Gross (2005), the number of breeding pairs of emperor penguins in the western Antarctic Peninsula plummeted from 300 to a mere 9. Similarly, polar bear populations have experienced notable declines along their southern range boundaries. In the southern reaches of Hudson Bay, a critical habitat for polar bears, both population numbers and average body weights are dwindling. As the presence of ice shelves is vital for their hunting grounds, the lengthening of ice-free periods in Hudson Bay due to climate change forces these bears to rely on their fat reserves, leading to starvation. (Barrett *et al.*, 2023).

Shifting Habitats of Birds

Beyond alterations in migration schedule, migratory birds may react to warming, especially when it impacts the spatial distribution of supplies and habitats. As temperatures rise, suitable breeding and staging habitats are probably going to move further north (Li. X *et al.*, 2022). Some bird species are thought to be seriously threatened by the changes in their environment brought on by global warming. Due to the extensive degradation of their appropriate habitats caused by escalating temperatures, certain prevalent avian species found in the high-altitude regions of the Peruvian mountains might experience a decline in population or face the risk of extinction (Li *et al.*, 2022). This alteration is also thought to be the main cause of the widespread extinction of high-altitude species in the tropical Andes, endangering bird populations in other tropical regions (Freeman *et al.*, 2018). Furthermore, research on climate change are causing temperatures to rise, which opens the door for viruses to spread to higher altitudes and infect nearby birds. According to a study, climate change is raising Hawaii's temperatures and expanding the places where diseases may thrive, which has led to an increase in the geographic area where avian malaria is spreading (Atkinson, C. T., and La Pointe, D. A. 2009). Under the scenario of sustained high greenhouse gas emissions, it is anticipated that the habitat of the White-tailed Ptarmigan on Vancouver Island will diminish by 25% in the 2040s, 44% in the 2050s, and 56% in the 2080s. Conversely, a more pronounced decline of 27%, 59%, and 74% is projected under conditions of elevated greenhouse gas emissions.

As temperatures continue to rise, the destruction of habitats is compelling a growing proportion of alpine bird species towards the brink of extinction. (Şekercioğlu et al., 2012).

Rising Sea Levels Are Submerging Turtle Nesting Sites

Highly correlated with the escalating global temperatures, the present mean sea level rise of 3.6 mm per year is anticipated to escalate by 0.5–2.0 meters by the year 2100. Freshwater reptile populations have borne significant consequences from the surging sea levels in various regions. For instance, the Roti Island snake-necked turtle (*Chelodina mccordi*) faces the grim prospect of extinction due to the encroaching water levels on the low-lying islands it inhabits (Bickford et al., 2010). Furthermore, it's predicted that up to 40% of the green turtle (*C. mydas*) nesting beaches in the northwest Hawaiian Islands could disappear due to a 0.9 m rise in sea level. Due to projected increases in sea levels, the primary nesting grounds of the world's largest green turtle population, situated in the northern region of Australia's Great Barrier Reef, could potentially diminish by up to 38%. Rising sea levels typically affect the availability of sea turtle nesting sites, pushing females to find other places to lay their eggs along neighbouring beaches. Due to coastal erosion, there are fewer potential nesting sites available, therefore numerous females may select the same area—a behaviour known as communal nesting. This higher nest density can also lead to increased embryo mortality and poorer overall reproductive success because of the increased danger of nest destruction by other nesting females. (Dayananda *et al.*, 2021).

MITIGATION

The primary strategies for mitigating the effects of climate change are evident: reducing greenhouse gas emissions substantially, halting deforestation, and expanding forest coverage. To curb emissions, a substantial decrease in the reliance on coal and other fossil fuels is imperative. Given the global nature of climate change, localized solutions from individual nations are insufficient, necessitating collaborative efforts to address it collectively. Numerous efforts have been made since many decades ago in an attempt to achieve these objectives globally. The majority of mitigation measures are discussed diplomatically between governments, nongovernmental groups, and states in international organizations (Shivanna *et al.*, 2022).

United Nations climate change conference: The United Nations Climate Change Conference (COP 21) convened in November 2015 in Paris, France, marking the first instance in the conference's history where a global consensus on mitigating climate change was achieved. The vast majority of nations expressed strong support for the agreement. The primary aim was to limit global warming to below 2°C compared to pre-industrial levels by the year 2100, as outlined in the UN Report of 2015. Additionally, the final version of the Paris Agreement, adopted in 2015, included a provision urging parties to strive towards limiting temperature rises to 1.5°C.

Conference of Parties (COP) The UN Framework Convention on Climate Change (UNFCCC) was established in 1992 to coordinate global responses to climate change. Its inaugural meeting (COP 1) convened in Berlin in March 1995, marking the start of yearly gatherings. The guidelines for implementing the Paris Agreement, aimed at capping global warming at 1.5°C, were established during COP 26. Major nations like the United States and the European Union have committed to achieving carbon neutrality by 2050, while China and India have set targets for net zero emissions by 2060 and 2070 respectively. India has also pledged to reduce its fossil fuel dependency by 40% by 2030. More than 100 countries have vowed to halt deforestation, aiming to slash global methane emissions by 30% by 2030 compared to 2020 levels. The US National Oceanic and Atmospheric Administration has observed a substantial rise in atmospheric methane levels, underscoring the urgency of fulfilling COP 26 commitments (Shivanna *et al.*, 2022).

Intergovernmental panel on climate change (IPCC): The United Nations Environmental Programme (UNEP) and the World Meteorological Organization collaborated to establish the Intergovernmental Panel on Climate Change (IPCC). Its primary mandate is to gather a comprehensive assessment of various aspects of climate change as analyzed by scientists. Three working groups are aiding the IPCC in crafting its fifth assessment report on climate change: Group I will focus on elucidating the scientific foundations of the issue, Group II will address impacts, adaptation, and vulnerability, and Group III will examine potential mitigation strategies.

The United Nations' framework convention on climate change (UNFCCC): Commencing on March 21, 1994, the UNFCCC was founded. The 195 countries that have in total approved the treaty are parties to it. The UNFCCC adopted the Rio Convention during the Rio Earth Summit in 1992. Reaching a stable concentration of greenhouse gases at which detrimental human interference with the climate system would not take place was the primary aim of the conference. As stipulated in the treaty, the developed nations must take the initiative. This is indicative since they are the main cause of greenhouse gas emissions from the past and present. Developed countries should endeavour to minimize their emissions (UNFCCC., 2015)

The Kyoto protocol: The Kyoto Protocol, which was adopted in Kyoto, Japan, in December 1997 and entered into force in February 2005, is connected to the United Nations Framework Convention on Climate Change. It established globally and nationally enforceable emission reduction targets. The entire set of guidelines for implementing the pact, approved during COP 7 in Marrakech, Morocco, in 2001, is known as the "Marrakesh Accords." It concluded in 2012, having begun in 2008. According to this treaty, all member countries must use internal resources to achieve their goals. The agreement also provides three additional market-based mechanisms—the Joint Implementation, International Emissions Trading, and Clean Development Mechanism—to help nations meet their goals.

Climate change and India's actions: According to estimates, the year 2007 saw 1331.6 million tons of greenhouse gas (GHG) emissions equivalent to carbon dioxide. India's share of global CO2 emissions was approximately 4%, and even less if historical concentrations are considered. This represents a yearly increment in emissions levels of 4.2% since the level of 1994. India is nevertheless aware of the worldwide threat posed by climate change. India has committed to carrying out its duties as a party to the United Nations Framework Convention on Climate Change (UNFCCC) by sharing information about how the convention is being implemented, taking into account its own national and regional development priorities, objectives, and circumstances in addition to its individual capacities and shared but separate responsibilities.

Indian perspective on the Paris 2015 international climate change accord: India pledges to actively participate in a range of UNFCCC Paris, 2015 negotiations in a way that is innovative, creative, and forward-thinking. Its principal goal is to create, within the framework of the UNFCCC, an unbiased, collaborative, and effective global planning system based on climate justice, equity, and separate but common responsibilities and capabilities.

India set a lofty goal to acquire 40% of its electricity from renewable sources by the year 2030.The Indian government recently introduced a solar power treaty with the goal of increasing solar power output in underdeveloped nations. India has also established a goal to increase its solar power production capacity to 100 GW by 2022, a significant increase over present levels (TIME, 2015).

The National action plan on climate change (NAPCC): In June 2008, the National Action Plan on Climate Change (NAPCC) unveiled a roster of initiatives and policies aimed at both adapting to and mitigating the effects of climate change for the first time. Oversight of strategizing and executing these measures in specific regions falls under the purview of nodal Ministries, primarily managed by the Ministry of Environment, Forest and Climate Change (MOEFCC). Several national missions have been established under the NAPCC, encompassing endeavors such as the National Solar Mission, the National Mission for Enhanced Energy Efficiency, the National Mission on Sustainable Habitat, the National Water Mission, the National Mission for Sustaining the Himalayan Ecosystem, the National Mission for a Green India, the National Mission for Sustainable Agriculture, and the National Mission on Strategic Knowledge for Climate Change. These missions also highlight ongoing significant projects in areas like clean energy and renewable energy geared towards fostering sustainable development.

Parliamentary forum on global warming and climate change: Since the Indian government established the forum in 2008, experts on climate change and global warming have had discussions with policymakers. The Indian government submitted a plan to the UNFCCC with the goal of lowering the emission intensity of India's GDP by 20–25% by 2020.

CONCLUSION

It is clear that climate change is causing biodiversity to disappear. Human activity is the main factor behind all of these environmental changes that are detrimental to biodiversity. The effects of increasing greenhouse gas emissions on the planet's temperature have an impact on biodiversity, ecological balance, and human populations. Even a small change in temperature has the potential to wipe out some delicate and endangered species. Changes in breeding and migration behaviour, changing distribution patterns, invasions by other species, an increase in insect assaults and forest fires, and other factors are some of the ways that climate change is affecting biodiversity. Facilitating the harmonious and efficient alignment of climate change and biodiversity efforts in India involves enacting environmentally sound legislation, integrating these concerns into national agendas, and implementing biodiversity-driven strategies for climate change mitigation. The United Nations Framework Convention on Climate Change (UNFCCC) has implemented specialized initiatives such as REDD (Reducing Emissions from Deforestation and Forest Degradation), which have progressively broadened to encompass endeavors focusing on forest conservation, sustainable forest management, and the augmentation of forest carbon reservoirs.

REFERENCES

Abrams, J., Davis, E. J., Moseley, C., & Nowell, B. (2017). Building practical authority for community forestry in and through networks – The role of community based organisations in the U.S. West. *Environmental Policy and Governance*, *27*(4), 285–297. doi:10.1002/eet.1765

Alongi, D. M. (2015). The impact of climate change on mangrove forests. *Current Climate Change Reports*, *1*(1), 30–39. doi:10.1007/s40641-015-0002-x

Atkinson, C. T., & La Pointe, D. A. (2009). Introduced avian diseases, climate change, and the future of Hawaiian honeycreepers. *Journal of Avian Medicine and Surgery*, *23*(1), 53–63. doi:10.1647/2008-059.1 PMID:19530408

Azam, C., Le Viol, I., Bas, Y., Zissis, G., Vernet, A., Julien, J. F., & Kerbiriou, C. (2018). Evidence for distance and illuminance thresholds in the effects of artificial lighting on bat activity. *Landscape and Urban Planning*, *175*, 123–135. doi:10.1016/j.landurbplan.2018.02.011

BarrettJ. E.AdamsB. J.DoranP. T.DuganH. A.MyersK. F.SalvatoreM.GooseffM. N. (2023). Weather whiplash in a terrestrial polar ecosystem following the March 2022 Antarctic weather anomaly. Authorea Preprints. doi:10.22541/essoar.170144175.54028237/v1

Brat, D., & Pathak, I. R. (2020). Impact of Pollution on Biodiversity: A Review. *IRE J.*, *4*(2), 11-15.

Cahill, A. E., Aiello-Lammens, M. E., Fisher-Reid, M. C., Hua, X., Karanewsky, C. J., Yeong Ryu, H., & Wiens, J. J. (2013). How does climate change cause extinction. *Proceedings of the Royal Society B: Biological Sciences, 280*(1750), 20121890. 10.1098/rspb.2012.1890

Davies, T. W., & Smyth, T. (2018). Why artificial light at night should be a focus for global change research in the 21st century. *Global Change Biology*, *24*(3), 872–882. doi:10.1111/gcb.13927 PMID:29124824

Dayananda, B., Bezeng, S. B., Karunarathna, S., & Jeffree, R. A. (2021). Climate change impacts on tropical reptiles: Likely effects and future research needs based on Sri Lankan perspectives. *Frontiers in Ecology and Evolution*, *9*, 688723. doi:10.3389/fevo.2021.688723

De Framond, L., & Brumm, H. (2022). Long-term effects of noise pollution on the avian dawn chorus: a natural experiment facilitated by the closure of an international airport. *Proceedings of the Royal Society B, 289*(1982), 20220906. IEEE. 10.1098/rspb.2022.0906

Dooling, R. J., & Leek, M. R. (2018). Communication masking by man-made noise. In H. Slabbekoorn, R. J. Dooling, A. N. Popper, & R. R. Fay (Eds.), *Effects of anthropogenic noise on animals* (pp. 23–46). Springer. doi:10.1007/978-1-4939-8574-6_2

Erbe, C., Dent, M. L., Gannon, W. L., McCauley, R. D., Römer, H., Southall, B. L., & Thomas, J. A. (2022). The effects of noise on animals. In Exploring animal behavior through sound: volume 1: methods (pp. 459-506). Cham: Springer International Publishing. doi:10.1007/978-3-030-97540-1_13

Falchi, F., Cinzano, P., Duriscoe, D., Kyba, C. C., Elvidge, C. D., Baugh, K., Portnov, B. A., Rybnikova, N. A., & Furgoni, R. (2016). The new world atlas of artificial night sky brightness. *Science Advances*, *2*(6), e1600377. doi:10.1126/sciadv.1600377 PMID:27386582

Freeman, B. G., Scholer, M. N., Ruiz-Gutierrez, V., & Fitzpatrick, J. W. (2018). Climate change causes upslope shifts and mountaintop extirpations in a tropical bird community. *Proceedings of the National Academy of Sciences of the United States of America*, *115*(47), 11982–11987. doi:10.1073/pnas.1804224115 PMID:30373825

Fuentes, M., Limpus, C., Hamann, M., & Dawson, J. (2010). Potential impacts of projected sea-level rise on sea turtle rookeries. Aquatic Conserv. Mar. *Aquatic Conservation*, *20*(2), 132–139. doi:10.1002/aqc.1088

Gaston, K. J., & Holt, L. A. (2018). Nature, extent and ecological impli- cations of night-time light from road vehicles. *Journal of Applied Ecology*, *55*(5), 2296–2307. doi:10.1111/1365-2664.13157 PMID:30147142

Hoegh-Guldberg, O., Poloczanska, E. S., Skirving, W., & Dove, S. (2017). Coral reef ecosystems under climate change and ocean acidification. *Frontiers in Marine Science*, *4*, 158. doi:10.3389/fmars.2017.00158

IPCC. (2007). An assessment of Intergovernmental Panel on climate change. Climate change synthesis report. Cambridge University Press, Cambridge.

IPCC. (2013). *Climate Change, the Physical Science Basis*. Cambridge University Press.

Kaye, T. N., Bahm, M. A., Thorpe, A. S., Gray, E. C., Pfingsten, I., & Waddell, C. (2019). Population extinctions driven by climate change, population size, and time since observation may make rare species databases inaccurate. *PLoS One*, *14*(10), e0210378. doi:10.1371/journal.pone.0210378 PMID:31622344

Li, X., Liu, Y., & Zhu, Y. (2022, April). The Effects of Climate Change on Birds and Approaches to Response. []. IOP Publishing.]. *IOP Conference Series. Earth and Environmental Science*, *1011*(1), 012054. doi:10.1088/1755-1315/1011/1/012054

McLaughlin, J. F., Hellmann, J. J., Boggs, C. L., & Ehrlich, P. R. (2002). Climate change hastens population extinctions. *Proceedings of the National Academy of Sciences of the United States of America*, *99*(9), 6070–6074. doi:10.1073/pnas.052131199 PMID:11972020

MoEFCC. (2014). *India's fifth national report to the convention on biological diversity 2014*. Ministry of Environment, Forests and Climate Change, New Delhi, India.

Nicholls, R. J., Marinova, N., Lowe, J. A., Brown, S., Vellinga, P., De Gusmao, D., Hinkel, J., & Tol, R. S. J. (2011). Sea-level rise and its possible impacts given a 'beyond 4_C 'world'in the twenty-first century. *Philosophical Transactions. Series A, Mathematical, Physical, and Engineering Sciences*, *369*(1934), 161–181. doi:10.1098/rsta.2010.0291 PMID:21115518

Owens, A. C., & Lewis, S. M. (2018). The impact of artificial light at night on nocturnal insects: A review and synthesis. *Ecology and Evolution*, *8*(22), 11337–11358. doi:10.1002/ece3.4557 PMID:30519447

Palita, S. K., Panda, D., & Nayak, J. K. (2023). Indigenous Communities And Biodiversity Conservation: An Indian Perspective. *SCIENCE AND CULTURE.*

Prakash, S., & Srivastava, S. (2019). Impact of Climate Change on Biodiversity: An Overview. *International Journal of Biological Innovations.*, *1*(2), 60–65. doi:10.46505/IJBI.2019.1205

Rajkhowa, R. (2014). Light pollution and impact of light pollution. [IJSR]. *International Journal of Scientific Research*, *3*(10), 861–867.

Şekercioğlu, Ç. H., Primack, R. B., & Wormworth, J. (2012). The effects of climate change on tropical birds. *Biological Conservation*, *148*(1), 1–18. doi:10.1016/j.biocon.2011.10.019

Shivanna, K. R. (2022). Climate change and its impact on biodiversity and human welfare. *Proceedings of the Indian National Science Academy. Part A, Physical Sciences*, *88*(2), 160–171. doi:10.1007/s43538-022-00073-6

Singh, S., Singh, P., Selvasembian, R., & Srivastava, K. K. (Eds.). (2021). *Global climate change*. Elsevier.

Srivastava, S., Shukla, S. N., & Singh, P. (2019). Climate change and Biodiversity management: A review. *International Journal of Environmental Sciences*, *10*(2), 71–75.

Thatoi, H. N. (2021). Biodiversity Conservation and Livelihood Management. Daya Publishing House, a division of Astral International Pvt. Limited.

UNFCCC. (2015). 19th Session of the Conference of the Parties to the UNFCCC. *International Institute for Sustainable Development*. UNFCCC.

UNFCCC (2015). *Adoption of the Paris agreement, Proposal by the President, Draft decision CP.21*. UNFCC.

Wang, Y., Liu, Y., & Gu, B. (2022). *COP26: Progress, challenges, and outlook.*

Warren, R., Price, J., Graham, E., Forstenhaeusler, N., & VanDerWal, J. (2018). The projected effect on insects, vertebrates, and plants of limiting global warming to 1.5 C rather than 2 C. *Science*, *360*(6390), 791–795. doi:10.1126/science.aar3646 PMID:29773751

Chapter 10
Biodiversity Loss Due to Forest Diversion

Vinaya Satyawan Tari
https://orcid.org/0000-0002-8352-252X
Universitas Airlangga, Indonesia

Vandana Gupta
B.K. Birla College (Autonomous), India

Ashutosh S. Joshi
University of Mumbai, India

Poonam Nirmalkumar Kurve
B.N. Bandodkar College, India

Naresh Chandra
B.K. Birla College of Arts, Science, and Commerce, India

ABSTRACT

The world is continuously progressing towards technological advancement. Eventually, global biodiversity is taking a consistent downfall. This is not only disrupting the ecosystem services and livelihood spatially but also impacts carbon offsetting globally in the present context. In the arena of biodiversity loss in the last three decades the problem remains unresolved, despite environmental regulations present at national and international levels to protect biodiversity. Forest diversion has been observed to be one of the major root causes for these losses which is further resulting in habitat deterioration and fragmentation. Numerous case studies have been reported that support biodiversity loss due to forest diversions. The conversion of forest land to commercial purposes such as plantations, infrastructural developments, and residential projects has shown a major impact on biodiversity loss, especially in the tropics.

DOI: 10.4018/979-8-3693-6950-0.ch010

Table 1. Global distribution of forests showing countries with the largest forest area, 2020

Country	Forest Area (in million hectares)	% of Worlds Forest
Russian Federation	815	20
Brazil	497	12
Canada	347	9
United States of America	310	8
China	220	5
Australia	134	3
Democratic Republic of the Congo	126	3
Indonesia	92	2
Peru	72	2
India	72	2
Rest of the World	1375	34

Source: Food And Agriculture Organization, 2020

INTRODUCTION

The term biodiversity or biological diversity refers to the variation of life right from genes to ecosystems on Earth. It comprises the evolutionary, ecological, and cultural processes that sustain life. Today's biodiversity is the result of the evolution of 4.5 billion years. Plants and oceans are important carbon absorbers. Forests store twice as much carbon as they emit. In contrast, an estimated 25% of global carbon emissions are sequestered alongside forests in other vegetative forms, such as grasslands or rangelands (fields, prairies, shrublands, etc.). Ecosystems and the biota inhabiting them face serious threats due to it. This is an attempt to highlight the specific case studies to envisage the effects of Forest Diversion on the biodiversity of India. This chapter also included some international scenarios through interesting case studies.

According to the report of UNEP on the state of the world's forests (2020), the global distribution of forests is shown in Table 1.

Unfortunately, the expanding population is demanding more space, hence leading to loss of forest and forest diversion. FAO (2020) mentions that the rate of net forest loss declined from 7.8 million ha per year in the decade 1990–2000 to 5.2 million ha in 2000–2010 and 4.7 million ha per year in 2010–2020. The rate of decline of net forest loss slowed in the most recent decade due to a reduction in the rate of forest expansion. According to the Policy papers of Analysis of Forest Diversion in India, 2019, the total forest cover in India is 7,08,273 Km2 which is 21.54% of its geographic area. In the year 2017 approvals were granted to 686 projects for diversion of 11,601.08 hectares of forest land for non-forestry uses. 43% of total forest land recommended for diversion falls in wildlife habitats and nearly 60% falls under the dense forest category.

About one-third of the greenhouse gas emissions reductions needed in the next decade could be achieved by improving nature's ability to absorb emissions. The government deals with climate change and biodiversity through two different international agreements – the UN Framework Convention on Climate Change (UNFCCC) and the UN Convention on Biological Diversity (CBD), both established at the 1992 Rio Earth Summit (Sage 2007). Similar to the historic Paris Agreement made in 2015 under the UNFCCC, parties to the Biodiversity Convention in December 2022 adopted an agreement for nature, known as the Kunming-Montreal Global Biodiversity Framework, which succeeds the Aichi Biodiversity Targets adopted in 2010. The framework includes wide-ranging steps to tackle the causes of biodiversity loss worldwide, including climate change and pollution.

During the presentation of the UN report in 2019, the Director-General of the United Nations Educational, Scientific and Cultural Organization (UNESCO), Audrey Azoulay, urged action with these words: "*We can no longer destroy the diversity of life. It is our responsibility to future generations*". As the planet, we are failing to meet all the targets set for slowing down biodiversity destruction by 2020. This is the devastating conclusion of the fifth Global Biodiversity Outlook report published in September 2020 by the Intergovernmental Science-Policy Platform on Biodiversity and Ecosystem Services (IPBES). The document not only warns of the alarming degradation of nature but points to it as a variable that increases the risk of future pandemics. In December 2022, governments met in Montreal, Canada to agree on a new framework to secure an ambitious and transformative global plan to set humanity on a path to living in harmony with nature.

Traditional Environmental, Social, and Governance (ESG) metrics have primarily focused on promoting sustainable finance, positive screening, and sustainability reporting. However, recent research highlights the urgency for greater accountability and action to counter species extinction. This chapter explores the potential of ESG frameworks in guiding corporate and managerial decision-making to address biodiversity loss. There is a need to call all stakeholders—business and policy decision-makers, conservationists, and environmental organizations—to formulate robust, inclusive, and ecologically sensitive strategies incorporating deep ecological perspectives. The findings of this study include recommendations for the Global Reporting Initiative (GRI). This chapter will also shed some light on the role of sustainable development goals viz. Decent Work and economic growth (SDG 8), Industry, innovation, and Infrastructure (SDG 9), Climate Action (SDG 13), Life below Water (SDG 14), and Life on Land (SDG 15) as a step towards biodiversity conservation.

IMPACT OF FOREST DIVERSION

The utilization of forest land and allied resources for purposes other than the forest is called 'Forest diversion'. India has a remarkable diversity of flora and fauna. Yet, for quite a few reasons this diversity has been impacted by forest diversion for various development projects such as expansion of agriculture, industrialization, infrastructural facilities, road construction, etc.

Biodiversity is an integral component of every ecosystem on this planet. It not only helps maintain the ecosystem structure and function but ensures an uninterrupted supply of ecosystem services as well (Costanza et al., 1997; Millennium Ecosystem Assessment (MEA), 2005; TEEB Synthesis, 2010). As per the Convention for Biological Diversity, over 4.3 billion people depend on biodiversity for their livelihoods, with 70% of the world's poor and vulnerable living in rural areas depending directly on it. Over 1.6 billion of these are completely or partially dependent on forests alone for their livelihood (CBD, 2017).

The impact of forest diversions on biodiversity loss is multifaceted and extends beyond the immediate conversion of land. Globally, the corporate industrial sector with approximately $44 trillion of financial revenue viz. half of the world's Gross Domestic Product (GDP) is moderately or highly dependent on nature and thus is endangered by biodiversity loss (The New Nature Economy Report, 2020). Looming threats from anthropogenic activities led by forest loss, fragmentation, and degradation (FAO, 2015) have already caused severe biodiversity decline, exotic invasion, and loss of habitat heterogeneity (Lindenmayer and Franklin 2002; Newbold et al., 2015; Van der Plas et al., 2016). According to data from Global Forest Watch (GFW), an online platform that invigilates global forest cover changes, the world lost an average of 10.1 million hectares of forest (including both natural forests and planted forests) per year between 2015 and 2020.

Regrettably, forest destruction and diversion have a long history in the world scenario. Most of the early civilizations across the world have been built and flourished by converting forest ecosystems into anthropogenic endeavors and infrastructure. Later, countries, particularly Western countries underwent an industrial revolution. Their very model of development was based on rapid urbanization and development at the cost of forests and other natural resources. These were the major culprits of forest diversion and resulting biodiversity loss in the Western world primarily. According to a study by Rober Neil and colleagues, the last 6,000 years, have seen more than a 50% reduction in forest habitats in European countries; mainly led by agricultural development and fuel wood (Robert et al, 2018). The international history of forest diversions for non-forest purposes spans centuries and has evolved in response to changing societal needs, economic dynamics, and environmental apathy influenced by the notion of progress.

Figure 1. Forest diversion and biodiversity

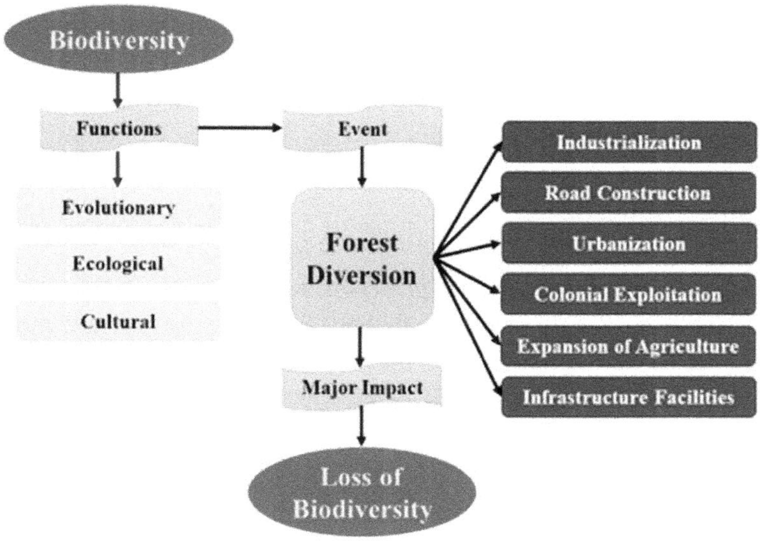

Four factors were majorly responsible for forest habitat destruction (Figure 1):

1. Agricultural Expansion

Agricultural expansion, driven by population growth and the need to feed growing populations, has been a major driver of forest diversions globally. Forests have been cleared to create farmland for crops, pastureland for livestock, Timberlands for infrastructure, and in later decades for agro-industrial plantations such as palm oil and soy.

2. Industrialization and Urbanization

The advent of industrialization and urbanization in the 19th and 20th centuries increased the demand for forest products and land for development (Norgaard R., 1988). Mass production made people realize that with a sufficient supply of raw materials, even everyone's demands can be satisfied. At the same time, rail and quicker transport facilities enabled long-distance export of goods commodities earning higher profits. Forests were extensively logged to meet the growing needs of industries such as mining, construction, and manufacturing, as well as to accommodate expanding urban populations.

3. Infrastructure Development

The construction of roads, railways, dams, and other infrastructure projects often required the diversion of forested land. Large-scale infrastructure developments, particularly in rapidly industrializing nations, contributed to widespread forest loss and ecosystem degradation.

4. Colonial Exploitation

During the colonial era, many European powers exploited forest resources in their colonies across the globe for timber, fuelwood, and other raw and new materials for industries. Forests were often diverted for other natural resources such as minerals, crude oil, local spices, cuisines, and gems which resulted in significant habitat destruction and species endangerment in resource-rich countries like Africa and South Asia.

On the other hand, as the industrial revolution hit the old world especially Africa, south and southeast Asia, a bit late this mass destruction of the forest was delayed. But after colonization and post-colonization, era vast stretches of forests were diverted for timber, industrial fuel, raw material, urbanization, rail, road construction, and other infrastructures.

Loss of Habitat

Habitats for a variety of plants and animals are being lost as the primary impact of forest diversion. Discontinuous diversion of forest leads to fragmentation of forest area and consequently, it separates populations from community leading to prevention of hereditary exchange which further may make the species prone to extinction (Figure 2).

- In India, there are a few species that are found only in particular ecosystems. These species are adapted to survive in specific conditions and are not found anywhere else in the world. These are known as 'Endemic' species. In the event of forest diversion, their habitat is threatened and consequently, these species progress towards extinction. Western Ghats once had a good number of Malabar Giant Squirrels but, there is a very small population of them surviving now (Mehta P.,2011).
- Whenever there is fragmentation of any habitat or forest, the populations get isolated leading to hindrances in the natural processes such as migration, genetic exchanges, patterns of movement, etc. The population follows inbreeding and there is no genetic diversity. This further makes the population

prone to diseases and its capability to adapt to changing environmental conditions. Wild cats and many other mammals in India are facing this challenge as they need large territorial areas for existence.

- As such, an ecosystem comprises varied forms of life ranging from microorganisms to large animals and plants. These living forms and the environment are deeply connected. In the case of forest diversion, these relationships (such as prey-predator, cross-pollination, and so on) among components of biodiversity are disturbed. When any coastal wetland is reclaimed, the plants in these areas along with various faunal species, their life processes, and habitats are affected causing a severe blow to the entire ecosystem.
- In many places, agricultural lands are close to forests. If these forests are diverted to develop agriculture, there is a detrimental impact on existing agriculture. The crops lose their capability to resist pests, and pathogenic invasions and further fail to cope with the changing climate. This not only impacts the farmer community but also affects the food security.
- Forest diversion predominantly involves ecosystems like wetlands, mudflats, grasslands in altitudes, and other coastal areas. It is because of such ecosystems that, various services such as flood control, carbon sequestration, and consequently water purification occur. For years together migratory birds visit specific locations for winter periods. These are usually wetlands or mudflats and such places are quite often reclaimed for agriculture or other development projects depriving the migratory birds of their winter migration locations. This reclamation of wetlands also switches off the natural water filtration process.

Figure 2. Biodiversity loss at a glance

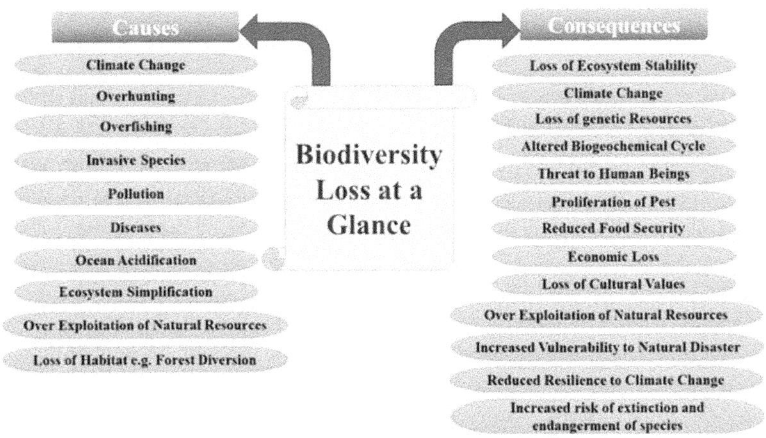

Disruption of Ecosystem Services

Any established ecosystem provides various services for biodiversity. These services are climate control, purification of water, soil retention, helping in pollination, etc. Diversion of forests affects some or all of these services leading to deleterious impacts on the floral and faunal diversity (Figure 2).

- When the direction or location of any water body is altered, the fertile soil near it is lost either due to exposure to the sun or due to submerging. This further results in a decline in soil fertility consequently, reducing in quality/ quantity of crops produced and eventually, the livelihood of the nearby human population adversely affected.
- The first and foremost effect of forest divergence is that the carbon sequestration process is hampered. This causes the accumulation of greenhouse gases that in turn result in global warming. Excessive rains causing floods, higher temperatures, and changes in precipitation are many such adversities which are observed in India because of above-mentioned conditions. As monsoon being an important source of water in many states of India, changes in precipitation are causing droughts in many parts of the country. This negatively affects agriculture as well as causes a shortage of water supply for normal consumption.
- Forests can hold the soil, thus helping to maintain the water table and river flow. When forests are destroyed for infrastructural development or for agriculture, cause disturbances in natural processes. The classic example is various rivers originating in the Western Ghats and supplying water to Southern India. Destruction of forests in the Western Ghats can be responsible for serious disruptions in water flows, and higher sedimentation rates. This would further affect not only the ecosystem but also the biota dependent upon these rivers.
- As mentioned above forests hold the soil and through cycling of various nutrients help retain its fertility. Deforestation leads to the degradation of soil nutrients and thus soil fertility is depleted. Indian economy is always said to be dependent mainly upon agriculture. So, deforestation and subsequently reduced soil fertility can have notable negative impacts on agriculture practices.
- The diverse flora and fauna are intimately related to an abiotic component of the ecosystem. Together they form a system that provides various services to human beings. Ecosystems with rich biodiversity are better adapted to cope with climate changes or similar disturbances. Forest diversion destroys biodiversity resulting in reduced resilience of the ecosystem making it

susceptible to infestation by pests, pathogenic invasions, and adverse impacts of climate change. This in turn affects adjacent human populations dependent upon ecosystem services for their existence.

- A large floral diversity is adapted for cross-pollination particularly Entomophily (insect pollination). Various faunal species like numerous insects, and a few birds serve as pollinating agents for this type of pollination. Forest is an ecosystem that supports the survival of many of these pollinator organisms. In the case of forest diversion such essential fauna is destroyed affecting the pollination process negatively and further the crop yield. Subsequently, food security and also farmers' livelihoods are impacted negatively.

Large trees in the forest are capable of absorbing considerable amounts of atmospheric CO_2. Urbanization, expansion of agriculture, and infrastructural development are the prime causes of deforestation. In the absence of these carbon sequesters, climate change caused by emissions from automobiles and industries directly affects human life.

Extinction and Endangerment of Species

Biodiversity loss refers to the decline or disappearance of biological diversity. Bandera, G. (2023) has enlisted the following causes for the loss of biodiversity viz. loss of habitat and fragmentation, climate change, over-exploitation of natural resources, overhunting, overfishing, invasive species, pollution, disease, ocean acidification, ecosystem simplification, etc (Bandera, G. 2023). Some of the consequences are viz. loss of ecosystem stability, the decline in ecosystem services, loss of genetic resources, altered biogeochemical cycles, increased risk of species extinction, threat to human beings, proliferation of pests, increase in CO2 emissions, etc (Figure 2). Apart from all the mentioned impacts, biodiversity loss can also affect humans in many ways viz. reduced food security, decline in human health, economic loss, loss of cultural values, increased vulnerability to natural disasters, reduced resilience to climate change, etc (Bandera, G. 2023).

- Destruction of habitats and subsequent effects of ecosystem services threaten the survival of a large number of species (Figure 2). In the Indian context where remarkable biodiversity is recorded and quite a few are known to be endemic species, habitat destruction has pushed many of the species towards endangerment or extinction. Specific habitats like the Andaman / Nicobar Islands, Western Ghats, and Eastern Himalayas have always been identified as unique ecosystems in India supporting many endemic species. Diversification of forests from these areas has challenged the survival of some of the species in these habitats. Purple frogs endemically recorded from the Western Ghats can be cited as an example of a seriously threatened species.

- Many ecosystems are interconnected with each other for various reasons and the biodiversity is also dependent upon their connectivity. Forest diversion often causes severing these connections thus, obliterating the mobility of several species. It particularly affects the species which require large expanses for mobility and migration in unfavorable seasons. Great Indian Bustard can be cited as a classic example of this.
- More than 1400 tree species are assessed as critically endangered and in urgent need of conservation action. Some 8 percent of assessed forest plants, 5 percent of forest animals, and 5 percent of fungi found in forests are currently listed as critically endangered (FAO 2020). Currently, there are more than 1,57,100 species on The IUCN Red List, with more than 44,000 species threatened with extinction, including 41% of amphibians, 37% of sharks and rays, 36% of reef-building corals, 34% of conifers, 26% of mammals and 12% of birds (IUCN 2013). In mid-2019, the United Nations (UN), in collaboration with the Intergovernmental Science-Policy Platform on Biodiversity and Ecosystem Services (IPBES), presented a report on biodiversity warning that out of a total of eight million, one million species are in danger of extinction. Some researchers even dared to talk about the sixth mass extinction in the history of the planet. The WWF's Living Planet Report 2022 finds wildlife populations have declined by an average of 69% in the past 50 years.

CASE STUDIES: IMPACT OF FOREST DIVERSIONS ON BIODIVERSITY LOSS

National Case Studies

Case Study One: Loss of Habitat

Location: Aravalli Range, Rajasthan, India

The mountain range called 'Aravalli Range' is in Northern-Western India, running around 670 km in a south-west direction, starting at Delhi, going through southern Haryana, Rajasthan, and finally ending in Ahmedabad Gujarat. The highest mountain peak in these ranges is 5650ft high. Luni Bansa, Sabarmati and Sakhi rivers originate in these mountains. The vegetation here is moist deciduous, dry deciduous, montane, tropical thorn, mixed deciduous and dry teak species. Various anthropogenic activities like mining, road building, agriculture, etc. in the Aravalli Ranges particularly in Rajasthan have caused extensive loss of habitat and fragmentation (Rathod, 2002). These ranges are exploited for Mining for minerals like marble, granite, and limestone leading to direct deforestation of large stretches in the Aravalli region. The area once known for its contiguous habitat has been subjected to fragmented patches of forests, affecting its rich biodiversity.

Impacts:

- Wildlife Isolation: Species like the Indian leopard, Indian pangolin, and various species of deer have been isolated in small populations due to fragmentation effects in the Aravalli forests. This isolation disrupts genetic exchange and further results into making them susceptible to local extinctions.
- Threat to Endemic Flora: several endemic plant species adapted to the unique climatic conditions is the unique features of The Aravalli Range. Destruction of habitat for varied reasons has pushed these plant species towards risk, thus creating an imbalance within biodiversity of the Aravalli ranges.
- Impact on Ecological Processes: the natural movement patterns of wildlife, processes like seed dispersal and pollination are seriously affected by fragmentation of forests in the Aravalli. Both flora and fauna of the entire ecosystem consequently faces cascading effects of this.

Case Study Two: Disruption of Ecosystem Services

Location: Yamuna River Basin, Northern India

The River Yamuna originates from the Yamunotri Glacier at about 14,800 ft[REMOVED HYPERLINK FIELD] on the southwestern slopes of Bandarpunch peaks of the Lower Himalaya in Uttarakhand, it travels 1,376 kilometers before reaching Prayagraj. There has been serious forest diversion for urbanization along The Yamuna River basin which has impacted the ecosystem services that were being provided by the surrounding forests (Pandey *et.al.* 2014).

Impacts

- Degradation of Water Quality: the Yamuna basin has faced severe soil erosion consequently causing a decline in the quality of water in the river. Prolonged sedimentation, as well as nutrient runoff from deforested areas, have resulted in water pollution. This has not only affected aquatic life but has also affected the supply of drinking water for human communities reliant on the river.
- Impact on Water Regulation: The natural forests in the Yamuna basin play a crucial role in regulating water flow. Hydrological patterns have undergone drastic changes due to deforestation. The frequent floods during monsoons and the shortage of water later on are the prime effects of this deforestation.

- Effect on Agriculture: Influence on water regime, and negative impact on soil fertility have resulted because of the disrupted ecosystem services. Agricultural productivity has gone down notably due to decreased soil fertility further affecting the livelihoods of farmers which for irrigation depend upon the Yamuna basin.

Case Study Three: Extinction / Endangerment of Species

Location: Silent Valley National Park, Kerala, India

In 1914, Silent Valley forests were declared as Reserved Forests. Initially, there was only 89.52 sq. km. area under the Division which forms the core zone of the National Park. In 2007, an area of 148 sq. km. was added to this Division as a buffer zone. Silent Valley National Park comprises 8,952 hectares of forestland on the Nilgiris plateau closed on all sides by mountains of the Western Ghats, some as high as 2000 meters. It is very well known for its rich biodiversity, quite a few of which are endemic. In the 1970s a serious threat was posed to it when a Pathrakkadavu Hydro-Electric Project (PHEP) was proposed. A major portion of the National Park was sure to be submerged after the construction of the proposed dam. The proposed dam would have submerged a significant portion of the park and thus many of the species would have either become extinct or would become endangered.

Impacts

- Lion-tailed Macaque (*Macaca Silenus*): One important species recorded at Silent Valley is the endangered Lion-tailed Macaque. The park flaunts a peculiar rainforest ecosystem housing these Lion Tailed Macaques. The proposed dam project if implemented would pose a threat to the existence of this species.
- Biodiversity Hotspot: Any kind of threat to the existence of single or more than once species in this area would certainly cause a global uproar. It is because of the presence of numerous ecologically important endemic species surviving in this area. The population of endemic species like the Nilgiri Wood Pigeon, Nilgiri Flycatcher, Nilgiri Langur, Nilgiri Tahr, and many other species will be threatened.
- Status of Conservation process: There was strong opposition by environmental activists and common people to this dam project and hence, the Silent Valley hydroelectric project was finally shelved. This is a classic example of how community involvement can play a significant role in protecting habitat and biodiversity.

Case Study Four: Forest Diversion for Mining in Hasdeo Arand, Chhattisgarh

Location: Chhattisgarh

The Hasdeo Arand forest region is spread across 1,70,000 hectares of Chhattisgarh and overlaps northern part of Jharkhand. It is a home for diverse flora and fauna, including several endangered species hence, known as a biodiversity hotspot. (Kundan Kumar, 2020). Substantial coal reserves have been found in this region thus, making it a prominent target for mining activities. The Hasdeo Arand coalfield is spread over an area of 1,879.6 km^2, and comprises 23 coal blocks. Hasdeo Arand is a large coalfield with 1.369 billion tonnes of proven coal reserves and 5.179 billion of estimated coal reserves. Parsa East and Kante Basan (PEKB) coal block in Hasdeo Arand has been allocated to Rajasthan Rajya Vidyut Utpadan Nigam Limited (RRVUNL) and Adani Enterprises Limited for coal extraction. Main purpose of the project was to supply coal to power plants in Rajasthan.

Impacts

- Deforestation and Habitat Loss: Extensive deforestation and habitat loss happened as a result of mining activities in the Hasdeo Arand region. Forest land measuring over 2,000 hectares was diverted for the PEKB coal block.
- Fragmentation of Ecosystems: Once a contiguous ecosystem but forest diversion led to fragmentation consequently, isolating populations of various species and disturbing natural ecological processes. Wildlife migration, genetic exchange, and the overall health of the ecosystem have been hugely affected by this fragmentation. This forest has been a migratory corridor for elephants.
- Loss of Biodiversity: Hasdeo Arand is known for its rich biodiversity, including various plant and animal species. several endemic and keystone species lost their natural habitats because of the mining project and there was huge loss of biodiversity. Animals like leopard, sloth bear, tiger and elephants have been known to inhabit these forests.
- Impact on Water Resources: Forests have always played a crucial role in regulation of water flow and maintenance of health of water sources like rivers and streams. Water regimes in Hasdeo Arand underwent drastic changes due to deforestation, affecting local water resources and quality.
- Displacement of Indigenous Communities: local indigenous communities, who have traditionally relied on the forest for their livelihoods got displaced by the mining project. Various communities were forcibly relocated disrupting their cultural practices and traditional ways of life.

- Effect on Livelihoods: livelihoods of local communities dependent on forest resources were impacted by forest diversion for mining. Traditional agriculture, hunting, and gathering have been disrupted, resulting in a generation of economic challenges for the affected populations.

Controversies and Opposition

I. Environmental Activism: There was strong opposition for the forest diversion for the PEKB coal block project from local communities, environmental activists, and non-governmental organizations (NGOs). Objections were raised for the irreparable damage caused to the unique ecosystems of Hasdeo Arand.
II. Legal Challenges: Writ petitions were filed by environmentalists and activists against the project forest diversion and the judiciary was requested to intervene for the protection of ecologically sensitive region.

To conclude this case study of forest diversion in Hasdeo Arand for mining. It highlights the complex trade-off between environmental conservation and economic development. Although mining projects contribute to energy production and economic growth, the biodiversity, ecosystems, and local communities need to be taken into consideration for long-term consequences. An appropriate balance needs to be maintained between development and environmental sustainability. This has always remained a challenge, and cases like Hasdeo Arand highlight the need for robust environmental impact assessments, sustainable practices, and community engagement in decision-making processes.

Case Study Five: Dams on Narmada River

Location: Madhya Pradesh

The Narmada River originates from the Amarkantak Plateau in Anuppur district in Madhya Pradesh. It makes the boundary between North and South India and flows westwards for 1,312 km before joining into the Arabian Sea through the Gulf of Khambhat, 30 km west of Bharuch city of Gujarat. The project called as Narmada Valley Project involves construction of large number of Dams on Narmada River which would benefit people from three states Gujarat, Maharashtra and Madhya Pradesh. Narmada Sagar Project in Madhya Pradesh and Sardar Sarovar Project in Gujarat are the two major parts of this project (Tanmayee Sahoo, 2014).

The land selected for the project involves forests with rich wildlife and unique vegetation. In the project, construction of Dams and submerging of land after construction of dam are two main factors responsible for posing a threat to the forests. Backwaters and canal building would be added to the issue. This project as a whole proves to be a classic example of forest divergence of a large magnitude.

This project plans to generate over 2450 megawatts of electricity and irrigate over 2 lakh hectares of land. The forest to be destroyed is a home to rare variety of Anjan tree and various Cheetals, Sambars, Neelgai and a few wildcats. The vegetation would be lost but attempts could be made to relocate the wildlife although the success in relocation is not guaranteed. A few tribes like Bhil and Talvadi would also be displaced and it is difficult to relocate them as they are dependent upon the forest resources for their survival. Relocation of affected families was also committed in the proposal. Yet, the data mentioned in the proposal and factual figures were found to be different and hence, there were huge protests against the project. Main leader of these protests was Ms. Medha Patkar (an alumnus of TISS, a politician and activist) who was demanding proper relocation of the people affected by the project. The protest later became a nationwide uproar and gathered good support from people.

Thus, Narmada Sagar and Sardar Sarovar projects would be example of forest diversion causing loss of habitat, extinction of species, loss of biodiversity and disruption of ecosystem services on a large scale. The project is planned for wellbeing of human population and it is indeed capable of helping human beings but, the impact on the ecosystem is of such a magnitude that one can't turn a blind eye to it.

INTERNATIONAL CASE STUDIES

In the 21st century, owing to global environmental awareness, multiple abiding and non-abiding treaties and protocols such Millenium Ecosystem Assessment and Sustainable Development Goals, there has been plethora of afforestation, restoration, and conservation programs that have been implemented by various governments and other agencies to mitigate forest destruction and species recovery. However, in reality for almost all developing and underdeveloped countries, it is a tough trade-off that they have to wager i.e. ecosystem conservation vs national development. It is almost impossible to achieve the perfect balance between the two and unfortunately, in most cases, governments tend to choose the latter. Let us examine some international case studies that shade light on this multifaced issue.

Case Study One: The Amazon Basin

The Amazon Forest of Amazon River in South America are undoubtedly the largest tracts of contiguous forest land even today occupying an area of 3,000,000 km^2. These tropical forests due to their unparallel species richness of over 3 million species, have rightly earned the title "Cradles of Biodiversity.

Before1970s, Amazon Forest were still largely intact and were safe due to impenetrability. The scenario changed after military power rise in especially in Brazil which expediated the forest diversion for new settlements, roads and highway construction such as Trans-Amazonian Highway in 1972 and industry led plantations. (Sandy M., 2019). These conditions started a progressive chain reaction to slash and burn further deeper and deeper into the forest in order to earn profits. Road, highways and infrastructural development chopped down old trees and forest patches for timber and other construction. Parallelly, in other parts of Amazon with low soil fertility, people turned to livestock keeping, cattle ranching and creating pasture lands. Reports by Amazon Environmental Research Institute iterate Brazil as largest meat exporter in the world. It had considerably more profit margins with minimal labour demand (Nepstad et al, 2008; Edorado L., 2019).

Another factor that led to habitat degradation and biodiversity loss was mechanized large scale farming of cash crops such as Soy, Oil palm and sugarcane etc. Soyabean was extensively produced for the production biodiesel and as alternative fodder livestock. Additionally, 2015 onwards illegal palm oil plantations have become widespread in many parts of Amazon. As of 2022, these once pristine habitats have lost 26% of its forest habitats (Quintanilla et al, 2022). In addition, small farmers who hold insignificant or infertile land also turned to hunting, capturing for pet trade and poaching for their livelihoods which directly affects species population.

The unobstructed forest diversion for these uses have periled many native species of plants and animals near to extinction. For example, Hyacinth Macaw a significant species in this forest which is threatened by its habitat destruction due to agricultural expansion and logging. This has greatly affected their population and is now categorized as "Vulnerable" from "Least Concern" by IUCN. Habitat destruction has affected populations of Various iconic species such as Jaguar, Harpy Eagle, Golden Lion Tamarin, Harlequin Poison Frog, Araripe Manakin, Yellow-footed Tortoise, Brazilian Tapir and many more are experiencing unprecedented population decline, range reduction due to logging and habitat destruction and overexploitation for meat and pet trade, and have become "critically endangered" as per IUCN (Bodmer et al, 1997; IUCN Red list v. 3.1). Aquatic species like Giant otter, Brazilian Merganser, Amazon River Dolphin, Amazonian Manatee etc. share no different story. Deforestation, dams and industry led agriculture deteriorated the water quality and polluted the river greatly which has affected the very home of these species pushing many of them towards extinction (Oliveira G. & Oren D., 2017; Peres C. & Lake. I., 2003; IUCN Red list v. 3.1).

These are just a few examples of species facing severe threats in the Amazon region. Unmanaged and unobstructed forest diversions and other human activities continue to pose significant challenges to the conservation of biodiversity in this one of kind ecosystem. Efforts are being taken to aware people to protect and restore habitats, combat illegal wildlife trade, and engage local communities in conservation which some key solutions for the survival of these and other threatened species in the Amazon.

Case Study Two: Southeast Asia

While the Amazon rainforest is often the most recited story in this context, other regions, such as Southeast (SE) Asia, also experience profound impacts. The case study is mainly focused on some particular key areas such as Sumatra and Indonesia within this widespread landscape. SE Asia harbors many diverse and unique ecosystems, including tropical rainforests, deciduous forests, mangroves, etc. Although ecosystems are different, the area has met with a similar fate as that of the Amazons.

Extensive deforestation for palm oil and pulpwood plantations, as well as illegal logging, has led to the loss and fragmentation of critical habitat for numerous species. SE Asia has one of the highest rates of forest conversion with 14.5% vegetation cleared in just 15 years (Miettinen et al, 2011). The major driver of forest diversion in this region is the Oil Palm and rubber plantations, and are one of the largest exporters in the world (Meijide et al, 2018). Pulp and paper industry plantations of Acacia and Eucalyptus is yet another rapidly spreading infestation in Indonesia and neighboring countries. This affects local and marginalized tribals as well and has threatened their livelihood and sustenance. Forest diversion often displaces indigenous and local communities that depend on forests for their livelihoods, food security, and cultural identity. Loss of access to traditional resources and disruption of cultural practices further exacerbate social inequalities and conflicts. It is this poverty that prompts them to switch to plantations which provides some financial stability.

Amidst of all this, habitats that were once vast free grounds for exceptionally diverse and endemic flora as well as fauna, have shrank into isolated pockets. Monocultures plantations although financially lucrative generally exert negative implications on the biodiversity. The loss of habitat and increased human-wildlife conflicts have resulted in population declines and local extinctions of many species. Another direct threat is the international trade in various endangered species. Data from CITES shows that, there has been considerable increase in the illegal trade in recent years. Several endemic species such as Pangolins, Pig nosed turtles are severely threatened by this new business.

Larger animals especially mammals require wide contiguous patches of forest with interconnectivity as their home ranges and thus are highly affected by habitat fragmentation. Endangered species such as the Sumatran tiger and the Sumatran orangutan are facing heightened threats by this habitat destruction (Gaveau et al., 2014). Moreover, endemism is proving to be curse to species with restricted ranges, such as the Sumatran rhinoceros, Asiatic elephants, marine turtles and innumerable coral species found in Apex Coral Triangle are particularly vulnerable to extinction, mainly due to illegal trade and habitat degradation (Meijaard et al., 2018).

These trends underscore the urgent need for conservation action to mitigate these impacts and safeguard the integrity of forest ecosystems. Strict Policies and statutory interventions that prioritize biodiversity protection from trade, sustainable land-use practices, and the rights of indigenous communities are key factors to focus in order to preserve Southeast Asia's irreplaceable biodiversity.

Case Study Three: Australia's Biodiversity Hotspot

Australia is renowned for its unique and diverse ecosystems, housing a vast array of plant and animal species found nowhere else on Earth. Australia's biodiversity hotspots encompass a variety of ecosystems, from tropical rainforests in the north to arid deserts in the interior and temperate forests in the south. Despite its vast and varied landscapes, Australia has one of the highest rates of mammal extinction in the world, with numerous species facing threats of extinction. It is an unsettling fact that more than 1100 species are posed with consistent annual drop in their population (WWF's Living planet Report WWF, 2022). Over the last 200 years Australia has suffered the largest documented decline in biodiversity of any continent. According to recent studies (e.g., Australian Government Department of the Environment and Energy report, 2018), millions of hectares of native forests have been diverted since European settlement, primarily for agriculture, logging for timber and infrastructure development.

Large-scale land clearing for agriculture, including cattle grazing and crop production, has been a major driver of forest diversion in regions such as Queensland and New South Wales (Pacheco et al., 2012). This has resulted in loss and fragmentation of critical habitat for numerous endemic plant and animal species. Uprooting of forests led to heavy soil erosion, leading to sedimentation of waterways and degradation of water quality (Vesk et al., 2020). Of its rich and varied habitats, Murray–Darling Basin has seen one of the largest denudations since European colonization in coastline areas (Lindenmayer et al., 2008). The Murray-Darling Basin, Australia's largest river system, has been heavily impacted by water extraction, river regulation, and pollution (Kingsford et al., 2011). Reduced water flows and deteriorating water quality have led to declines in fish populations, waterbird numbers, and wetland habitats. One of the well noted example is - Spotted Handfish. This critically endangered marine fish species is found in coastal waters of Tasmania. Its number is dwindling mainly owing to habitat degradation, pollution, and invasive species threaten its shallow-water habitats, including seagrass beds and rocky reefs.

In addition to this, rapid urbanization and population growth have led to the diversion of the forests for residential, commercial, and industrial purposes, particularly in semi-urban areas (Laurance et al., 2015). Construction of roads, highways, and other infrastructure projects often necessitates vegetation clearing, leading to habitat fragmentation and loss (Laurance et al., 2002). The Daintree Rainforest in Queensland, one of the oldest tropical rainforests in the world, is under threat from deforestation, land clearing, and fragmentation. Loss of habitat has resulted in declines in endemic species such as the southern cassowary and the Bennett's tree-kangaroo (Goosem & Tucker, 2013). Southern Cassowary, a large, flightless bird native to the tropical rainforest has been listed as "Endangered" (EPBC Act, 2022). Leadbeater's possum have also been classified as "Critically Endangered" after a sudden dip in their population in the mountain ash forests of Victoria (IUCN Red list, 3.1). Apart from this, of 660 floral plant species categorized as "Critically Endangered or Endangered" under the Environment Protection and Biodiversity Conservation Act, 62 species have less than 50 individuals, and other 300 species have less than 250 individuals surviving in the wild (Silcock et al. 2020).

Forest diversion in Australia poses significant challenges to biodiversity conservation, ecosystem resilience, and sustainable development. In such conditions, effective policies and management strategies are need of the hour to attenuate factors of forest diversions and promote sustainable land use practices that balance environmental conservation with socio-economic development.

Case Study Four: The African Congo

The Congo Basin or Congo is the sedimentary basin of the Congo River in equatorial Central Africa. It contains some of the largest tropical rainforests in the world and is an important source of water used in agriculture and energy generation. The rainforest in the Congo Basin is the largest rainforest in Africa and second only to the Amazon rainforest in size with 300 million hectares (FAO, 2011). However, this virgin ecosystem is degrading at an alarming rate in recent years. Rate of tree felling in Congo basin was observed to be doubled between 2015 to 2019 (Kinver M., 2019). In fact, according to Global Forest Atlas, logging industry spans over 44 to 66 million hectares of woodland vegetation.

The forest diversion situation at Congo is interlinked and has compounding effect on the habitat. Multiple studies have through years have pointed agricultural expansion, demand for firewood, ill-managed logging and infrastructure construction as prime drivers of deforestation (Wilkie D. et al, 2000; Iloweka Ernest, 2004; United Nations Environmental Programme, 2011) The most pressing concern regarding forest diversion among these, as identified by the United Nations Environment Programme, is the method of Slash-and-Burn agriculture, posing significant environmental challenges. Given that a considerable proportion of population relies on this intensive farming, the issue is serious and widespread. This approach ultimately results in soil degradation and desertification. This forces poor farmers to clear further deep into the forest for fertile land thereby exacerbating habitat fragmentation (Iloweka Ernest, 2004; United Nations Environmental Programme, 2011). Collection of fuelwood and charcoal stands as a secondary driver of deforestation, as it fulfils 95% of the population's energy requirements. This leads to the formation of circular fragmented forest patches around urban zones, necessitating progressively inner forest ventures for fuelwood collection (Wolfire D., 1998). Both mining and logging activities also contribute to secondary deforestation by constructing roads. Logging operations enforced by companies, often built new road to access previously untouched forest areas, facilitating the conversion of these logged forests into agricultural land. Artisanal and industrial logging is another major threat to these habitats and species that is conducted in unrestricted manner till date. (Witte J. 1992).

Some of the most significant and flagship species such as African forest elephant, Chimpanzee, Bonobo, Okapi, pygmy hippopotamus, Leopard and plenty others that inhabit this pristine ecosystem, are now imperilled by forest diversion and habitat deterioration. For example, Western lowland gorillas are "Endangered" due to habitat degradation and may become extinct in next decade as per UNEP. (UNEP, 2011; Morgan et al., 2019)

The African forest elephant is a distinct species from the savanna elephant found primarily in the forests of the Congo Basin. Habitat loss due to logging, agriculture, construction development, and poaching for ivory have led to a significant decline in forest elephant populations (Maisels et al., 2013). Endemic to Congo Basin forests, the Okapi or "Forest Giraffe" faces threats from habitat loss and illegal hunting (IUCN SSC Giraffe and Okapi Specialist Group, 2013). Logging and agricultural expansion have significantly reduced their numbers. The Congo peafowl is a large, ground-dwelling bird endemic to the Congo Basin. Forest fragmentation, habitat destruction and anthropogenic disturbance have threatened their survival, as they rely on dense forest habitats for shelter and foraging (BirdLife International, 2018). These are just a few examples of the many species threatened by deforestation in the Congo Basin. Conservation efforts focused on habitat protection, sustainable land use practices, and addressing the root causes of deforestation are crucial for safeguarding the biodiversity of this critical ecosystem.

SUSTAINABLE BUSINESS OPERATION

It is very important to adapt sustainable business operations to avoid environmental loss especially where forest diversion or forest products are being used. A report

Figure 3. Three pillars of sustainable business operation

by Russo (2020) says that over half of the global GDP of the world is directly or indirectly dependent on nature. According to the report of UNEP on the state of the world's forests (2020), forests provide more than 86 million green jobs and support the livelihoods of many more people. Here, we have discussed briefly Environmental Social Governance (ESG) for sustainable development, sustainable strategy, corporate sustainability area, etc.

Environmental Social Governance for Sustainable Development

Today, globally people are aware that activities undertaken by different sectors like industries, commercial, and residential sectors are the root cause of environmental degradation and climate change. To address the issue, Sustainable Development Goals (SDGs), the blueprint to achieve a better and more sustainable future for all was agreed in 2015 in which 17 goals are framed which is to be met by 2030 (David, 2018).

The SDGs establish universal goals that provide a roadmap for sustainability in business in target areas such as poverty, inequality, environmental degradation, and climate change. One way to meet these goals is sustainability in different businesses. Sustainability in business means strategy and actions taken by the organization to reduce negative environmental and social effects resulting from its operations. Earlier, there were efforts made in businesses to tackle the above problems through Corporate Social Responsibility (CSR). It has been a precursor to ESG with a more sustainable approach (Figure 3) through ESG (Senadheera, S. S. et al 2021).

Those paving the way in sustainable business practices are embracing new business models to win customers, increase brand loyalty, and uncover new opportunities to lower costs. Many of the world's top economies have or are developing corporate disclosure requirements around environmental impact, driving businesses to curb GHG emissions. Adopting an ESG policy enhances the innovation capacity, innovative activities, value creation, and financial performance of businesses. Overall, social and environmental performance demonstrated a significantly positive relationship with business sustainability, indicating that business economy and creating value for society are mutually dependent (Ahmad, 2024).

Some of the ways by which sustainability in business can be achieved are:

- Switching to alternative energy resources to improve energy management efficiency.
- Replacing infrastructure with those that reduce GHG emissions, preserve water resources, and eliminate waste.
- Empower circular economy through operating dynamic and efficient supply chains.
- Encourage reuse, design out waste, promote sustainable consumption, and protect natural resources.
- Enabling sustainable development by assessing risks and improving resiliency while adhering to external regulations and development goals.

In recent years, business investments and sustainability have become major concerns for investors. Investors evaluate corporate social and environmental reactions based on ESG performance because firms with low scores face high ESG disputes. Governments and governing bodies also pressurize nonsocial and environmental companies that have failed to adopt ESG policies. Today, investors are more interested in the stable processes of a firm than in their operational and financial benefits. Firms that ignore ESG features and do not integrate them into their businesses face unnecessary consequences for investors (Shakil, 2020; Ahmad et al., 2024). The GRI Sustainable Reporting Guidelines promote sustainable reporting practices at a level comparable to financial reporting. The GRI guidelines recommend assessing ESG disclosure (Asante-Appiah, 2020).

To summarize, we recognize that environmentally friendly processes and CSR systems are receiving greater attention to find sustainable competing businesses in the future. On the other hand, shareholders, managers, and human rights play vital roles in determining ESG performance in developing countries. Therefore, managers should focus on these indicators to improve the efficiency and effectiveness of the firms. The sustainability points provided by various reference providers should be considered for a comparative analysis in future studies. Industrial segregation can lead to highly detailed results, another avenue for future research. Test methods can be used to collect baseline data from industries, and current research can be considered in the future to compare similar results.

CONCLUSION AND OUTLOOK

The case studies discussed above reflect that biodiversity loss directed by forest diversion is an extremely intricate and multifaceted issue. Thus, conservation strategies should be long term and tailormade considering local interventions and their implication in the future. Apart from direct biodiversity losses, other negative impacts of forest diversions may get ignored while focusing on larger perils. Local extirpation occurs when a species is sufficiently common within protected areas while is rare outside it. It may not be considered under threat right away, but in the long term may risk the survival of the population due to inbreeding depression. Continual deforestation may alter the course of the season and local climate patterns which may transform the vegetation composition in the remaining forest, which impacts species that are directly or indirectly dependent on a particular vegetation composition. Moreover, habitat degradation has globally shrunk the ranges and modified distribution patterns and migratory routes of numerous species across the globe.

UNFCCC and CBD strongly advocate the concept of nature-based solutions which has gained traction as a medium to harness natural processes to tackle the climate crisis and mitigate biodiversity loss, while also delivering supplementary benefits for nature's contributions to human welfare. Forests are our natural heritage that is invaluable to sustaining biodiversity, delivering ecosystem services, supporting livelihoods, and nourishing a plethora of communities. Thus, their sustenance and well-being are of paramount importance for attenuating greenhouse emissions and combating the climate change crisis.

Environmental insights are transforming business and society, leading to the development of sustainable business. Sustainability in business will continue to have a significant impact on companies' competitiveness and even their survival in the marketplace. Leaders are looking to leverage data, artificial intelligence, and blockchain to mitigate climate and environmental risks, optimize asset and resource performance, decarbonize, and build sustainable supply chains.

REMEDIAL MEASURES: COMPLEMENTARY AFFORESTATION

Legal provisions have been made to overcome this issue. Whenever any forest land is acquired for some development activity, double to size of the land will have to be reforested and will then be declared as a Reserved / Protected Forest. The land required for this type or afforestation will be acquired from revenue lands/zudpi jungle/ jharka jungle/jungle-jhari land/civil-soyam/orange lands or any other such forest lands that are not under administrative control of the Forest Department of the State/UT.

REFERENCES

Ahmad, H., Yaqub, M., & Lee, S. H. (2024). Environmental-, social-, and governance-related factors for business investment and sustainability: A scientometric review of global trends. *Environment, Development and Sustainability*, *26*(2), 2965–2987. doi:10.1007/s10668-023-02921-x PMID:36714213

Asante-Appiah, B. (2020). Does the severity of a client's negative environmental, social and governance reputation affect audit effort and audit quality? *Journal of Accounting and Public Policy*, *39*(3), 106713. doi:10.1016/j.jaccpubpol.2019.106713

Bandera, G. (2023). *10 Causes of Biodiversity Loss and Its Effect on The Environment*. Fair Planet. https://www.fairplanet.org/story/causes-effects-biodiversity-loss/

BirdLife International. (2018). Afropavo congensis. *The IUCN Red List of Threatened Species*. BirdLife International.

Bodmer, R. E., Eisenberg, J. F., & Redford, K. H. (1997). Hunting and the likelihood of extinction of Amazonian mammals. *Conservation Biology*, *11*(2), 460–466. doi:10.1046/j.1523-1739.1997.96022.x

Conservation Society. (2013). The plight of forest elephants. *Wildlife Journal*, *15*(3), 45-52.

Convention on Biological Diversity. (2017). *Biodiversity and the 2030 agenda for sustainable development policy brief*. Convention on Biological Diversity Technical Note.

Costanza, R., d'Arge, R., de Groot, R., Farber, S., Grasso, M., Hannon, B., Limburg, K., Naeem, S., O'Neill, R. V., Paruelo, J., Raskin, R. G., Sutton, P., & van den Belt, M. (1997). The value of the world's ecosystem services and natural capital. *Nature*, *387*(6630), 253–260. doi:10.1038/387253a0

David, M. P. (2018). Sustainable Development Goals (SDGs)- Challenges for India. *Indian Journal of Public Health Research & Development*, *9*(3), 1. doi:10.5958/0976-5506.2018.00172.9

Edorado, L. (2019). The impact of cattle ranching on Amazon deforestation: An evidence-based review. *Journal of Agricultural Economics*, *72*(1), 3–21.

FAO. (2011). The State of Forests in the Amazon basin and Southeast Asian, Brazzaville, Republic of Congo. Food and Agriculture Organization of the United Nations.

FAO. (2015). *Global forest resources assessment 2015: How are the world's forests changing?* Food and Agriculture Organization of the United Nations.

FAO. (2020). *Global Forest Resources Assessment 2020 – Key findings*. FAO. https://doi.org/ doi:10.4060/ca8753en

Gaveau, D. L., Sloan, S., Molidena, E., Yaen, H., Sheil, D., Abram, N. K., Ancrenaz, M., Nasi, R., Quinones, M., Wielaard, N., & Meijaard, E. (2014). Four decades of forest persistence, clearance and logging on Borneo. *PLoS One*, *9*(7), e101654. doi:10.1371/journal.pone.0101654 PMID:25029192

Goosem, M., & Tucker, N. I. J. (2013). Repairing the rainforest: a biodiversity vision for the Wet

IUCN SSC Giraffe and Okapi Specialist Group. (2013). Okapia johnstoni. *The IUCN Red List of Threatened Species 2013*. IUCN.

Kingsford, R. T., Biggs, H. C., & Pollard, S. R. (2011). Strategic adaptive management in freshwater protected areas and their rivers. *Biological Conservation*, *144*(5), 1194–1203. doi:10.1016/j.biocon.2010.09.022

Kinver, M. (2019). *World 'losing battle against deforestation*. BBC News.

Koh, L. P., & Wilcove, D. S. (2008). Is oil palm agriculture really destroying tropical biodiversity? *Conservation Letters*, *1*(2), 60–64. doi:10.1111/j.1755-263X.2008.00011.x

Kundan, K. (2020). *Allocation of forest land for mining in the state of Chhattisgarh: A brief analysis of Hasdeo Arand stretch*. Department of Social Work, University of Delhi.

Laurance, W., Peletier-Jellema, A., Geenen, B., Koster, H., Verweij, P., Van Dijck, P., Lovejoy, T. E., Schleicher, J., & Van Kuijk, M. (2015). Reducing the global environmental impacts of rapid infrastructure expansion. *Current Biology*, *25*(7), 25. doi:10.1016/j.cub.2015.02.050 PMID:25754645

Laurance, W. F., Lovejoy, T. E., Vasconcelos, H. L., Bruna, E. M., Didham, R. K., Stouffer, P. C., Gascon, C., Bierregaard, R. O., Laurance, S. G., & Sampaio, E. (2002). Ecosystem Decay of Amazonian Forest Fragments: A 22-Year Investigation. *Conservation Biology*, *16*(3), 605–618. doi:10.1046/j.1523-1739.2002.01025.x

Lindenmayer, D., Hobbs, R. J., Montague-Drake, R., Alexandra, J., Bennett, A., Burgman, M., Cale, P., Calhoun, A., Cramer, V., Cullen, P., Driscoll, D., Fahrig, L., Fischer, J., Franklin, J., Haila, Y., Hunter, M., Gibbons, P., Lake, S., Luck, G., & Zavaleta, E. (2008). A checklist for ecological management of landscapes for conservation. *Ecology Letters*, *11*(1), 78–91. doi:10.1111/j.1461-0248.2007.01114.x PMID:17927771

Lindenmayer, D. B., & Franklin, J. F. (2002). *Conserving forest biodiversity: a comprehensive multi-scaled approach*. Island Press.

Manganda, I. E. (2004). The Deforestation of Rural Areas in the Lower Congo Province. *Environmental Monitoring and Assessment*, *99*(1–3), 245–250. PMID:15641386

MEA. (2005). *Ecosystems and Human Well-being: Biodiversity Synthesis*. World Resources Institute.

Mehta, P. (2011) *An Assessment of Distribution, Status and Threats to Endemic Malabar Giant Squirrel (Ratufa indica) in Northern Western Ghats, India*. Rufford. https://www.rufford.org/projects/prachi-mehta/an-assessment-of-distribution-status-and-threats-to-endemic-malabar-giant-squirrel-ratufa-indica-in-northern-western-ghats-india/#:~:text=Although%20giant%20 squirrels%20are%20widely,the%20 squirrel%20by%20local%20people. 19.03.2024

Meijaard, E. (2018). Sumatran orangutan (Pongo abelii). In J. Kingdon & M. Hoffmann (Eds.), *The IUCN Red List of Threatened Species 2018: e.T121097935A123797042*. IUCN.

Meijide, A., Badu, C., Moyano, F., Tiralla, N., Gunawan, D., & Knohl, A. (2018). Impact of forest conversion to oil palm and rubber plantations on microclimate and the role of the 2015 ENSO event. *Agricultural and Forest Meteorology*, *252*, 208–219. doi:10.1016/j.agrformet.2018.01.013

Miettinen, J., Shi, C., & Liew, S. (2011). Deforestation rates in insular Southeast Asia between 2000 and 2010. *Global Change Biology*, *17*(7), 2261–2270. doi:10.1111/j.1365-2486.2011.02398.x

Millennium Ecosystem Assessment. (2005). *Ecosystems and human well-being: synthesis*. Island Press.

Morgan, D., Strindberg, S., Winston, W., Stephens, C. R., Traub, C., Ayina, C. E., Ndolo Ebika, S. T., Mayoukou, W., Koni, D., Iyenguet, F., & Sanz, C. M. (2019). Impacts of Selective Logging and Associated Anthropogenic Disturbance on Intact Forest Landscapes and Apes of Northern Congo. *Frontiers in Forests and Global Change*, *2*, 448311. doi:10.3389/ffgc.2019.00028

Nepstad, D. C., Stickler, C. M., Filho, B. S., & Merry, F. (2008). Interactions among Amazon land use, forests and climate: Prospects for a near-term forest tipping point. *Philosophical Transactions of the Royal Society of London. Series B, Biological Sciences*, *363*(1498), 1737–1746. doi:10.1098/rstb.2007.0036 PMID:18267897

Newbold, T., Hudson, L. N., Hill, S. L., Contu, S., Lysenko, I., Senior, R. A., ... Borger, L. (2015). Global effects of land use on local terrestrial biodiversity. *Nature*, *520*(7545), 45–50. doi:10.1038/nature14324 PMID:25832402

Norgaard, R. B. (1988). Sustainable development: A co-evolutionary view. *Futures*, *20*(6), 606–620. doi:10.1016/0016-3287(88)90003-1

Oliveira, G., & Oren, D. C. (2017). Exploring the long-term impacts of anthropogenic disturbances in the Amazon basin: A review of vegetation and carbon stock changes in the Brazilian state of Roraima. *Anthropocene*, *17*, 13–29.

Pacheco, P., Mo, K., Dudley, N., Shapiro, A., Aguilar-Amuchastegui, N., Ling, P. Y., Anderson, C., & Marx, A. (2021). *Deforestation fronts: Drivers and responses in a changing world*. WWF.

Pandey, K. (2014, January 25). Climate footprints in the Late Quaternary–Holocene landforms of Dun Valley, NW Himalaya, India. *Current Science*, *106*(2), 245–253.

Peres, C. A., & Lake, I. R. (2003). Extent of nontimber resource extraction in tropical forests: Accessibility to game vertebrates by hunters in the Amazon basin. *Conservation Biology*, *17*(2), 521–535. doi:10.1046/j.1523-1739.2003.01413.x

Quintanilla, M., León, A. G., & Josse, C. (2022). *The Amazon against the clock: a Regional Assessment on Where and How to protect 80% by 2025*. WWF.

Rathod, N. (2002). *Deforestation in great Aravalli Mountain Region of India. AARS (ACRS 2002) report*. Hazard Mitigation and Disaster Management.

Sage, C. (2007). 7 - The scope for North–south co-operation. *Environmental Policy in an International Context, 3*, 167–196. doi:10.1016/S1874-7043(03)80026-4

Sandy, M. (2019). The Amazon Rain Forest Is Nearly Gone. We Went to the Front Lines to See If It Could Be Saved. *Time Magazine*.

Senadheera, S. S., Withana, P. A., Dissanayake, P. D., Sarkar, B., Chopra, S. S., Rhee, J., & Ok, Y. (2021). Scoring environment pillar in environmental, social, and governance (ESG) assessment. *Sustainable Environment*, *7*(1), 1–7. doi:10.1080/27658511.2021.1960097

Shakil, M. H. (2020). Environmental, social and governance performance and stock price volatility: A moderating role of firm size. *Journal of Public Affairs*, *22*(3), e2574. doi:10.1002/pa.2574

Silcock, J. L., Fensham, R. J., & Lewis, T. (2020). Involving Indigenous Landowners in Feral Animal Management in Remote Northern Australia: A Case Study of Wunambal Gaambera Country, Kimberley, Western Australia. *Environmental Management*, *65*(5), 568–579.

Synthesis, T. E. E. B. (2010). *The Economics of Ecosystems and Biodiversity: Mainstreaming the Economics of Nature: A synthesis of the approach, conclusions and recommendations of TEEB*. European Commission.

The New Nature Economy Report. (2020). Geneva: World Economic Forum.

The State of the Worlds Forest. (2020). *Food and Agriculture Organization of the United Nations*. FAO. https://www.fao.org/state-of-forests/en/

United Nations Environment Programme. (2009). *From Conflict to Peacebuilding: The Democratic Republic of the Congo*. Nairobi, Kenya: United Nations Environment Programme.

United Nations Environment Programme. (2010). *Global Forest Resources Assessment 2010: Main Report*. Food and Agriculture Organization of the United Nations.

Van der Plas, F., Manning, P., Allan, E., Scherer-Lorenzen, M., & Fischer, M. (2016). Biodiversity effect on ecosystem functioning stronger in more variable environments. *Ecology Letters*, *19*(4), 391–399.

Vesk, P. A., Morris, W. K., Neal, W. C., Mokany, K., & Pollock, L. J. (2020). Transferability of trait-based species distribution models. *Ecography*, *44*(1), 134–147. doi:10.1111/ecog.05179

Wilkie, D., Shaw, E., Rotberg, F., Morelli, G., & Auzel, P. (2000). Roads, development, and conservation in the Congo Basin. *Conservation Biology, 14*(6), 1614–1622. doi:10.1111/j.1523-1739.2000.99102.x PMID:35701921

Witte, J. (1992). Deforestation in Zaire: Logging and Landlessness. *The Ecologist, 22*(2), 58–64.

Wolfire, D., Brunner, J., & Sizer, N. (1998). *Forests and the Democratic Republic of Congo*. World Resources Institute.

WWF's Living planet Report. (2022). WWF.

KEY TERMS AND DEFINITION

Biodiversity Loss: The decline or disappearance of species.

ESG: Environmental and social Governance metrics are used to analyse an organization's sustainability practices.

Forest Diversion: The utilization of forest land and allied resources for purposes other than the forest is called 'Forest diversion'

Habitat Loss: Discontinuous diversion of forest leads to fragmentation of forest area and consequently, it separates populations from community leading to prevention of hereditary exchange which further may make the species prone to extinction.

Chapter 11
Biodiversity Loss Due to Mining Activities

Shanmukha N. T.
https://orcid.org/0009-0001-9988-2490
Government Engineering College, Krishnarajpet, India

Vinayaka M.
University BDT College of Engineering, Davangere, India

Lokeshappa B.
University BDT College of Engineering, Davangere, India

Shanwaj Nadaf
University BDT College of Engineering, Davangere, India

ABSTRACT

The mining sector, despite its significant role in the climate and biodiversity crises, has not garnered substantial attention in global discussions on these matters. COP15 presents an additional chance to integrate biodiversity and forests into standard practices within the mining sector. In 2019, a UN report highlighted a global biodiversity crisis characterized by an alarming pace of animal and plant extinctions, putting one million species at risk of vanishing. Surprisingly, the biodiversity crisis doesn't garner the same level of media and political focus as the climate crisis, despite the imperative to address both simultaneously. Preserving habitats such as forests serves a dual purpose: safeguarding species diversity while also serving as a solution to mitigate greenhouse gas emissions. Forests act as vital carbon sinks, absorbing and storing atmospheric carbon as biomass. Mining activities, essential for meeting global resource demands, have become a pervasive threat to biodiversity, leading to widespread ecological disruption and biodiversity loss.

DOI: 10.4018/979-8-3693-6950-0.ch011

INTRODUCTION TO MINING

One of the main forces behind both direct and indirect economic growth is still the extraction of mineral resources from the planet. The total production value of minerals (excluding coal) produced by China, Australia, Brazil, Russia, Chile, USA, South Africa, Canada, and India in 2012 was USD 123 billion, USD 109 billion, USD 66 billion, USD 53 billion, USD 42 billion, USD 39 billion, USD 32 billion, and USD 27 billion, respectively (1). Zambia (31%), Chile (17%), Peru (14%), Ukraine (9%), Uzbekistan (9%), South Africa (9%), Australia (7%), and Kazakhstan (7%) are among the countries whose mining contributed more than 5% of their GDP in 2010 (1). The Gross Value Added (GVA) at current prices for the fiscal year 2021–2022 showed that the mining estimated at Rs.327984 crores in India. A significant portion of the mining sector in India consists of small, active mines. In India, there were 1319 mines that reported producing minerals in 2021–2022 (not including fuel, atomic, or minor minerals), down from 1375 the year before. Madhya Pradesh accounted for 263 of the 1319 reported mines, with Gujarat (147), Karnataka (132), Odisha (128), Chhattisgarh (114), Andhra Pradesh (108), Rajasthan (90), Tamil Nadu (88), Maharashtra (73), Jharkhand (45), and Telangana (39) following, in order of decreasing number. In the years 2021–2022, these 11 States collectively accounted for 93% of all mines in the nation.

MINING OPERATIONS

Significant effects of mining operations are seen on biodiversity over a range of spatial scales, from the local site to the larger landscape, regional, and even global levels. These effects result from direct processes like the actual extraction of minerals as well as indirect ones like the actions of external stakeholders who get access to places rich in biodiversity as a result of mining operations and the industries that support mining operations. The majority of research to date has been devoted to comprehending the effects of mining at the site level. It seems sense to place such focus on site preparation for waste management and mining expansion, as these activities can be quite damaging and change the abiotic and biotic environments. These changes frequently lead to habitat deterioration and loss, which can have serious repercussions for biodiversity. Such site-level effects have been demonstrated to cause extensive reductions in rare and threatened species and ecosystems throughout entire regions in some situations.

MINING AND ITS TYPES OF MINING ACTIVITIES

Mining is the process of removing valuable minerals or geological features from the surface of the Earth, such as reefs, lodes, veins, seams, and ore bodies. This extraction process includes a wide range of non-renewable natural resources, including water, natural gas, and petroleum. The features of the ore body are usually the determining factor in the mining method selection process, which involves ranking and evaluating the numerical values associated with these features to identify the best course of action.

Modern mining plans encompass searching for minerals, measuring the turnover potential of proposed mines, extracting desired materials, and ultimately regaining the land post-closure. However, mining actions can have important environmental influences, both during operations and for years afterward. So, many countries globally have applied regulations aimed at mitigating the contrary effects of mining processes.

WATER POLLUTION

One of the main issues with mining operations is water pollution. (Sumi and Thomsen, 2001; Hudson, 2012; Dasgupta, 2012). Water quality is lowered when hazardous chemical-containing effluent spills or leaks, when leachate from mine waste is released, or when surface runoff from overburden dumps occurs. (Hudson, 2012; Karmakar and Das, 2012).

The mining process negatively affects the environment (Sumi and Thomsen, 2001; Dasgupta, 2012). Over a sizable area of surface or groundwater, the outcome may be an abnormally high concentration of various substances, such as arsenic, sulfuric acid, and mercury (Karmakar and Das, 2012). The several chemicals employed in the mining process, along with the potentially harmful substance and the metal extracted from the ore, present the possibility of significant pollution of the environment around mines (Hudson, 2012). The risk of these compounds contaminating ground and surface water grows with the volume of water produced by mining activities such as aqueous extraction, mine cooling, and drainage. (Dasgupta, 2012).

Spills And Tailing

Accidental spills of toxic chemicals from storage or processing facilities are potential concerns where ever they are used (Sumi and Thomsen, 2001; Dasgupta, 2012; Hudson, 2012). Some leaching chemicals may be left behind after the metals have been removed, mines commonly employ a combination of rinsing, physical isolation, and detoxification of heap leach pads before they are reclaimed (Hudson, 2012).

Erosion

Erosion of waste materials, especially waste rock and tailing, can affect surface water quality. Major sources of erosion at mining sites can include open pit areas, heap and dump leaches, waste rock and overburden piles, tailings piles and dams, haul roads and access roads, ore stockpiles, vehicle and equipment maintenance areas, exploration areas, and reclamation areas. Waste rock disposal areas are located as close to mines as possible to minimize haulage costs, but if they are not properly sited, metal bearing element can be eroded into streams or other bodies of water. When this occurs, the material may react with water and oxygen to release metals and other potentially undesirable elements into streams. Metals that become dissolved in surface water in this way are more

Acid Drainage

Acid drainage is a major environmental problem facing the mining industry is the formation of acid drainage and the associated mobilization of contaminants. Commonly called acid mine drainage (AMD) or acid rock drainage (ARD), acid drainage primarily depends on the mineralogy of the rock material and the availability of water and oxygen. Acid drainage is generated at both abandoned and active mine sites. This acidic water is able to dissolve-toxic metals, such as copper, aluminum, cadmium, arsenic, lead and mercury, from the surrounding rock. These metals, particularly the iron, may coat the stream bottom with an orange-red colored slime called yellow boy.

Groundwater Quality

Groundwater can also be impacted by acid runoff and mining operations. To stop acidic drainage seepage, an impermeable barrier must typically be erected at the tailing pond's base. Groundwater contamination has resulted from acidic runoff created by older tailing ponds built without impermeable bases. Acidic water seeps into groundwater, where it ultimately loses its acidity, dissolves heavy metals found in rocks, and contaminates the quality of groundwater.

Groundwater Level

The process of mining alters the drainage system, the surface contour, and lowers the water table by extracting groundwater for mining operations, which also modifies the way groundwater migrates beneath the surface (Jhariya and Chourasia, 2010; Hudson, 2012; Karmakar and Das, 2012). According to Chauhan's (2010) research, the sandstone used in the Bijolia Mine produces dust particles made solely of silica, which is heavy and insoluble, and settles at the bottom of reservoirs like ponds and wells. This effect lowers the water level by recharging the ground water's capacity rather than affecting the mine's portability.

Dewatering Mine

Groundwater enters an open pit when it intersects with the water table. Mine water pumping and discharge has a special set of effects on the environment. In order to minimize infiltration and groundwater level, a cone-shaped depression in the groundwater table is created during mining operations when groundwater is pumped from nearby wells.

AIR POLLUTION

Massive amounts of dust of all sizes are produced by opencast mining operations, and this dust enters the transportation system and disperses a large amount of gaseous pollutants and suspended particulate matter (SPM) into the atmosphere. The surrounding communities, livestock, and agricultural crops are all impacted by these contaminants in addition to the mine workers. RSPM and SPM have been observed to be the main sources of emissions from Jhansi's open-pit granite mining during the field investigation. The range of values for RSPM and SPM are, respectively, 155µg m-3 to 234µg m-3 and 393µg m-3 to 541µg m-3. The severity of the negative effects is mostly determined by the local weather conditions. Depending on their size, dispersed particulate matter can lead to: (a) Human and animal respiratory disorders.

SOIL DEGRADATION

Large amounts of overloaded waste are produced by mining, particularly open cast mining, and are disposed of in overburden dumps. This damages the land's soil quality in addition to taking up a lot of space. Major nutrients are lacking in the top material of the overburden dump, which makes plantations unsupportable. Additionally, the microbial community in my soil is impacted, which indirectly affects the health of the soil, due to its unstable structure and low nutrient content.

After mining, soil serves as the foundation and guides post-reclamation land use as we strive for vegetation restoration. Although reclamation can be evaluated in a variety of methods, the most important factor in creating a healthy environment is the quality of the soil, as mining.

DEFORESTATION

Depending on the mineral and the nation, mining has varying effects on deforestation. 60 percent of Guyana's total deforestation between 2000 and 2009 was brought on by gold mining. Nine thousand hectares of forests were lost as a result in 2009, making up 90% of all deforestation. Coal mining in the US has been connected to the yearly degradation of 2300 km of water streams. Furthermore, it has been discovered that, in comparison to pristine streams, US coal mining reduces the taxonomic richness of fish, salamanders, and invertebrates by 32% and their overall number by 35% in damaged streams. The tropics are where most of the present deforestation is taking place. Approximately 14,000 km^2 of forest land were lawfully cleared in India.

SOCIAL IMPACTS

Numerous social effects are possible for mining operations on nearby towns and cultures. Among the principal societal effects are:

Displacement and Resettlement: Land is frequently needed for mining projects, which may force local residents living in or near the mining location to relocate. Loss of identity and a breakdown in community cohesion can result from displacement, which can also upset social networks, livelihoods, and cultural customs.

Health Impacts: Mining operations can contaminate the air, water, and soil, putting the local population at risk for illnesses, respiratory disorders, and contaminated water supplies. For those who are impacted, this may result in long-term health problems and a lower quality of life.

Economic Impacts: Although mining can benefit local communities economically by creating jobs and improving infrastructure, it can also make a person financially dependent on the mining sector. Commodity price fluctuations have the potential to cause boom-bust cycles, making communities susceptible to unstable economies.

Social Conflict: Conflicts over resource ownership, land rights, and benefit sharing can arise from mining operations. Conflicts between governments, local populations, mining firms, and indigenous people can result in social unrest, demonstrations, and even acts of violence.

Cultural Impacts: Mining operations have the potential to upend indigenous peoples' and the local populations' traditional ways of life, cultural customs, and spiritual beliefs. Cultural assets, traditional knowledge, and sacred locations are vulnerable to destruction or attack, which can lead to a loss of social cohesiveness and cultural identity.

Infrastructure and Services: Mining operations have the potential to upend indigenous peoples' and the local populations' traditional ways of life, cultural customs, and spiritual beliefs. Cultural assets, traditional knowledge, and sacred locations are vulnerable to destruction or attack, which can lead to a loss of social cohesiveness and cultural identity.

NOISE POLLUTION

Blasting is a typical procedure in open-pit mining that generates a lot of noise. The gradual but progressive degradation of the inner ear's neuro-sensorial cells is the cause of deafness. Furthermore, it is well recognized that noisy work environments at mining sites affect miners' ability to communicate, interfere with their ability to complete tasks, disrupt their sleep, alter their personal behavior, etc. Furthermore, noise has negative impacts on one's health, impairs productivity at work, and hinders communication. Furthermore, noise has an impact on the wildlife in the forests and other areas near mines and industrial complexes. It is commonly held that wildlife is more sensitive to noise and vibrations than people are. The granite quarries' operating zones have rather high noise levels due to drilling, blasting and the mine service stations.

ECOSYSTEM

The increasing loss of biodiversity in the Bundelkhand region is one of the primary issues facing the region today. The region's ecological variety has been gradually declining due to large-scale mining operations, either directly or indirectly. There are several locations where vegetative covers are opened up to speed up ore excavation, infrastructure development, and overburden disposal. When vegetative cover is removed, significant soil erosion and siltation frequently occur.

Direct effects on living things in mining sites can include everything from plant and animal deaths brought on by mining operations or contact with toxic wastes and mine drainages to disruption of wildlife habitats by heavy machinery and blasting. Modifications in the cycle of nutrients, disturbances.

BIODIVERSITY AND MINING ACTIVITIES

Mining operations have the potential to significantly affect biodiversity by causing species extinction, habitat degradation, and ecosystem disruption. The following are a few ways that mining contributes to the loss of biodiversity:

Importance of Biodiversity for Ecosystem Functioning and Services

Ecosystem services—the advantages that humans receive from nature—and the proper operation of ecosystems depend on biodiversity. The following are some main arguments in biodiversity's significance for ecosystem services and functioning:

- **Resilience and Stability**: Ecosystems that are rich in biodiversity are typically more resilient to external shocks like disease outbreaks, natural disasters, and climate change. The ecosystem's redundancy and diversity of species, which act as a buffer against the extinction of individual species and preserve ecological services in the face of change, are the source of this resilience.
- **Nutrient Cycling:** The cycling of nutrients like carbon, nitrogen, and phosphorus through the soil, water, and atmosphere is greatly aided by biodiverse ecosystems. Through activities including decomposition, nitrogen fixation, and nutrient uptake, various species contribute to nutrient cycle, which guarantees the supply of vital nutrients for plant development and ecosystem productivity.
- **Pollination and Seed Dispersal:** Pollinators like bees, butterflies, and birds, as well as seed dispersers like mammals and birds, are essential to the provision of many ecological services, such as crop pollination and forest regeneration. By offering a wide variety of species with various pollination and dispersion techniques, biodiversity supports these services by guaranteeing plant reproduction and dissemination across a range of habitats.
- **Pest Control:** By providing habitat for diseases, parasites, and predators that help control insect and herbivore populations, biodiversity promotes natural pest control. In addition to promoting sustainable forestry and agriculture practices, biological pest management lowers the need for chemical pesticides.
- **Climate Regulation:** Through mechanisms including carbon sequestration, evapotranspiration, and cloud formation, ecosystems with high biodiversity

contribute to the regulation of regional and global climate patterns. As carbon sinks, wetlands, forests, and other biodiverse ecosystems store vast volumes of carbon dioxide and so reduce greenhouse gas emissions, thereby moderating climate change.

- **Water Purification:** Wetlands, mangroves, and other highly diverse aquatic ecosystems are vital for water purification because they collect sediments, break down organic waste, and filter contaminants. Aquatic ecosystems depend on biodiversity to stay healthy and function properly, providing clean water for drinking, irrigation, and sanitary purposes.
- **Cultural and Aesthetic Values:** Through its cultural and aesthetic benefits—such as leisure activities, spiritual meaning, and aesthetic enjoyment—biodiversity enhances human well-being. Because of their aesthetic appeal, cultural significance, and diversity, biodiverse landscapes and wildlife habitats enhance human experiences and foster a sense of interconnectedness with the natural world.

Impacts of Mining on Biodiversity

The following are some of the ways that mining can negatively effect species, ecosystems, and ecological processes:

Habitat Destruction and Fragmentation

The two most important effects of mining on biodiversity are habitat degradation and fragmentation. This is how mining fuels these occurrences:

Clearing of Vegetation: Large tracts of land must frequently be cleared for mining activities in order to reach mineral resources. Usually, this clearance entails removing vegetation, such as meadows, woods, and other natural habitats. Consequently, the destruction of habitat that supports a multitude of plant and animal species results in the direct loss of biodiversity.

Surface Mining: In order to access mineral resources, surface mining techniques like open-pit mining and strip mining require removing rock and soil layers. These techniques have the potential to completely remove topsoil and plants, changing the terrain and damaging natural habitats in the process.

Blasting and Excavation: Mining-related blasting and excavation operations have the potential to physically harm neighbouring habitats by destroying flora, compacting the soil, and changing the topography. These disruptions have the potential to upend ecosystems, drive animals out, and worsen habitat degradation.

Infrastructure Development: Building roads, railroads, processing plants, and other infrastructure is necessary for mining operations in order to facilitate extraction processes. This infrastructural expansion may cause natural habitats to become fragmented, obstructing the movement of wildlife and destroying the ecological linkages between various locations. Fragmentation can lead to species isolation, a decrease in genetic diversity, and a higher chance of local extinctions.

Waterway Alteration: Dredging, dam building, and sedimentation are examples of mining operations that can change river flows and harm aquatic environments. Freshwater ecosystems and the species that rely on them can be negatively impacted by changes in river flow, sedimentation levels, and water quality. This can result in the fragmentation and destruction of aquatic organisms' habitats.

Indirect Impacts: Mining can have indirect effects on nearby ecosystems, such as changes in water supply, soil erosion, and pollution, in addition to directly destroying habitats. These unintentional effects have the potential to worsen habitat fragmentation by accelerating habitat degradation and upsetting biological systems that sustain biodiversity.

Loss of Keystone Species

Keystone species are vital to the stability and biodiversity of ecosystems, and their extinction can be caused by mining activities. For instance, removing particular plant species may have an impact on the cycling of nutrients in the soil and soil stability, which may worsen habitat deterioration.

Direct Mortality of Fauna and Flora

Mining operations have the potential to directly cause the extinction of specific creatures, including plants and animals. Blasting-related explosives have the potential to kill nearby species, such as birds, animals, and reptiles. When plants and animals are crushed or moved by heavy machinery and equipment used in mining operations, rapid death results. Further adding to direct death is the habitat degradation brought on by mining, which eliminates many species' food sources, nesting places, and hiding places.

Example: In order to extract coal resources, mountaintop removal mining is a type of surface mining that is particularly common in the Appalachian region of the United States. In order to access coal seams, this mining technique entails clearing forests and blasting through mountaintops, burying streams and causing biodiversity loss and ecological deterioration. Because mountaintop removal mining disrupts and destroys habitats, it has a direct impact on bat and bird populations. The direct destruction of small mammal species' habitats during mountaintop removal mining frequently has an impact on these species, including rats, shrews, and rabbits. Mountaintop removal mining also affects the amphibians and reptiles that live in the Appalachian region's woods and streams. Animals such as snakes, turtles, and salamanders

Disturbance of Wildlife

Significant noise, vibrations, and environmental disturbances are produced by mining operations, and these factors have the potential to affect biological processes and wildlife behavior. When mining-related disturbances occur, animals may move differently or be forced out of their natural habitats. The distribution of species and the composition of communities may alter as a result of sensitive species, for instance, avoiding areas close to mining activities. Important ecological processes like seed dispersal, pollination, and predator-prey relationships can all be upset by disturbance, which lowers the ecosystem's total biodiversity.

Example: Rising gold prices and technological developments have led to a considerable increase in gold mining in the Peruvian Amazon in recent years. In one of the biodiverse areas of the planet, this has resulted in significant deforestation and habitat destruction. In the Peruvian Amazon, gold mining has a wide range of effects on wildlife, affecting many species belonging to diverse taxonomic groupings. These are a few instances of impacted species.

Mammals: Tapir (*Tapirus terrestris*), Jaguar (*Panthera onca*), Giant Anteater (*Myrmecophaga tridactyla*), Spider Monkey (*Ateles spp.*), and Sloth (*Bradypus spp.*)

Birds: Harpy Eagle (*Harpia harpyja*), Scarlet Macaw (*Ara macao*), Toucans (*Ramphastidae family*), Hummingbirds (*Trochilidae family*), and Tinamous (*Tinamidae family*)

Reptiles and Amphibians: Green Anaconda (*Eunectes murinus*), Poison Dart Frogs (*Dendrobatidae family*), Caiman (*Caiman crocodilus*), Boa Constrictor (*Boa constrictor*), and Spectacled Caiman (*Caiman crocodilus*)

Fish: Piranhas (*Serrasalmidae family*), Electric Eel (*Electrophorus electricus*), Arapaima (*Arapaima gigas*), Catfish (*Pimelodidae family*), and Tetras (*Characidae family*)

Invertebrates: Butterflies (Various species), Beetles (Various species), Ants (Various species), Dragonflies (Various species), and Moths (Various species)

Plants: Orchids (Various species), Bromeliads (Various species), Epiphytes (Various species), Trees (Various species), and Medicinal plants (Various species)

These are just a few examples of the diverse wildlife affected by gold mining in the Peruvian Amazon (Asner et al., 2015).

Disruption of Breeding and Reproduction

The breeding and reproduction cycles of animal species can be disturbed by mining activities, especially those that depend on certain habitats or environmental conditions for effective reproduction. Reductions in reproductive success and population levels can result from animals being discouraged from breeding or nesting in areas affected by noise, vibrations, and habitat loss. Breeding behaviour disruptions can worsen biodiversity loss by having long-term effects on genetic diversity and population viability.

Impact on Endangered Species

Mining operations frequently take place in areas where threatened or endangered animals are also present. These already endangered species may face serious dangers to their survival from direct death and disturbance caused by mining operations. Populations of endangered species may be disproportionately affected by even minor perturbations, which could drive the species closer to extinction.

Example: As an illustration, consider the red-cockaded woodpecker, an endangered species native to the Southeast region of the United States that lives in old-growth pine woods. Coal mining operations also target these forests, resulting in habitat degradation and loss for the woodpecker population.

Indirect Effects on Ecosystem Functioning

Mining-related direct death and disturbance of creatures can have a domino impact on ecosystem services and functioning. For instance, plant reproductive cycles can be disturbed and plant diversity decreased by the loss of important pollinators or seed dispersers. Similar to how changes in predator populations can result in an increase in prey species, community dynamics and ecosystem structure can also be affected. These unintended consequences may worsen the decline of biodiversity in areas where mining operations are taking place.

CASE STUDIES

Mining in the Western Ghats

The Western Ghats, an ecological hotspot and UNESCO World Heritage Site, are under tremendous human pressure due to deforestation, population growth, and irresponsible mining. These pressures are made worse by issues with governance and competing conservation plans put forth by committees chaired by Dr. Gadgil and Dr.Kasturirangan. Using mining as a case study, this paper explores the challenges of managing the Ghats and emphasizes the critical necessity for coordinated conservation and resource management plans.

Overview of the Location

The Western Ghats extend over six Indian states, from Gujarat to Tamil Nadu. They are roughly 160,000 km^2 in area, 1600 km in length, and have heights between 300 and 2700 meters. Similar to the northern Himalayas, they have a major impact on Peninsular India's biogeography and ecology.

One of the world's hotspots for biodiversity, the Western Ghats are home to a wide variety of habitats and species, including more than 400 biological species and a large number of endangered plants and animals. It has high levels of endemism, sustains priceless genetic resources, and is home to 39 national parks and wildlife sanctuaries. Its ecological significance in India is highlighted by its impact on rainfall patterns and carbon sequestration, which neutralizes about 4 million tonnes of carbon dioxide annually.

Mining in the Ghats and its Impacts in Biodiversity

Economic development is inextricably linked to mining and mineral extraction, which is utilized in a wide range of sectors including transportation, energy, defense, construction, agriculture, and ICTs. The heavily mineral-rich six coastal states that make up the Ghats contribute significantly to India's total mining output. Major haematite iron ore reserves, for example, are found in Karnataka (11% of total reserves) and Goa (5%).15 The greatest reserves of magnetite iron ore in India are found in Karnataka (72%), followed by Tamil Nadu (5%), and Goa (2%).16 Due to the Western Ghats' abundant supply of these resources, mining has been widespread in the area for a lengthy amount of time.

Five taluks in Goa's Western Ghats region—Sattari Darbandora, Sanguem, Quepem, and Canacona—are heavily mined for late rite basalt, iron, and manganese.

To facilitate comprehension, the following infographic illustrates the amount of mining in the next five states:

LOSS OF BIODIVERSITY

Numerous plant and animal species have disappeared as a result of the substantial negative effects that mining operations have had on the Western Ghats' biodiversity.

An extensive report on a few of the species impacted by mining in the area may be found here:

Endangered Mammals

Numerous endangered animal species are in danger due to habitat degradation and fragmentation caused by mining operations in the Western Ghats. Famous animals like the Indian elephant (Elephas maximus indicus), Bengal tiger (Panthera Tigris), and Indian gaur (Bos gaurus) can be found in the Western Ghats. For these massive beasts to survive, their habitat must span enormous contiguous expanses. However, mining operations cause biological corridors to be disrupted and forests to be destroyed, isolating communities and making it harder for them to find food and mates.

Endemic Amphibians

There are numerous amphibian species in the Western Ghats that are unique to this part of the world. However, these amphibians have suffered greatly from pollution and habitat devastation brought on by mining. For instance, the loss of their forest habitat and the pollution of water bodies by mining effluents are causing population decreases in species such as the Malabar torrent toad (Ghatophryne ornata) and the Amboli bush frog (Pseudophilautus amboli).

Endemic Birds

The Western Ghats are home to many endemic bird species, making them a hotspot for avian diversity. However, significant bird habitats including evergreen woods and grasslands are being destroyed as a result of mining activities. Population decreases are being caused by mining operations that disrupt and destroy habitat for species like the Malabar parakeet (Psittacula columboides) and the Nilgiri wood pigeon (Columba elphinstonii).

Endemic Plants

Many of the endemic plant species found in the Western Ghats are under threat from mining operations. These plants may be directly impacted by the clearing of forests for mining operations and related activities like blasting and road construction. Because of habitat loss brought on by mining, some species, including the once-a-decade blooming Neelakurinji (Strobilanthes kunthiana), have seen population decreases and habitat degradation.

Freshwater Fishes

Many endemic fish species can be found in the rivers and streams of the Western Ghats. Fish populations have been negatively impacted by the pollution and sedimentation of water bodies caused by mining activity. Certain species, such as the indigenous Mahseer (Tor spp.), are more susceptible to pollution and habitat destruction, which can result in population decreases.

How Effect the Critical Endangered Species

The Western Ghats, a globally recognized biodiversity hotspot, is home to several critically endangered species that are at risk of extinction due to mining activities. Here's a detailed report some of the critically endangered species impacted by mining in the region:

- **Lion-tailed Macaque (*Macaca silenus*):** A flagship species of the Western Ghats, the lion-tailed macaque is among the world's most critically endangered monkeys. Its already limited habitat has been severely diminished by habitat degradation and fragmentation caused by mining. Population losses and an increase in conflict between humans and wildlife have resulted from the mining industry's destruction of dense evergreen forests.
- **Malabar Civet (*Viverra civettina*):** A nocturnal carnivore native to the Western Ghats is the Malabar civet. Its range has been reduced and its population number has decreased due to mining activities that have destroyed its woodland habitat.
- **Nilgiri Tahr (*Nilgiritragus hylocrius*):** The Western Ghats' high-altitude grasslands are home to the famous Nilgiri tahr - mountain goat species. Due to competition from cattle, hunting, and habitat loss, it is listed as severely endangered. Its habitat has been invaded by mining operations, which has caused habitat fragmentation and population decreases.
- **Great Indian Bustard (*Ardeotis nigriceps*):** In the semi-arid grasslands and scrublands of the Western Ghats, there exists a critically endangered bird species called the Great Indian bustard. Its habitat has been invaded by mining operations, which has resulted in habitat loss, disturbance, and mortality from collisions with mining infrastructure. Reduced grassland habitats have led to population decreases and heightened extinction susceptibility.
- **Wayanad Laughingthrush (*Montecincla wayanadensis*):** Native to the Western Ghats, the Wayanad laughingthrush is a severely endangered species of bird. It is extremely susceptible to habitat fragmentation and loss brought on by mining operations. Population decreases and greater isolation of the

surviving populations are the results of habitat destruction in the montane evergreen forest.

Impact of Iron Ore Mining on Avian Diversity in Bellary District

This study evaluated the impact of iron ore mining on bird diversity, and it was carried out in the Bellary area of Karnataka. It discovered that habitat damage and pollution brought on by mining operations had resulted in notable drops in bird populations and species diversity.

Affected Species: Various bird species including raptors, waterfowl, and passerines.

Effects of Granite Quarrying on Amphibian Communities in Mysore District

This study looked at how amphibian populations were affected by granite mining in the Karnataka district of Mysore. It discovered that the loss and degradation of amphibian habitats led to a drop in the diversity and abundance of species.

Affected Species: Various amphibian species including frogs, toads, and caecilians.

Impact on Aquatic Species

Impact of Sand Mining on Riverine Ecosystems in Dakshina Kannada District

This study looked into how sand mining affected riverine ecosystems in the Dakshina Kannada district of Karnataka. Fish and aquatic invertebrate population decreases were shown to be caused by habitat degradation, sedimentation, and changes in water quality.

Affected Species: Various fish species including Mahseer and indigenous freshwater fish.

Effects of Limestone Mining on Freshwater Fauna in Tumkur District

This study looked at how mining for limestone affected freshwater animals in the Tumkur area of Karnataka. It discovered notable changes in flow patterns and water quality that resulted in population decreases of fish and macroinvertebrates.

Affected Species: Native fish species, freshwater crabs, mollusks, and aquatic insects (Manjunath, 2015).

Impact of Iron Ore Mining on Riverine Ecosystems in Bellary District

This study evaluated the impact of iron ore mining on riverine ecosystems in the Bellary area of Karnataka. It showed how heavy metal pollution, habitat deterioration, and sedimentation caused fish and macroinvertebrate communities to decline. Affected Species: Indigenous fish species, freshwater crabs, aquatic insects, and other benthic organisms (Girisha and Sridhar, 2017).

Mitigative Measures for Biodiversity Losses Due to Mining Activities

A variety of tactics are included in mitigation efforts for biodiversity losses brought on by mining operations with the goal of lessening the detrimental effects on ecosystems and species.

Pre-Mining Biodiversity Assessment

To find rare species, vulnerable habitats, and ecological hotspots, thorough biodiversity evaluations must be carried out prior to mining operations. These evaluations aid in the development of practical mitigation plans that reduce the loss of biodiversity. For instance, before beginning mining operations in the Pilbara region of Western Australia, Rio Tinto carried out comprehensive biodiversity surveys to identify important ecosystems and species, enabling the implementation of focused conservation measures.

Habitat Conservation and Restoration

In order to mitigate the loss of biodiversity, strategies for habitat conservation and restoration must be implemented. This include designating protected areas, creating corridors for species, and working to restore habitat after mining. The conservation program of Richards Bay Minerals (RBM) in South Africa serves as an illustration; it involves the restoration of wetlands and the rehabilitation of mined areas in order to promote local biodiversity.

Adoption of Best Available Technologies

Mitigating habitat disturbance and environmental implications can be achieved by implementing advanced mining technology and procedures. For example, in order to decrease habitat fragmentation and reduce surface disturbance, corporations involved in oil sands mining activities in Alberta, Canada, have invested in new technology like steam-assisted gravity drainage (SAGD).

Biodiversity Offsetting

By utilizing conservation or restoration initiatives elsewhere, biodiversity offsetting attempts to make up for losses in biodiversity brought on by mining operations. One instance is the Biodiversity Offsets Program in Queensland, Australia, which mandates that mining corporations fund conservation projects like habitat restoration or species reintroduction efforts in order to balance their impacts on biodiversity.

Community Engagement and Stakeholder Collaboration

For effective biodiversity conservation in mining sites, interaction with stakeholders, local communities, and indigenous groups is essential. Partnerships between local organizations and mining firms, as well as community-led conservation programs, are examples of collaborative approaches that can help increase support for biodiversity conservation efforts. One such instance is the cooperative conservation projects that are carried out in Peru as part of the partnership between Newmont Corporation and Conservation International to safeguard biodiversity and assist local populations.

Long-Term Monitoring and Adaptive Management

By putting in place long-term monitoring programs, it is possible to assess the success of mitigation strategies and spot new risks to biodiversity. Mining businesses can modify their operations in response to stakeholder feedback and monitoring data by implementing adaptive management approaches. For instance, the Biodiversity Management Committees in the Indian state of Madhya Pradesh coordinate the monitoring of biodiversity in mining sites and support adaptive management techniques aimed at reducing impacts.

Pollution Control

It is necessary to reevaluate the pollution management methods now in use. The "polluter pays" idea must be put into practice, with fines commensurate with the extent of environmental damage and favourable rewards for excellent work. Pollution level monitoring must be done in a completely transparent manner and involve private organizations, such as science and engineering college laboratories. In this setting, new easily observable biodiversity indicators must be used to complement physical and chemical data. Establishing appropriate double-blind protocols akin to drug testing for athletes is necessary to guarantee total dependability and public confidence in the pollution monitoring system. In order to allow NGOs to participate in the pollution monitoring process as committee members, we must methodically establish.

From the standpoint of biodiversity conservation and sustainable use, the new Biological Diversity Act offers great opportunities for reforming the current legislative and policy frameworks of civic bodies ranging from zilla parishads to municipalities, gram sabhas, gram panchayats, and municipal corporations. It asks for the creation of biodiversity management institutions at the state, district, municipality, and panchayat levels. To address the new issues, the Karnataka government should take advantage of its ability to foster new, broad-based institutions with the involvement of the civil society.

The case study analysis highlights the significant impact of mining activities on biodiversity loss, emphasizing the urgent need for comprehensive measures to mitigate these adverse effects. From habitat destruction and fragmentation to pollution and altered ecosystems, mining operations have been shown to have far-reaching consequences on both terrestrial and aquatic biodiversity. Throughout the analysis, it became evident that the current regulatory frameworks often fall short in adequately safeguarding biodiversity against the onslaught of mining activities. Weak enforcement, lack of stringent environmental assessments, and insufficient rehabilitation efforts exacerbate the negative impact on ecosystems and species diversity.

Furthermore, the case study underscores the importance of adopting a holistic approach that integrates ecological considerations into mining planning and decision-making processes. This approach should prioritize conservation strategies, restoration efforts, and sustainable practices to minimize biodiversity loss while meeting the demands for mineral resources. Findings of this case study underscore the critical need for collaborative efforts among governments, mining companies, conservation organizations, and local communities to address biodiversity loss caused by mining activities effectively. By implementing robust regulations, promoting responsible mining practices, and investing in biodiversity conservation and restoration initiatives, we can strive towards achieving a balance between resource extraction and environmental preservation, ensuring the long-term health and resilience of our planet's ecosystems.

LEGAL INTERVENTIONS: ANALYSIS AND REFLECTIONS FOR THE FUTURE

The past and future of mining in the Ghats is riddled with policy challenges. Forests and wildlife being under concurrent jurisdiction as per the Constitution of India has led to the fragmentation in the legislative and administrative state control. Not only do the laws vary within and across states, the Government of India has also tipped the scale towards the developmental activities despite mounting evidence of it putting pressure on the environment. Given the ecological sensitivity and significance of the Ghats and the intricate nature of its geology, the Ministry of Environment and Forests established Western Ghats Ecology Expert Panel (WGEEP) headed by Prof. Madhav Gadgil in 2010. The Panel was approached to recognize Ecological Sensitive Areas (ESAs) along the Western Ghats, and to recommend the manner in which they could be managed and regulated. Based on extensive survey and consolidation of data, the WGEEP recommended 64% of the Western Ghats as 'ecologically sensitive', and dissected the entire region into three classes of Ecologically Sensitive Areas, in the order of varying degrees of ecological affectability. For mining, the Committee recommended an "indefinite moratorium" on new environmental clearance applications in ESZ 1 & ESZ 2. Further, it advised for the phasing out of mining from ESZ 1 by 2016, as well as all existing mining activities in ESZ 2 to be under strict regulations as well as subject to regular social audit. However, the Gadgil Report was labelled too 'environmentally-friendly' and deaf to the cause of economic development. Subsequently, the Ministry of Environment and Forest constituted a High-Level Working Group (HLWG) under Dr. K. Kasturirangan, which was mandated to receive inputs upon the ESZs demarcated by the Gadgil report from all the relevant states as well as to formulate its implementation. Submitted in 2013, the Kasturirangan Report tagged 37% of the Ghats as an ESA. With respect to mining, the Panel still recommended a total ban upon such activities, and granted existing operations leeway to continue operating until their lease expired. No leases in the ESA, however, would be renewed.

This HLWG Report became the basis for a draft notification released by the Ministry of Environment, Forest and Climate Change which seeks to declare one-third of the Ghats as ecologically sensitive. Although the Kasturirangan Report had a similar perspective upon the negative impacts of mining and quarrying in the Western Ghats, yet the whittling down of the total area declared as ESA only grants an illusory protection to the Ghats from the extractive industries. Environmentalists regard the Report to be "biased and pandered to entrenched lobbies and dubious elements", which was drafted after primarily relying upon satellite pictures, as opposed to the years of fieldwork undertaken by the WGEEP. Despite prohibitions, reports show that state governments are permitting private miners to operate.

The conservation debate upon the management of the Ghats continues with no consensus in sight, while the noises of mining and quarrying continue to echo in the mountains. The regard with which the WGEEP sought to protect the Ghats rightly acknowledged its fragility and sensitivity. Irrespective of how fragmented the Ghats currently are, they have been proven to have the capacity to support the rich repertoire of species and their habitat. The conservation is more urgent in the case of endemic species, as they can only be protected through in situ planning and conservation. Continuation of mining may cause contamination of air and water bodies, and its cessation may increase the quality of water as well as biodiversity improvements. The importance of the Ghats puts immense legislative pressure to effectively ban ecologically-destructive activities such as mining, as well as to revive the landscapes which have already damaged through it. The Gadgil Report must be revisited again to understand the environmental consequences of mining in order to minimize the impending hazards, which will get exacerbated by climate change. As highlighted by the Hon'ble Madras High Court in S. Kumar v. The District Collector, the precautionary principle, as "one of the pillars of foundation upon which the edifice of Environmental Law is built", should be applied while granting mining licenses and clearances. The extent of lawlessness that prevails in India's mining industry also cannot be exaggerated, and illegal mining is also major cause of concern. This alludes to situations wherein miners extract resources that they have no legitimate authority to or consent for. Bellary, Karnataka, for instance, was once treated as a hotspot for mining, yet the flagrant violations of the MMDR Act 1957, the EPA 1986 and the Forest Conservation Act 1980 lead to the cancellation of mining activities in the environmentally ravaged region. However, what drew attention to this predicament was the connivance of officials at all levels, regardless of their associations with "police, mining, forest, revenue, weights and measurements, commercial taxes, labour and the Karnataka State Pollution Control Board". Parallels can also be drawn in the case of illegal mining in Goa. These events necessitate the policy to decentralize as well as establish a stronghold on ground to check instances of flouting of rules and regulations, internally as well as externally. Mining projects are required to mandatorily assess the ecological and social risks of their proposed and ongoing operations. However, past instances have demonstrated that the inputs received during public hearings of such assessments regularly ignored or the information is not made freely accessible to the populace in the relevant vernacular, prompting significant levels of social disappointment and disagreement. A case in point made by the Gadgil report is of the unanimous opposition by the Gram Panchayat of Sindhudurg to a proposed mining project, due to the water contamination of the Kalane river, which was overlooked by the Maharashtra Government to grant clearance to the mine. Such failure to ensure an inclusive, transparent and fair decision-making process runs counter-intuitive to the goals of the Environmental Clearance process.

CONCLUSION

There is a need for the mining sector, which is often demonized in this respect, to coordinate with all the relevant stakeholders to ensure satisfactory legal and social approval. Lastly, it must also be borne in mind that the Ghats' ecological significance coincides with human settlements, where the people are profoundly dependent upon its biological resources for subsistence and livelihood. This makes distributional equity and environmental justice to be significant concerns in this debate, whereby the interests of the locals cannot invariably be presumed to be morally inferior to the urgency of conservation – highlighting the need for environmental protection to itself be more socially and economically inclusive. A key way to release from the cycle of 'development by imposition' and 'conservation by imposition' is to increase public participation within conservation strategies, not only to grant democratic legitimacy to environmental decisions but also to improve the quality and efficacy of such decisions. In the case of the Ghats, this necessitates recognition of both formal and informal modes of protection, "so that all stakeholders can participate actively in the endeavour". Carvalho (2017) suggests engaging the local community "in reforestation and remediation efforts-parallel with mining activity". Another instance is of recognizing sacred groves and sacred kan forests under the Biodiversity Act, 2002, that have historically and extensively been a traditional form of conservation in the Western Ghats which, if provided due acknowledgement, would not only lead to biodiversity preservation but also decentralize and democratize conservation planning.

REFERENCES

Asner, G. P., Llactayo, W., Tupayachi, R., Luna, E. R., & Heyenga, S. (2013). Elevated rates of gold mining in the Amazon revealed through high-resolution monitoring. *Proceedings of the National Academy of Sciences of the United States of America*, *110*(46), 18454–18459. doi:10.1073/pnas.1318271110 PMID:24167281

Carvalho, F. P. (2017). Mining industry and sustainable development: Time for change. *Food and Energy Security*, *6*(2), 61–77. doi:10.1002/fes3.109

Dasgupta, A. (2012). Impact of mining on rural environmental and economy. A case study, Kota district, Rajasthan. *International Journal of Remote sensing and Geoscience (IJRSG), 2*, 21-26.

Girisha, S. K., & Sridhar, K. R. (2017). Impact of Iron Ore Mining on Macroinvertebrate Communities in Two Streams of the Sandur Hills, Bellary District, Karnataka, India. *Journal of Entomology and Zoology Studies*, *5*(1), 95–100.

Jhariya, D., & Choraisia, K. (2010) Impact of mineral production on water resources AIVAM water conservation. Ground Water Augmentation AIVAM Conservation of Water Resources.

Karmakar, H. N., & Das, K. P. (2012). *Impact of Mining on Ground & Surface water*. International Mine Water Association. www. IMWA.info

Sumi, L., & Thomsen, S. (2001). *Mining in Remote Areas, Issues and Impacts. MinindWatch Canada/ Mines Alerte. Hudson, T. (2012).* Living with Earth, An Introduction to Environmental Geology. PHI Learning Private Limited.

ADDITIONAL READING

Balakrishnan, R., Rajkumar, R., & Induchoodan, V. (2017). Impact of mining on floral diversity in Western Ghats, India. *International Journal of Environment, Agriculture and Biotechnology*, *2*(5), 2291–2298.

Chapin, F. S. III, Carpenter, S. R., Kofinas, G. P., Folke, C., Abel, N., Clark, W. C., & Leemans, R. (2010). Ecosystem stewardship: Sustainability strategies for a rapidly changing planet. *Trends in Ecology & Evolution*, *25*(4), 241–249. doi:10.1016/j.tree.2009.10.008 PMID:19923035

Costa, R. G., Hooper, R. G., & Barrett, R. H. (2002). Red-cockaded woodpeckers in mined areas: Factors influencing habitat quality. *The Journal of Wildlife Management*, *66*(2), 502–513.

D'Souza, N. J., Gurumurthy, G. P., & Mamatha, P. (2015). Impact of mining activities on land use/land cover in Bicholim taluk, Goa. *International Journal of Research in Engineering and Technology*, *4*(10), 213–219.

Dahanukar, N., Raghavan, R., Ali, A., & Abraham, R. (2011). The status and distribution of freshwater fishes of the Western Ghats. In R. T. Watson, T. E. M. Brooks, & N. D. Burgess (Eds.), The Status and Distribution of Freshwater Biodiversity in the Western Ghats, India (pp. 21-36). IUCN.

Dahanukar, N., Raut, R., Bhat, A., & Molur, S. (2016). Endemic Amphibians of the Western Ghats. In S. Molur, K. Smith, B. A. Kutty, S. M. J. Walker, & P. Priya (Eds.), The Status and Distribution of Freshwater Biodiversity in the Western Ghats, India (pp. 23-34). IUCN.

Deb, M., Tiwari, G., & Lahiri-Dutt, K. (2008). Artisanal and small scale mining in India: Selected studies and an overview of the issues. *International Journal of Mining, Reclamation and Environment, 22*(3), 194–209. doi:10.1080/17480930701679574

Ganzhorn, J. U., Goodman, S. M., & Vincelette, M. (2007). *Biodiversity, ecology and conservation of littoral ecosystems in Southeastern Madagascar* (A. Alonso, Ed.). Smithsonian Institution.

Government of Alberta. (2021). *Oil Sands Mining*. Government of Alberta. https://www.alberta.ca/oil-sands-mining.aspx

Government of Madhya Pradesh. (2022). *Madhya Pradesh Biodiversity Board*. Government of Madhya Pradesh. http://www.mpsbb.nic.in/

Gupta, V., Singh, R. P., & Banerjee, A. (2020). Impact of mining on tree diversity and population structure in iron ore and associated forest ecosystem of Goa, India. *Environmental Monitoring and Assessment, 192*(7), 465. PMID:32601744

Indian Bureau of Mines. (2015a). *Indian Minerals Yearbook 2014 (Part-III: Mineral Reviews)*. Government of India, Ministry of Mines, Indian Bureau of Mines. ⟨https://ibm.gov.in/writereaddata/files/ 12172015172623IMYB- 2014_Bauxite(Adv).pdf⟩. https://mines.gov.in/webportal/nationalmineralscenario

Jacobi, C. M., do Carmo, F. F., Vincent, R. C., & Stehmann, J. R. (2007). Plant communities on ironstone outcrops: A diverse and endangered Brazilian ecosystem. *Biodiversity and Conservation, 16*(7), 2185–2200. doi:10.1007/s10531-007-9156-8

Jathanna, D., Karanth, K. U., & Johnsingh, A. J. T. (2008). Estimation of large carnivore densities in the tropical forests of Bhadra Wildlife Sanctuary, India. *Journal of Zoology, 274*(2), 188–198.

Johnsingh, A. J. T. (2001). Conservation status of tigers in India: A summary. In A. Seidensticker, S. Christie, & P. Jackson (Eds.), Riding the Tiger: Tiger Conservation in Human-Dominated Landscapes (pp. 24-34). Cambridge University Press.

Kulkarni, C., Nayak, G. K., & Nayak, B. (2017). Impact of iron ore mining on suspended sediment response in a tropical catchment in Kudremukh, Western Ghats, India. *Environmental Earth Sciences, 76*(1), 37.

Kumar, M. A., Shashikumar, M. C., & Karanth, K. K. (2014). Occupancy and density of diurnal primates in logged and unlogged forests of the Western Ghats, India. *International Journal of Primatology, 35*(5), 927–937.

Kumar, S., Majumder, J., & Baskaran, N. (2019). Distribution and status of Nilgiri tahr (Nilgiritragus hylocrius) in the Western Ghats, India. *Journal of Threatened Taxa, 11*(5), 13504–13512.

Manjunath, M. T., & Sreekantha, B. (2015). Impact of Limestone Mining on the Aquatic Fauna of River Shimsha, Karnataka, India. *Journal of Applied and Natural Science, 7*(1), 1–6.

Nandagopal, S., & Vasudeva, R. (2016). Study of Biodiversity of Amphibians and Reptiles in Mining Area of Southern India. *Journal of Fisheries & Livestock Production, 4*(1), 1–5.

Newmont Corporation. (n.d.). *Partnerships for Conservation*. Newmont Corporation. https://www.newmont.com/sustainability/partnerships-for-conservation

Palmer, M. A., Bernhardt, E. S., Schlesinger, W. H., Eshleman, K. N., Foufoula-Georgiou, E., Hendryx, M. S., Lemly, A. D., Likens, G. E., Loucks, O. L., Power, M. E., White, P. S., & Wilcock, P. R. (2010). Mountaintop mining consequences. *Science, 327*(5962), 148–149. doi:10.1126/science.1180543 PMID:20056876

Queensland Government. (2022). *Biodiversity Offsets Program*. Queensland Government. https://www.qld.gov.au/environment/land/management/biodiversity-offsets-program

Raman, T. R. S. (2006). Mining and biodiversity. *Current Science, 90*(2), 161–162.

Ramesh, B. R., Vyas, V., Venkatesha, M. T., & Krishnan, S. (2016). Land use dynamics and their ecological consequences in the Western Ghats of Karnataka, India. *Environmental Monitoring and Assessment, 188*(2), 94. PMID:26780413

Richards Bay Minerals. (n.d.). *Environment and Community*. Richards Bay Minerals. https://www.rbm.co.za/sustainability/environment-and-community/

Robin, V. V., Katti, M., Purushotham, C. B., & Sreekar, R. (2018). Disappearing acts: Evidence of habitat and species loss from a tropical biodiversity hotspot. *Biological Conservation, 221*, 320–326.

Ruiz, O., Thomson, B., & Cerrato, J. (2016). Investigation of in situ leach (ISL) mining of uranium in New Mexico and post-mining reclamation. *New Mexico Geology, 38*, 77–85.

Sankaran, R. (2016). Impact of mining on the distribution, habitat use and conservation status of the Great Indian Bustard in Karnataka, India. *Oryx*, *50*(3), 442–448.

Singh, G. (2010). Assessment of environmental impacts by mining activities: A case study from Jhansi open cast mining site - Uttar Pradesh, India. *Journal of Experimental Sciences, 1*(1), 09- 13.

Singh, M., & Tare, V. (2017). Impact of Sand Mining on Aquatic Species in Two Rivers of Dakshina Kannada District, Karnataka, India. *Current World Environment*, *12*(3), 939–947.

Sreedharan, R. (2018). Environmental Impact Assessment of Iron Ore Mining in Goa. *International Journal of Research in Environmental Studies*, *5*(2), 17–24.

Srinivasulu, B., Srinivasulu, C., Venkateshwarlu, M., & Srinivasulu, A. (2019). Impact of Iron Ore Mining on Avifauna of Sandur Hills, Bellary District, Karnataka, India. *Journal of Threatened Taxa*, *11*(5), 13504–13512.

Tinto, R. (2020). *Biodiversity*. Riotinto. https://www.riotinto.com/sustainability/environment/biodiversity

Wickham, J., Wood, P. B., Nicholson, M. C., Jenkins, W., Druckenbrod, D., Suter, G. W., Strager, M. P., Mazzarella, C., Galloway, W., & Amos, J. (2013). The overlooked terrestrial impacts of mountaintop mining. *Bioscience*, *63*(5), 335–348. doi:10.1525/bio.2013.63.5.7

Chapter 12
Plastic Pollution and Its Impact on Biodiversity

Pulak Kumar Patra
https://orcid.org/0000-0002-8773-1978
Visva-Bharati, India

Ankita Chatterjee
Visva-Bharati, India

ABSTRACT

Biodiversity loss is one of the most critical environmental problems nowadays. Several natural and anthropogenic causes are known to trigger it. Environmental pollution, including land degradation, has been long considered a significant contributor to biodiversity loss. However, the role of plastic pollution, especially microplastics, in biodiversity loss has been recognised relatively recently. This chapter delves into the complex relationship between biodiversity and plastic pollution, assessing how plastic degrades natural habitats, distorts the ecosystem, and threatens the survival of countless species, eventually disrupting the ecosystem's delicate balance. Through a review of scientific literature, the chapter probes the sources of plastic pollution, its various forms and the impacts of plastic on marine and terrestrial ecosystems. It examines how plastic waste ingestion, entanglement, and chemical contamination affect individual species, populations, and ecological communities. Lastly, the chapter discusses measures to manage plastic pollution and safeguard biodiversity.

DOI: 10.4018/979-8-3693-6950-0.ch012

INTRODUCTION

Biodiversity has a great significance from ecosystem functioning to human wellbeing. It promotes ecosystem stability and functioning. Diverse ecosystems are more flexible to environmental changes such as climate change and natural disasters. Each species within an ecosystem plays a unique role or niche, and their interaction helps in nutrient cycling, pollination, water purification, and other essential ecosystem processes. The utility of rich biodiversity can be grouped into economic benefits and eecosystem services. Biodiversity come up with a wide range of economic benefits, including the medicines, provision of food and raw materials for industries such as agriculture, biotechnology and drug industry. It also provides many ecosystem services which are beneficial to human wellbeing such as clean air and water, soil fertility, flood control, carbon sequestration, and climate regulation. Biodiversity also provides us a variety of genetic resources which helps in biotechnology innovation, medical research and crop improvement. Considering the accelerating loss of biodiversity due to human activities such as habitat destruction, pollution especially plastic pollution, overexploitation of resources, and climate change, it is becoming crucial to emphasis on conservation efforts and to adopt sustainable practices.

Biodiversity loss refers to the decrease or disappearance of specific species in a particular area. Exacerbated by various human activities, it is a pressing environmental challenge that has far-reaching consequences for ecosystems and the planet's well-being. Human activities like deforestation, urbanisation, and agricultural expansion destroy and fragment natural habitats, losing numerous plant and animal species. Environmental pollution, including air and water pollution and climate change driven by human activities, poses significant threats to biodiversity by disrupting ecosystems, altering temperature patterns, and impacting the survival of various species. A critical contributor to this issue is plastic pollution, which has emerged as a pervasive and detrimental force affecting the biodiversity of both terrestrial and marine ecosystems. The use and disposal of plastics is a transboundary threat to biodiversity and natural ecosystems due to their low degradation rate and unsustainable production. Global plastic production has reached 368 million metric tons (Mt) annually and is estimated to double within 20 years (Lebreton and Andrady, 2019). The article explores the intricate relationship between plastic pollution and biodiversity loss. It highlights the diverse processes and pathways through which plastics impact the ecosystems, leading to declining biodiversity. The paper also deals with the latest technology for effectively managing plastic pollution, including monitoring through remote sensing.

Causes of Biodiversity Loss

The significant driver of biodiversity loss can be of two types- natural and human-driven. Natural ecological disturbances can include- wildfires, floods, and volcanic eruptions, which can alter biological diversity by reducing the local population of some species or transforming the whole biological community. Contrary to that, human-induced biodiversity loss will likely be harsh and deep-rooted. Some of them are: habitat loss, invasive species, overexploitation of natural resources, pollution, climate change resulting from global warming, and overpopulation. They are discussed below.

- Habitat fragmentation: Human activities such as urbanisation, deforestation, mining and infrastructure lead to the reduction of food resources and living space, resulting in loss of species' natural habitat.
- Overexploitation: Overutilisation of natural resources such as marine fish through overfishing and grazing results in the depletion of some species. This ultimately disrupts the intricate balance of the ecosystem and can have a ripple effect on other species within the food web.
- Pollution: Pollution from various sources, including industrial runoff, agricultural chemicals, air pollutants, and plastic waste, poses a significant threat to biodiversity. This can degrade their habitat as well as their lives.
- Climate change: Global warming-induced climate change is another reason for biodiversity loss. Climate change alternates the precipitation and temperature pattern, catalyses habitat loss, changes in species dispersion and disrupts the ecological process of migration and reproduction. The risk of species loss increases with every degree of warming. According to the United Nations Climate Change, mammals will have a 4% habitat loss with a 1.5°C rise in temperature and an 8% habitat loss with a 2°C increase.
- Land Use Change: Land use change often results in habitat loss, degradation, and fragmentation. The continuous alteration of natural habitats into urban areas, roadways, agricultural land reduces the chances of availability of suitable habitats for species.
- Population Growth and Overconsumption: The increasing population and consumption level aggravate natural resources and ecosystems. The growing demand for food, water, energy, and housing creates many opportunities for biodiversity loss.
- Invasive species: Introducing non-native species into an ecosystem sometimes disturb the ecological balance. These new species can outcompete the others in many ways (diseases, prey on them), leading to the extinction of the previous one.

Out of all the factors mentioned above, the accumulation of plastic waste here and there poses a significant menace to biodiversity nowadays and can be considered an emerging threat. The main question that comes to mind is why plastic is becoming the most important environmental hazard. Plastics are not natural products. They are produced synthetically from fossil fuels (coal and petroleum products) in industries. Plastics are long-chain polymers and cannot be degraded or decomposed by natural biological processes. Thus, they persist and accumulate in the ecosystem once used and discarded as waste. The widespread use of plastics has had detrimental effects on terrestrial and aquatic ecosystems. From the disposal sites, plastics move by wind and water to other parts of the ecosystem, contaminating the soil and water bodies. The toxic chemicals released from them are also transported around the globe through air and water, leading to health hazards. Plastic wastes are disintegrated into finer particulates, now identified as microplastics. These invisible plastics roam in the air, water and soil and then can enter the biosphere through the food chain and inhalation. It intimidates ecosystems animal and plant species, hindering their capacity to deliver essential services to humanity. To understand the impact of plastics on an ecosystem, we must know the nature of plastic materials.

Components of Plastic and its Sources

Plastic is a ubiquitous resource regarded as a marvel of modern convenience. Its adaptability and durability have modified industries and consumer behaviour in recent times. Plastics, derived from fossil fuels, have universal uses, from packaging to construction materials, from healthcare to electronics, and their uses are multifarious and seemingly endless. Yet, its durability poses significant environmental risks. Plastic is an artificial material derived from petroleum or natural gas polymers. The versatility and adaptability of plastics make them highly multifaceted and flexible for various uses. These polymers can be moulded or shaped into different shapes.

The primary substance of plastic (Beaumont et al., 2019) include-

- Polyethylene (PE): These are the most common forms of plastic in plastic bags, bottles, and containers. They can be found in both high-density and low-density polyethene forms.
- Polypropylene (PP): It is a firm and flexible type of plastic used in packaging, textiles, automotive parts, and household items like containers and appliances.
- Polyvinyl Chloride (PVC): It is durable and resistant to weathering and flame. It is used in construction materials, pipes, flooring, medical devices, and various consumer products.
- Polystyrene (PS): These are lightweight plastics used in packaging, disposable cups and utensils, insulation, and consumer electronics.

Other forms are Polyethylene Terephthalate (PET) and Polyurethane (PU). Many plastics harbour hazardous substances, including bisphenol A (BPA), phthalates, styrene, vinyl chloride, and flame retardants, which can leach into the surroundings, posing health risks to humans and wildlife. Furthermore, the degradation of more oversized plastic items through exposure to environmental factors such as mechanical breakdown, abrasion, biological and chemical degradation, and polymer weathering can lead to the formation of microplastics (particles less than 5 mm) and nano-plastics (particles less than 1 micron), compounding environmental pollution. Each of the forms of plastic has unique properties and usefulness. Despite their functionalities, rigorous production and disposal of plastic have induced significant ecological challenges, including pollution, habitat destruction, and menace to wildlife and ecosystems.

The primary sources of plastic include:

- Industry: Plastic is produced by polymerisation using petroleum and natural gas as raw materials. The wide range of production of plastic includes packaging, construction, automotive, electronic goods, and consumer goods.
- Packaging: Plastic bags, bottles, containers, and wrappers are the best examples. This packaging versatility has led to massive adoption in the food, beverage, and retail sectors.
- Consumer Goods: Due to their durability, affordability, and lightweight nature, household items, such as toys, electronics, and clothing, incorporate plastic components or packaging.
- Single-Use Items: Single-use plastics have a short lifespan despite their extensive use in daily life. These include straws, utensils, cups, and take-out containers (Prata et al., 2019).
- Construction: The construction industry uses plastics for items such as pipes, insulation, flooring, windows, and siding widely for its affordability.
- Medical and Healthcare: Plastic materials that are sterile and lightweight, such as syringes, IV bags, medical packaging, and equipment components, are made mainly of plastics.
- Agriculture: In agriculture, plastics are used to improve production, preserve water, and check weeds. Plastic films, mulches, and irrigation pipes are used.

Apart from the procedures mentioned earlier, inappropriate waste disposal practices on land, littering, and illegal dumping bring plastic waste into various niches, impacting soil quality and water bodies. Plastic wastes are now pervasive from the poles to the most isolated islands and from the sea's surface to its lowest depth.

Plastic debris mainly disseminates from land and accumulates in the oceans, representing the major part of marine litter. (Lartaud et al., 2020). Microplastics, derived from personal care products like exfoliating scrubs and toothpaste, further amplify the problem. According to a recent study, Microplastics can cause harm to the UN Sustainability Development Goals to improve atmospheric, water and soil quality and maintain biodiversity (Agathokleous et al., 2021; Davies et al., 2022). Sewage emerges as a significant contributor to the dispersion of microplastics, with a substantial portion—approximately 80 to 90 per cent—of plastic particles in sewage persisting in the sludge. The widespread dispersal of microplastics in our soils results from sewage sludge extensively utilised as fertiliser. Maritime activities, including oil spills and ship litter, significantly contribute to plastic pollution in aquatic environments. Human and industrial waste is to the ocean in developing countries, including micro and macro plastic (Woodall et al., 2014). Inadequacies of waste management infrastructure become apparent in poorly managed landfills. Inefficient containment leads to the leakage of plastic pollutants into the surrounding soil and water, contributing to permeating environmental pollution. Furthermore, the open burning of plastic waste amplifies the issue, releasing pollutants into the air and further degrading the air quality. Plastic pollution has multifaceted impacts on biodiversity. From marine ecosystems to terrestrial habitats, the consequences are far-reaching.

Microplastic and its Effect on Biodiversity

Microplastic are plastic that are less than 5 mm of size. Their type and source can be categorized in to two types- primary microplastic which are manufactured for various purposes, including personal care products, microbeads (<1mm), industrial abrasives, and microfiber textiles and secondary microplastic which originate from the fragmentation and degradation of larger plastic items, such as plastic bags, bottles, fishing gear, and packaging materials. This breakdown of microplastic includes the process of chemical degradation, UV radiation and mechanical process. (Wu et al, 2017).

The effects of micro-plastics are as follows:

1. Growth and reproduction: the impact of polyvinyl chloride (PVC) microplastics on the growth, oxidative damage, oviposition, and biological enzyme activity of gilthead seabream have been demonstrated. (Hu et al, 2019). Micro-plastics can increase the permeability of sediments and change the heat transfer which further affect temperature-related biological development and growth processes.
2. Disruption of ecosystem: Microplastics affect ecosystem dynamics by changing the behaviour, physiology, and reproductive success of organisms.

For example, studies have shown that microplastics can impede the ability of fish to detect predators or prey, change feeding behaviour of marine invertebrates, and affect the development of larvae. (Alberghini et al, 2022).

3. Chemical contamination: Microplastics can absorb and concentrate toxic chemical such as persistent organic pollutants (POPs) such as polychlorinated biphenyls (PCBs) and pesticides. When organisms ingest these chemical, this eventually cause toxicity in their body and disrupt ecosystem.
4. Affect on soil: Microplastics can change the physicochemical properties of soil, e.g., nutrient cycling and soil structure, eventually influencing growth of plant and soil-dwelling organisms. (Kumar et al, 2023).
5. Affect on aquatic animals: Aquatic organisms, such as fish, shellfish, and plankton, can ingest microplastics which affect their feeding behaviour, and lead to physical blockage, disrupt their growth and reproductive process. (Kumar et al, 2022). It can also lead to health issues such as oxidative stress, immune system alterations, neurotoxicity, and hormonal disruption in aquatic animals. (Dhar et al, 2023).
6. Climatic changes: with the degradation of plastic, many harmful greenhouse gases such as Furans, Mercury, Dioxins, and Polychlorinated Biphenyls are produced (Verma et al, 2016). Airborne microplastic can affect climate by absorbing solar and terrestrial radiation resulting into atmospheric warming which in due course cause global warming.

Effects of Plastic Pollution on Terrestrial Biodiversity

Many terrestrial organisms mistakenly consume plastic items, resulting in issues like ingestion, digestive complications, and disturbances in their natural habitats. This disturbance extends to soil fertility, nutrient cycles, and the overall terrestrial food web, intensifying the gravity of the problem. Plastics, through their interaction with soil, have the potential to release harmful chemicals that impact the well-being of both plants and animals. Accumulated plastic waste on the soil's surface can potentially discharge detrimental chemicals, eventually influencing soil quality. The presence of sizable plastic items obstruct trees' growth and nutrient absorption capabilities, leading to inhibited growth, reduced productivity, and heightened vulnerability to diseases. Microplastics may gather in plant root zones, potentially impacting nutrient absorption. Plastic debris on the soil's surface can hinder water infiltration and disturb natural drainage systems. The leaching of additives and chemicals into the soil poses toxicity risks to plants, ultimately resulting in food chain contamination. The enduring presence of these chemicals may give rise to enduring ecological consequences, posing a potential threat to biodiversity across different trophic levels. Species unable to adapt to or withstand the encroachment of plastic waste may witness declines in their populations or even local extinctions. Large plastic debris, a prevalent form of pollution, can obstruct terrestrial wildlife's typical movements and migration patterns. This interference results in the isolation of populations and hampers crucial behaviours such as breeding and foraging, further exacerbating the challenges faced by terrestrial ecosystems. In terrestrial ecosystems, plastic wastes affect the biota mainly through the destruction of their habitats and through entanglement and ingestion. Toxic chemical contaminations also affect the terrestrial biodiversity. Terrestrial soil is one of the significant zones of accumulation of plastic debris. Plastic litter is a considerable source of pollution in urban and rural soil. Plastic can release harmful chemicals and additives such as BPA (Bisphenol A) and phthalates, which can leach into the soil, leading to degraded soil health. Other causes include road runoff (including tyre wear particles) and atmospheric deposition of micro- and nano-plastic particles (Blasing & Amelung., 2018). A study reveals that the microplastics in agricultural soils affect soil health and productivity (Ragusa et al., 2021). In a study published in Environmental Pollution in 2020, the agrarian soil containing plastic residue impacts the soil microbial community and alters nutrient cycling, ultimately demeaning plant health and ecosystem quality. Research has shown that microplastics can change soil structure and degrade soil fertility. On solid ground, wildlife mistakenly eats up plastic items. This can lead to issues like ingestion, entanglement, suffocation, digestive complications, even death and disturbances in their natural habitats. Not only that, plastic wastes are breeding grounds for disease vectors, leading to the spread of infectious diseases (Anunobi, 2022). There have been reports that tiny plastic particles have been shown to penetrate biological membranes (Nizzetto et al., 2016). A case study of the Taiwan forest in Japan reveals that plastic flakes and tiny particles have been found in bear faeces. Another case study in January 2018 showed a 20-year-old wild elephant in Periyar, India, died from plastic consumption from waste left behind by the numerous Sabarimala pilgrims. Investigations reported vast quantities of plastic in its intestines, leading to internal bleeding and organ failure, ultimately resulting in its death (Geneva Environment Network, 2024).

According to the Humane Society of the United States, plastics retard the flying and hunting capabilities of birds. Species like raccoons often get stuck in plastic ring beverage holders, which causes deep cuts on their bodies. According to a report published by UN Environment Programme in 2019, Marine ducks, divers, penguins, albatrosses, petrels, grebes, pelicans, gannets and boobies, gulls, terns, auks, as well as tropic birds are particularly in danger. Plastic ingestion can kill them or cause severe injuries, and the digestive tract can get damaged by the accumulation of plastic, leading to malnutrition and starvation due to a feeling of full stomach.

Chemicals from plastic are found in birds' eggs in remote environments such as the Canadian Arctic (Tekman et al., 2022).

Plastic also pervasive affects plant growth by contaminating soil, creating physical barriers, changing habitats, introducing chemical additives in nature, and disrupting niche interactions. Plastic pollution can affect terrestrial habitats by changing soil composition and deteriorating plant growth. The reduced fertility due to plastic contamination in the soil can constrain root growth, seed germination, and nutritional availability, eventually reducing plant diversity in the ecosystem. Additionally, large plastic items such as bottles and carry bags can hinder light penetration into soil and air circulation, essential for plant growth. This eventually can disrupt the natural structure and composition of the ecosystem and degrade the habitat. A study examines the effects of microplastics on soil properties and plant growth using a greenhouse experiment. It suggests that microplastics can change soil pH, moisture holding capacity, and nutrient availability, affecting plant production and ecosystem dynamics (de Souza et al., 2019). Some researchers suggest microplastics can hinder root elongation, reduce shoot biomass, and modify plant nutrient uptake. Then, they highlighted the necessity of better waste management practices (Wu et al., 2020).

Though microplastics in marine and aquatic ecosystems research has received much attention by researchers until now, terrestrial ecosystems are estimated to receive 4-23 times more microplastic waste annually (Xu et al., 2020). The detection, occurrence, characterisation, and toxicology of microplastics in soil ecosystems have recently drawn considerable attention. Microplastics have been shown to affect soil physical and chemical properties, microbial and enzyme activities, and plant growth. It has also adverse ecotoxicological effects on soil fauna (Guo et al., 2020). Microplastic inputs to terrestrial ecosystems have been linked to poor waste management (leakage) application of biosolids to agricultural soils (Crossman et al., 2020; Rolsky et al., 2020). Atmospheric microplastic deposition has also been reported from many pristine and remote protected areas, such as the Pyrenees mountain catchments in France (Allen et al., 2019) and Rocky Mountain National Park in the USA (Brahney et al., 2020).

Microclimatic conditions in terrestrial ecosystems, such as surface albedo, temperature, and moisture levels, can be altered by plastic debris. Dark-coloured plastics can absorb solar radiation and increase surface temperatures, affecting microclimate and vegetation growth. Microclimatic changes can change the habitats of plants and animals, conclusively modifying species distributions and community composition.

Effects of Plastic Pollution on Marine Biodiversity

According to a report by IUCN, 2021 In 2021, global plastic production reached 390.7 million tons, a 4% increase, with 14 million tons going into the ocean yearly, making up 80% of marine debris. Plastic debris has multifaceted effects on marine biodiversity, putting light on the entanglement, ingestion, habitat destruction, microplastic proliferation, and disruption of ecosystem services that disrupt the balance of the ecosystem. A study quantifies global plastic waste inputs into the ocean, providing crucial insights into the scale of plastic pollution and its impact on marine ecosystems (Jambeck et al., 2015). Another study reveals the effect of plastic pollution on freshwater biodiversity. It states that plastic can be dangerous for freshwater species like invertebrates to mammals, leading to the risk of death and birds losing their ability to move, feed and breed; plastic can also be dangerous as it changes the overall quality of water, alters habitats, food availability and the overall freshwater ecosystem (Valter et al., 2021).

Marine animals like turtles, seals, dolphins, and seabirds often wrap themselves around or get entangled in plastic debris such as ropes, nets, traps and monofilament lines from abandoned, lost or discarded fishing gear. This eventually causes strangulation, wounds, restricted movement and death. According to a report by WWF, Germany, Fishing lines entangled 65% of coral colonies in Oahu, Hawaii, and in the deep Arctic sea. Plastic has been entangled in nearly 20% of the sponge colonies.

Many sea animals mistakenly swallow plastic debris as food, ultimately leading to death. This ingestion can block the digestive system and lead to malnutrition and internal injuries. The sea turtle is an ancient mariner whose very existence has been irreversibly altered by plastic debris. These majestic creatures face dire consequences by ingesting plastic items, mainly bags, under the illusion of jellyfish. Similarly, microplastics embed themselves in the tissues of whales and dolphins, posing a pervasive threat to these intelligent marine mammals. The impacts ripple through the depths of our oceans, affecting the intricate balance of life within the world's coral reefs, home to many marine species. Laboratory experiments have shown reduced growth in fish when their food is contaminated by high volumes of microplastics (Naidoo & Glassom., 2019). Another event shows the death of whales in Thailand by ingestion of plastic drinking straws and rupture of the digestive system, resulting in death (Haetrakul et al., 2009). Plastic debris is causing irreversible damage to sea turtles and whales, causing them to ingest plastic items and microplastics, which are embedded in their tissues. Plastic debris can encroach on sea turtles, affecting their movement, breeding and feeding behaviours. These impacts affect life balance within coral reefs, home to numerous marine species, and the ocean's depths. Some studies indicate altered food uptake, negative impacts on growth, immune response, fertility, reproduction, and modified cell functions in impacted species (Prokic et al., 2019). Another study explores the physical effects of microplastics on marine organisms, including ingestion, abrasion and entanglement, and the probable ecological and physiological impacts on aquatic biodiversity (Wright et al., 2013). In the case of fish, both small and large, microplastics can accumulate in the digestive system, causing reduced nutrient absorption, inflammation, and altered feeding behaviours. Microplastics cause toxic effects, reduced food consumption, slow growth, abnormal behaviour, and oxidative damage, and act on fishes at the molecular level (Davies & Anwuri., 2022). Microplastics can change assemblage composition and structure, impacting nutrient cycling and primary productivity. Microplastics have the potential to spread antimicrobial resistance by acting as a carrier for antimicrobial-resistant bacteria (Daniel & Green., 2020).

Persistent organic pollutants, like polychlorinated biphenyls (PCBs), have long-lasting effects on organisms and environmental health due to their non-degradable nature, which can be transported over long distances by wind and water. Animals' consumption of plastic can lead to the leaching of those toxic chemicals into their tissues, causing bioaccumulation and biomagnification and exposing predators at the top of the food chain, including humans, to high levels.

Plastic debris can alter mangrove habitats, coral reefs and seagrass beds. Accumulation of plastic in these habitats can smother and damage organisms, disrupt nutrient cycling, and inhibit gas exchange by depleting light, food, and oxygen in sediments, particularly for bottom-dwelling animals, corals, sponges, which leads to sediment oxygen deficiency and reduced organism numbers (Green et al., 2015).

In coral ecosystems, plastic can cause harmful effects, causing physical damage to coral reefs, leading to coral bleaching, reduced growth, and increased vulnerability to diseases. The toxic chemicals from plastic can also have pervasive effects on coral, disrupting their reproductive and developmental processes (Lartaud et al., 2020). Plastic can also disrupt the essential ecosystem services the marine ecosystem provides, such as nutrient cycling, carbon sequestration, fisheries and shoreline protection. However, habitat degradation and biodiversity decline can decrease their pliability to environmental stressors.

Management of Plastic Waste

A three-pronged strategy is adopted to manage plastic waste. The first and best method is to reduce the production and use of plastic materials. The second approach is to manage plastic waste scientifically, reducing its adverse effects. Plastic production and single-use plastics have led to a noteworthy waste management issue. While reducing plastic waste generation is essential, managing existing plastic waste efficiently is equally crucial for environmental protection. The third approach is to recycle and reuse plastic products. Despite recycling being a standard solution, only 14% of plastic waste is currently collected for recycling. Critics argue that recycling is insufficient, emphasising the toxic chemical recycling procedure. They highlight a comprehensive approach needed to address the plastic crisis. There are many sustainable strategies involved in the management of plastic pollution that are faced nowadays. Plastic pollution management requires a multidisciplinary approach, including control of production and consumption, eco-design, and increased demand for recycled plastic (Prata et al., 2019).

Strategies like policy initiatives and technological innovations aim to reduce plastic pollution and protect biodiversity, with the "reduce, reuse, and recycle" approach being a sustainable approach (Davies & Anwuri., 2022; Atoi et al., 2023)). A recent study says the solution to this problem lies in the following 3Rs: reduce, reuse and recycle. Recycled waste plastic can be used in several ways, including constructing roads. (Gupta., 2017). Reduction is the finest way to control waste (Siddique et al., 2008).

The initial step includes – lesser production and consumption of single-use plastics, banning or taxing plastic bags (Dan et al., 2012), utensils, and straws, and encouraging the business to adopt a plastic-free substitute or eco-friendly alternatives of plastic (Beaumont et al., 2019), promoting sustainable consumer behaviour and lifestyle by making people aware about the pervasive effect of plastic on both human as well as biodiversity, more innovation and research on biodegradable, compostable, or easily recyclable alternatives. Reuse is another approach that lessens the demand for new plastic, saves money, and saves energy and resources (Dan et al., 2012). The reuse concept emphasises reusable containers, packaging, and bags, which helps reduce the creation of new plastic waste. It also promotes the adoption of durable, long-lasting products and encourages businesses to implement take-back or refill programs for products like electronics, beverage containers, and packaging materials. Further, recycling plays a pivotal role in the management of plastic. The proposed measures include improving recycling infrastructure, educating consumers about proper waste segregation and recycling practices, and investing in advanced recycling technologies like pyrolysis and chemical recycling. These measures aim to increase plastic waste collection, sorting, and processing, especially in regions with limited recycling capabilities. This comprehensive strategy promotes responsible and environmentally conscious handling of plastic materials.

Apart from this method, some more integrated and holistic mitigation measures include- removing microbeads from personal care products, developing clean-up and bioremediation techniques for betterment of wastewater treatment efficiency (Davies et al., 2022), organising educational campaigns to make people aware of the impacts of plastic pollution, make consumer aware about their purchasing habits, Establish plastic-free zones in public spaces, events, and institutions to reduce the prevalence of plastic pollution, The initiative aims to encourage entrepreneurship and innovation in the green economy, particularly in start-ups and businesses that create sustainable solutions for plastic waste management, recycling, and upcycling. The government should also take up legislative measures, incentives for compliance and enforcement mechanisms, and implement local, regional and national policy regulation. There should be enough coordination among stakeholder engagement and government agencies and transparency in decision-making processes. Recent research explores the prospective for sustainable plastic pollution management, emphasising the need for policy changes, technological advancements and public awareness campaigns to lessen its impact on marine biodiversity (Kibria et al., 2023).

Another recent research explores some other mitigate approaches, which include plastic leakage management through proper control strategies by defining clear guidelines, standards, and protocols to guide ground-based practices, Zero landfilling policies and mainly some technologies to control plastic in solid waste for ex-mechanical recycling which produces recycled material of degraded quality, chemical recycling or feedstock recycling which produces syngas, plastic monomers and incineration for recovery of energy (Nikiema & Aseidu, 2022). Thermochemical conversion technologies like pyrolysis, liquefaction and gasification are all helpful in transforming plastic waste into valuable resources (Atoi et al., 2023).

The first step in managing plastic waste is to monitor the accumulation, movement and fate of these waste products in the ecosystem. Life cycle assessment is necessary for evaluating the environmental impact of plastic throughout its whole life cycle i.e. from production to its disposal. The production, use and disposal of every goods possess a potential health and environmental impact. Life cycle assessment helps in understanding and managing that environmental implications. Life cycle of a plastic involved different stages-

1. Raw material extraction (natural gas or crude oil extraction)
2. Manufacturing (processing of raw materials through the process of polymerisation, molding, extrusion etc)
3. Distribution (distribution to consumers, retailers, wholesaler etc)
4. Use phase (the actual use of plastic by the consumer which mainly has the environmental impacts)
5. Disposal (at this stage plastic waste goes through different methods of disposal such as incineration, landfilling, composting, recycling etc.)

The disposal stage of plastic waste has the most environmental impacts. Landfilling cause leaching of toxic chemicals into soil and water. Incineration process can emit greenhouse gas such as methane into air, causing air pollution. Composting process mainly uses biodegradable plastic, which breakdown the plastic into organic matter through effective process.

Researchers found that recycling the most effective process that reduces the environmental impact of plastic waste. Recycling can notably save energy and terminate harmful emissions of various carcinogens and GHGs into the environment (Gandhi et al, 2021). Recycling has the least environmental impact on total energy impact and global warming potential. While involving community citizens in collecting data related to pollution is encouraging, earth observation from space-based remote-sensing techniques provides quick and affordable monitoring. Due to its easy availability and applicability, remote sensing can be used to locate and quantify plastic debris in terrestrial and marine environments. For example, remote sensing data have been used to evaluate plastic waste in agriculture (Lanorte et al., 2017; Zhang et al., 2022) and monitor plasticulture in agricultural landscapes (Levin et al., 2007). Novelli and Tarantino (2015) utilised spectral bands to differentiate plastic debris from vegetation, sand, and water within terrestrial environments with the help of Landsat 8 OLI/TIRS spectral bands. In aquatic environments, floating plastic debris on water surfaces was detected using remote sensing techniques. For example, (Zhang et al., 2016) utilised high-resolution satellite imagery to identify and quantify floating plastic waste in the Bohai Sea in China. Hu (2021) evaluated the spectral characteristics of different bands of moderate spatial resolution Sentinel-2 satellite data in the visible (VIS) and near-infrared (NIR) spectra to detect and differentiate various forms of plastic debris in the marine environment. Biermann et al.(2020) highlight that the remarkable high spatial resolution, capable of providing details as small as 10 m × 10 m, proves invaluable in identifying and tracking plastic litter in marine environments. In the coming days, with the development of drone-based sensors, remote sensing is expected to be widely utilised to monitor plastic litter and its impact on biodiversity.

CONCLUSION

Biodiversity has an ecological and commercial value that ensures human well-being. The production, overconsumption and improper plastic disposal contaminate terrestrial and marine biodiversity habitats. The disruption of land and marine biodiversity through human-induced methods, mainly plastic hazards, threatens the ecosystem. From marine ecosystems to terrestrial habitats, the consequences are far-reaching. That disruption results in endangered species, disruption of the biogeochemical cycle, and altered food chain. The surge in plastic production and single-use plastics has turned plastic into a significant waste management issue. There is a need for concerted action at local, regional and global levels to protect the ecosystem from the pervasive effect of plastic. Previously, there was a ban on plastic in developing countries like India, but that was unsuccessful due to a lack of policy awareness, poor governance, weak institutional capacity, and a lack of affordable alternatives to plastic products. Now, to ensure the protection of biodiversity and ecosystems, there is an urgent need to adopt sustainable approaches to plastic use and disposal. Sustainable management of plastic pollution involves a holistic and integrated approach that addresses the entire lifecycle of plastic, from production to disposal, while promoting environmental sustainability, resource efficiency, and circularity. Proper management strategies, better waste management practices, recycling, enhanced public awareness and education and, more research and development in alternate materials, better collaboration among governments, industries, non-governmental organisations, and communities are the best approaches to reduce the prevalent effect of plastic in the environment and to protect the integrity and resilience of Earth's diverse ecosystems. The use of high-resolution satellite imagery not only enhances the accuracy of plastic detection but also provides a nuanced understanding of the spatial distribution and dynamic patterns of plastic pollution. Insights that are crucial for formulating targeted strategies and interventions to combat plastic pollution can be gained by employing such advanced technologies. The integration of these precise satellite observations presents a significant advancement in our ability to monitor and address plastic pollution on a global scale comprehensively.

The exploration of the effects of plastic pollution on biodiversity reveals a narrative of ecological vulnerability and human responsibility. From the oceans teeming with marine life to the diverse tapestry of terrestrial ecosystems, the insidious infiltration of plastics disrupts the delicate balance of nature. However, amid the harsh realities, glimpses of hope emerge through innovative solutions, technological advancements, and growing global awareness. The commitment to reduce, reuse, and recycle, sustainable alternatives, and responsible waste management represents a collective lifeline to safeguard our planet's biodiversity. Through concerted efforts, we can journey towards a future where biodiversity flourishes.

REFERENCES

Agathokleous, E., Iavicoli, I., Barceló, D., & Calabrese, E. J. (2021). Ecological risks in a 'plastic' world: A threat to biological diversity? *Journal of Hazardous Materials*, *417*, 126035. doi:10.1016/j.jhazmat.2021.126035 PMID:33992919

Alberghini, L., Truant, A., Santonicola, S., Colavita, G., & Giaccone, V. (2022). Microplastics in Fish and Fishery Products and Risks for Human Health: A Review. *International Journal of Environmental Research and Public Health*, *20*(1), 789. doi:10.3390/ijerph20010789 PMID:36613111

Allen, S., Allen, D., Phoenix, V. R., Le Roux, G., Jiménez, P. D., Simonneau, A., Binet, S., & Galop, D. (2019). Atmospheric transport and deposition of microplastics in a remote mountain catchment. *Nature Geoscience*, *12*(5), 339–344. doi:10.1038/s41561-019-0335-5

Anunobi, T. J. (2022). Hazardous effects of plastic wastes on land biodiversity: A review. *The Zoologist*, *20*(1), 80–86. doi:10.4314/tzool.v20i1.10

Beaumont, N. J., Aanesen, M., Austen, M. C., Börger, T., Clark, J. R., Cole, M., Hooper, T., Lindeque, P. K., Pascoe, C., & Wyles, K. J. (2019). Global ecological, social and economic impacts of marine plastic. *Marine Pollution Bulletin*, *142*, 189–195. doi:10.1016/j.marpolbul.2019.03.022 PMID:31232294

Biermann, L., Clewley, D., Martinez-Vicente, V., & Topouzelis, K. (2020). Finding Plastic Patches in Coastal Waters using Optical Satellite Data. *Scientific Reports*, *10*(1), 5364. doi:10.1038/s41598-020-62298-z PMID:32327674

Bläsing, M., & Amelung, W. (2018). Plastics in soil: Analytical methods and possible sources. *The Science of the Total Environment*, *612*, 422–435. doi:10.1016/j.scitotenv.2017.08.086 PMID:28863373

Brahney, J., Hallerud, M., Heim, E., Hahnenberger, M., & Sukumaran, S. (2020). Plastic rain in protected areas of the United States. *Science*, *368*(6496), 1257–1260. doi:10.1126/science.aaz5819 PMID:32527833

Bubu-Davies, O., & Anwuri, P. (2022). Microplastics: Potential impacts on aquatic biodiversity. *Tropical Freshwater Biology*, *31*(1), 45–60. doi:10.4314/tfb.v31i1.4

Crossman, J., Hurley, R. R., Futter, M., & Nizzetto, L. (2020). Transfer and transport of microplastics from biosolids to agricultural soils and the wider environment. *The Science of the Total Environment*, *724*, 138334. doi:10.1016/j.scitotenv.2020.138334 PMID:32408466

Dan, C., Xingyuan, H., & Peng, W. (2012). Recycling effective ways of waste plastics. *Engineering Plastics Application.*, *40*, 92–94.

de Souza Machado, A. A., Lau, C. W., Kloas, W., Bergmann, J., Bachelier, J. B., Faltin, E., Becker, R., Görlich, A. S., & Rillig, M. C. (2019). Microplastics Can Change Soil Properties and Affect Plant Performance. *Environmental Science & Technology*, *53*(10), 6044–6052. doi:10.1021/acs.est.9b01339 PMID:31021077

Dhar, M., Jasrotia, R., Langer, S., & Suwartiningsih, N. (2023). Impact of Microplastics on Reproductive and Physiological Aspects of Aquatic Inhabitants. In M. A. Rather, A. Amin, Y. A. Hajam, A. Jamwal, & I. Ahmad (Eds.), *Xenobiotics in Aquatic Animals*. Springer. doi:10.1007/978-981-99-1214-8_6

Gandhi, N. R., Farfaras, N., Wang, N. L., & Chen, W. (2021). Life Cycle Assessment of Recycling High-Density Polyethylene Plastic Waste. *Journal of Renewable Materials*, *9*(8), 1463–1483. doi:10.32604/jrm.2021.015529

Green, D. S. (2020). Biological and Ecological Impacts of Plastic Debris in Aquatic Ecosystems. In F. Stock, G. Reifferscheid, N. Brennholt, & E. Kostianaia (Eds.), *Plastics in the Aquatic Environment - Part I. The Handbook of Environmental Chemistry* (Vol. 111). Springer. doi:10.1007/698_2020_509

Green, D. S., Boots, B., Blockley, D. J., Rocha, C., & Thompson, R. (2015). Impacts of discarded plastic bags on marine assemblages and ecosystem functioning. *Environmental Science & Technology*, *49*(9), 5380–5389. doi:10.1021/acs.est.5b00277 PMID:25822754

Gregory, M. R., & Andrady, A. L. (2003). Plastics in the marine environment. In A. L. Andrady (Ed.), *Plastics and the environment* (pp. 379–401). John Wiley and Sons, Inc. doi:10.1002/0471721557.ch10

Guo, J. J., Huang, X. P., Xiang, L., Wang, Y. Z., Li, Y. W., Li, H., Cai, Q. Y., Mo, C. H., & Wong, M. H. (2020). Source, migration and toxicology of microplastics in soil. *Environment International*, *137*, 105263. doi:10.1016/j.envint.2019.105263 PMID:32087481

Gupta, P. (2017). Management of plastic waste: A step towards clean environment. *International Journal of Renewable Energy Technology*, *8*(3/4), 387. doi:10.1504/IJRET.2017.088986

Haetrakul, T., Munanansup, S., Assawawong-kasem, N., & Chansue, N. (2009). A case report: Stomach foreign object in whaleshark (Rhincodon typus) stranded in Thailand. *Proceedings of the 4th International Symposium on Seastar 2000 and Asian Bio-Logging Science*, (pp. 83–85). Geneca Environment Network. https://www.genevaenvironmentnetwork.org/resources/updates/plastics-and-biodiversity/-

Hu, D., Shen, M., Zhang, Y., Li, H., & Zeng, G. (2019). Microplastics and nanoplastics: Would they affect global biodiversity change? *Environmental Science and Pollution Research International, 26*(19), 19997–20002. doi:10.1007/s11356-019-05414-5 PMID:31102222

Jambeck, J. R., Geyer, R., Wilcox, C., Siegler, T. R., Perryman, M., Andrady, A., & Law, K. L. (2015). Plastic waste inputs from land into the ocean. *Science, 347*(6223), 768–771. doi:10.1126/science.1260352 PMID:25678662

Jatoi, A. S., Ahmed, J., Mubarak, N. M., Shaikh, M. S., Kumar, S., Ahmed, M. I., & Karri, R. R. (2023). A Comprehensive Review on Handling of Plastic Waste for Energy Generation—Current Status and Future Challenges. *Water, Air, and Soil Pollution, 234*(12), 773. doi:10.1007/s11270-023-06784-w

Kibria, M. G., Masuk, N. I., Safayet, R., Nguyen, H. Q., & Mourshed, M. (2023). Plastic Waste: Challenges and Opportunities to Mitigate Pollution and Effective Management. *International Journal of Environmental Research, 17*(1), 20. doi:10.1007/s41742-023-00507-z PMID:36711426

Kumar, A., Mishra, S., Pandey, R., Yu, Z.-G., Kumar, M., Khoo, K. S., Thakur, T. K., & Show, P. L. (2023). Microplastics in terrestrial ecosystems: Un-ignorable impacts on soil characterises, nutrient storage and its cycling. *Trends in Analytical Chemistry, 158*, 116869. doi:10.1016/j.trac.2022.116869

Kumar, A., Upadhyay, P., & Prajapati, S. K. (2022). Impact of microplastics on riverine greenhouse gas emissions- a view point. *Environmental Science and Pollution Research International, 29*(54), 107300–107303. doi:10.1007/s11356-022-23929-2 PMID:36336740

Lanorte, A., Santis, F. D., Nolè, G., Blanco, I., Loisi, R. V., Schettini, E., & Vox, G. (2017). Agricultural plastic waste spatial estimation by Landsat 8 satellite images. *Computers and Electronics in Agriculture, 141*, 35–45. doi:10.1016/j.compag.2017.07.003

Lartaud, F., Meistertzheim, A. L., Reichert, J., Ziegler, M., Peru, E., & Ghiglione, J. F. (2020). Plastics: An Additional Threat for Coral Ecosystems. In S. Rossi & L. Bramanti (Eds.), *Perspectives on the Marine Animal Forests of the World*. Springer. doi:10.1007/978-3-030-57054-5_14

Lebreton, L., & Andrady, A. (2019). Future scenarios of global plastic waste generation and disposal. *Palgrave Communications, 5*(1), 1–11. doi:10.1057/s41599-018-0212-7

Levin, N., Lugassi, R., Ramon, U., Braun, O., & Ben-Dor, E. (2007). Remote sensing as a tool for monitoring plasticulture in agricultural landscapes. *International Journal of Remote Sensing*, *28*(1), 183–202. doi:10.1080/01431160600658156

Naidoo, T., & Glassom, D. (2019). Decreased growth and survival in small juvenile fish, after chronic exposure to environmentally relevant concentrations of microplastic. *Marine Pollution Bulletin*, *145*, 254–259. doi:10.1016/j.marpolbul.2019.02.037 PMID:31590784

Nikiema, J., & Asiedu, Z. (2022). A review of the cost and effectiveness of solutions to address plastic pollution. *Environmental Science and Pollution Research International*, *29*(17), 24547–24573. doi:10.1007/s11356-021-18038-5 PMID:35066854

Nizzetto, L., Futter, M., & Langaas, S. (2016). Are agricultural soils dumps for microplastics of urban origin? *Environmental Science & Technology*, *50*(20), 10777–10779. doi:10.1021/acs.est.6b04140 PMID:27682621

Novelli, A., & Tarantino, E. (2015). Combining ad hoc spectral indices based on LANDSAT-8 OLI/TIRS sensor data for the detection of plastic cover vineyard. *Remote Sensing Letters*, *6*(12), 933–941. doi:10.1080/2150704X.2015.1093186

Pandey, P., Dhiman, M., Kansal, A., & Subudhi, S. P. (2023). Plastic waste management for sustainable environment: Techniques and approaches. *Waste Disposal & Sustainable Energy*, *5*(2), 205–222. doi:10.1007/s42768-023-00134-6 PMID:37359812

Prata, J. C., Silva, A. L. P., da Costa, J. P., Mouneyrac, C., Walker, T. R., Duarte, A. C., & Rocha-Santos, T. (2019). Solutions and integrated strategies for the control and mitigation of plastic and microplastic pollution. *International Journal of Environmental Research and Public Health*, *16*(13), 2411. doi:10.3390/ijerph16132411 PMID:31284627

Prokic, M. D., Radovanovic, T. B., Gavric, J. P., & Faggio, C. (2019). Ecotoxicological Effects of Microplastics: Examination of Biomarkers, Current State and Future Perspectives. *Trends in Analytical Chemistry*, *111*, 37–46. doi:10.1016/j.trac.2018.12.001

Ragusa, A., Svelato, A., Santacroce, C., Catalano, P., Notarstefano, V., Carnevali, O., Papa, F., Rongioletti, M. C. A., Baiocco, F., Draghi, S., D'Amore, E., Rinaldo, D., Matta, M., & Giorgini, E. (2021). Plasticenta: First evidence of microplastics in human placenta. *Environment International*, *146*, 106274. doi:10.1016/j.envint.2020.106274 PMID:33395930

Rolsky, C., Kelkar, V., Driver, E., & Halden, R. U. (2020). Municipal sewage sludge as a source of microplastics in the environment. *Current Opinion in Environmental Science & Health, 14*, 16–22. doi:10.1016/j.coesh.2019.12.001

Siddique, R., Khatib, J., & Kaur, I. (2008). Use of recycled plastic in concrete: A review. *Waste Management (New York, N.Y.), 28*(10), 1835–1852. doi:10.1016/j.wasman.2007.09.011 PMID:17981022

TekmanM. B.WaltherB. A.PeterC.GutowL.BergmannM. (2022). Impacts of plastic pollution in the oceans on marine species, B*iodiversity and Ecosystems*, 1–221, WWF Germany, Berlin. Doi: doi:10.5281/zenodo.5898684

Valter, M. (2021). Plastic pollution: A focus on freshwater biodiversity. *Ambio, 50*(7), 1313–1324. doi:10.1007/s13280-020-01496-5 PMID:33543362

Verma, R., Vinoda, K. S., Papireddy, M., & Gowda, A. N. S. (2016). Toxic pollutants from plastic waste-a review. *Procedia Environmental Sciences, 35*, 701–708. doi:10.1016/j.proenv.2016.07.069

Woodall, L. C., Sanchez-Vidal, A., Canals, M., Paterson, G. L., Coppock, R., Sleight, V., Calafat, A., Rogers, A. D., Narayanaswamy, B. E., & Thompson, R. C. (2014). The deep sea is a major sink for microplastic debris. *Royal Society Open Science, 140317*(4), 140317. doi:10.1098/rsos.140317 PMID:26064573

Wright, S. L., Thompson, R. C., & Galloway, T. S. (2013). The physical impacts of microplastics on marine organisms: a review. *Environmental pollution (Barking, Essex: 1987), 178*, 483–492. doi:10.1016/j.envpol.2013.02.031

Wu, B., Liu, M., & Xu, Z. (2020). Accumulation of Microplastics in Soil and Their Potential Effects on Plant Growth. *The Science of the Total Environment, 715*, 136952.

Wu, W. M., Yang, J., & Criddle, C. S. (2017). Microplastics pollution and reduction strategies. *Frontiers of Environmental Science & Engineering, 11*(1), 6. doi:10.1007/s11783-017-0897-7

Xu, B., Liu, F., Cryder, Z., Huang, D., Lu, Z., He, Y., Wang, H., Lu, Z., Brookes, P. C., Tang, C., Gan, J., & Xu, J. (2020). Microplastics in the soil envi ronment: Occurrence, risks, interactions and fate–a review. *Critical Reviews in Environmental Science and Technology, 50*(21), 2175–2222. doi:10.1080/10643389.2019.1694822

Zhang, J., Ren, S., Xu, W., Liang, C., Li, J., Zhang, H., Li, Y., Liu, X., Jones, D. L., Chadwick, D. R., Zhang, F., & Wang, K. (2022). Effects of plastic residues and microplastics on soil ecosystems: A global meta-analysis. *Journal of Hazardous Materials, 435*, 129065. doi:10.1016/j.jhazmat.2022.129065 PMID:35650746

Zhang, W., Zhang, S., Zhao, Q., Qu, L., Ma, D., & Wang, J. (2020). Spatio-temporal distribution of plastic and microplastic debris in the surface water of the Bohai Sea, China. *Marine Pollution Bulletin, 158*, 111343. doi:10.1016/j.marpolbul.2020.111343 PMID:32753167

ADDITIONAL READING

Kumar, R., Verma, A., Shome, A., Sinha, R., Sinha, S., Jha, P. K., Kumar, R., Das, S., Sharma, P., & Prasad, P. V. V. (2021). Impacts of Plastic Pollution on Ecosystem Services, Sustainable Development Goals, and Need to Focus on Circular Economy and Policy Interventions. *Sustainability (Basel), 13*(17), 1–41. doi:10.3390/su13179963

Nollet, L. M. L., & Siddiqi, K. S. (Eds.). (2020). *Analysis of Nanoplastics and Microplastics in Food* (1st ed.). CRC Press. doi:10.1201/9780429469596

Pawar, P. R., Shirgaonkar, S. S., & Patil, R. B. (2016). Plastic marine debris: Sources, distribution and impacts on coastal and ocean biodiversity. *PENCIL Publication of Biological Sciences, 3*(1), 40–54.

Singh, P., & Sharma, V. P. (2016). Integrated Plastic Waste Management: Environmental and Improved Health Approaches. *Procedia Environmental Sciences, 35*, 692–700. doi:10.1016/j.proenv.2016.07.068

Welden, N. A. (2020). The environmental impacts of plastic pollution. In *Plastic waste and recycling* (pp. 195–222). Academic Press. doi:10.1016/B978-0-12-817880-5.00008-6

KEY TERMS AND DEFINITIONS

Biodiversity: It refers to the variability on Earth, which includes the species diversity, i.e. variety and abundance of species in the ecosystem; ecosystem diversity, i.e. the variety of species in different ecosystems in a particular region; and genetic diversity, i.e. array of genes within a species or population. Biodiversity is crucial for the human population's survival and maintaining the balance of ecosystems.

Entanglement: Plastic entanglement is a widespread issue resulting from plastic pollution in aquatic and land biodiversity. It refers to animals trapped in abandoned plastic debris, such as fishing nets, lines, ropes, and packaging materials. It results in physical injury, strangulation, and immobilisation. Many aquatic animals like seals, sea lions, turtles and whales get entangled in fishing nets or other plastic debris, leading to the inability to swim, fly, or move freely. Which eventually leads to death.

Ingestion: Plastic ingestion is a widespread issue affecting many marine and terrestrial species worldwide. It is mainly profound in aquatic ecosystems where plastic can be found everywhere. Ingestion refers to the situation when animals mistakenly consume plastic materials, thinking plastic to be a food item. Ingestion of plastic can lead to serious health hazards like- blockage of the digestive tract, leading to starvation, internal injuries, and death. The harmful chemicals released from plastic can harm the tissue of animals and cause organ damage, reproductive issues.

Microplastics: Microplastic are small plastics that have a particle size of 5 mm or less that. They can originate from primary sources (for ex., microbeads in personal care products or industrial waste) and secondary sources (when larger plastic breaks down into smaller particles due to weathering and erosion over time). Like larger plastic, microplastics are ubiquitous and profound in oceans, lakes, forests, and urban areas.

Plastic Pollution: It refers to the assemblage of plastic waste in the environment, mainly in marine (lakes, ocean, ponds) and terrestrial (land, soil, forests, urban) ecosystems. Plastic pollution takes place from various types of plastic, the most menancing is the single use plastics. Plastic pollution have a colossal consequence on the environment, human health and society.

Recycle: It is a crucial aspect of plastic mitigation strategies aimed at reducing the environmental impact of plastic waste. It involves collecting, processing, and converting used plastic materials into new products or materials, extending their lifecycle and reducing the need for new plastic production. There are some processes involved in this recycling, i.e. collection and sorting of used plastic, processing and treatment for further use, and conversion into another material. Recycling is considered a great approach in a circular economy where there is a minimum waste of resources.

Reuse: This concept involves using plastic items multiple times to extend their lifespan and reduce the amount of plastic waste generated over time. This mitigation strategy needs to emphasise alternatives to single-use plastic products, behavioural changes in consumers, new research and innovation towards eco-friendly alternative products, government initiatives and policy regulations. By practising reuse as part of plastic mitigation strategies, single-use plastic demand can be reduced, and a sustainable environment can thrive.

Single-Use Plastic: It refers to those types of plastic used for once before being thrown away or recycled. It is mainly made to attain convenience and for short-term usage. Single-use plastics include plastic bags, straws, water bottles, disposable cutlery, food packaging, and packaging materials like bubble wrap. Like microplastics, single-use plastics can survive hundreds of years without degrading themselves. Thus, they contribute to pollution and distress to the biodiversity.

Species Extinction: Species extinction due to plastic pollution occurs when plastic debris in natural environments directly or indirectly leads to the decline or loss of species. This can impact biodiversity in various ways, such as entanglement, ingestion, abrasion, habitat degradation. All these factors eventually lead to species extinction.

Sustainable Management: It refers to adopting strategies and practices that aim to reduce the environmental impact of plastic throughout its lifecycle, from production to disposal. It involves minimising the use of plastic, proper waste management strategies, promoting alternative, new innovative technology, better collaboration between government, industries, and societies, and adequate policy implementation by policy makers.

Chapter 13
Carbon Sequestration by Terestrial, Aquatic, and Marine Biodiversity

Pradosh Kumar Parida
Tamil Nadu Agricultural University, Coimbatore, India

G. Asha Selvi
https://orcid.org/0009-0009-3699-5437
Tamil Nadu Agricultural University, Coimbatore, India

S. M. Vinodhini
https://orcid.org/0009-0005-7079-2167
Tamil Nadu Agricultural University, Coimbatore, India

Maram Bhargav Reddy
SNRM, College of Postgraduate Studies in Agricultural Science, Umiam, India

ABSTRACT

The concentration of atmospheric CO2 is currently at around 415 ppm and is increasing by approximately 2-3 ppm annually. This rise is mainly caused by human activities, particularly the burning of fossil fuels for energy production, industrial processes, and changes in land use. Although annual emission rates have fluctuated, they generally remain high, with recent estimates indicating around 40-45 GtCO2 being emitted each year. This chapter examines the crucial involvement of terrestrial, aquatic, and marine ecosystems in carbon sequestration, a key process in contending climate change. By investigating various mechanisms such as terrestrial vegetation absorption of carbon dioxide through photosynthesis and the deposition of organic material in marine sediments, the chapter elucidates the intricate connections between these ecosystems and their impact on the global carbon cycle and regulating climate and stresses the vital requirement for cooperative efforts to ensure their ability to sequester carbon for the benefit of future generations.

DOI: 10.4018/979-8-3693-6950-0.ch013

INTRODUCTION

In the face of climate change, there is a growing urgency to explore and implement strategies that mitigate its effects. One such strategy is carbon sequestration, which holds promise to reduce the concentration of greenhouse gases in the atmosphere. Additionally, biodiversity plays a crucial role in climate mitigation efforts due to its ability to enhance the resilience of ecosystems and contribute to carbon sequestration (Farrelly *et al.,* 2015). In this discussion, we will delve into the definitions of carbon sequestration and biodiversity, as well as the significance of biodiversity in climate mitigation.

DEFINING CARBON SEQUESTRATION

Carbon sequestration refers to the process of capturing carbon dioxide (CO_2) from the atmosphere or other sources and storing it in long-term reservoirs to prevent its release into the atmosphere. The primary goal of carbon sequestration is to mitigate climate change by reducing the concentration of greenhouse gases in the atmosphere. This process involves various natural and artificial methods, including forestation, soil management, carbon capture and storage (CCS), and direct air capture (DAC). Carbon sequestration plays a crucial role in efforts to combat global warming and its associated impacts on the environment and society (Lal, 2008).

Worldwide, around 60 GtC y^{-1} are temporarily stored by land vegetation through Net Primary Productivity (NPP). However, this carbon is eventually released back into the atmosphere as vegetation dies and decomposes, a process known as heterotrophic respiration (Rh). In a balanced state, the rates of carbon component deaths match their decomposition rates, resulting in a net land-atmosphere carbon flux that approaches zero (Fig 1).

Significance of Biodiversity in Climate Mitigation

Biodiversity, the variety of life forms present in an ecosystem, plays a crucial role in climate mitigation efforts. One of the primary ways biodiversity contributes to climate mitigation is through carbon sequestration. Healthy ecosystems with high biodiversity tend to be more efficient at capturing and storing carbon dioxide. Forests, for example, act as important carbon sinks, absorbing CO_2 from the atmosphere through photosynthesis and storing it in biomass and soil organic matter. The greater the diversity of plant and animal species within a forest ecosystem, the more resilient and productive it tends to be, enhancing its capacity for carbon sequestration.

Figure 1. Pools and fluxes of the global carbon cycle
Courtesy: *Zeng, (2008)*

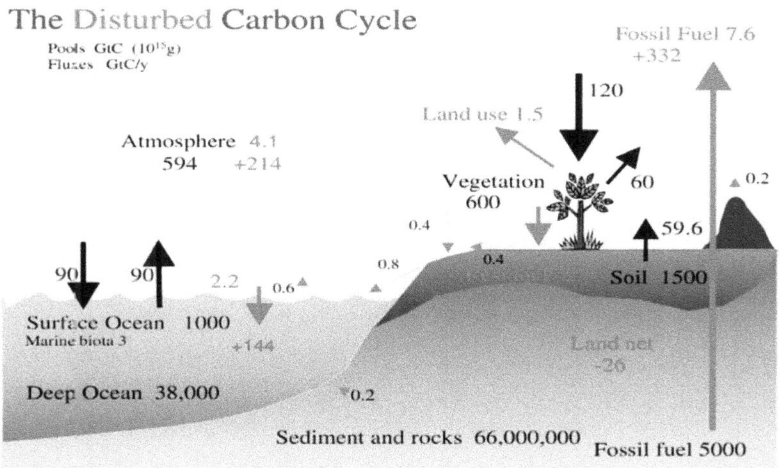

Furthermore, biodiversity helps to enhance the resilience of ecosystems to climate change impacts. Diverse ecosystems are better able to adapt to changing environmental conditions, such as temperature extremes and precipitation patterns, which are becoming more frequent and severe due to climate change. By preserving biodiversity and restoring degraded ecosystems, we can improve their ability to withstand and recover from the adverse effects of climate change, thus contributing to overall climate resilience. Carbon sequestration and biodiversity are interconnected aspects of climate mitigation efforts. Understanding and harnessing the role of biodiversity in enhancing carbon sequestration not only helps to mitigate climate change but also supports the conservation and sustainable management of ecosystems for the benefit of both present and future generations.

Terrestrial Ecosystems: Diverse Guardians of Carbon

Terrestrial ecosystems play a crucial role in the global carbon cycle, acting as diverse guardians of carbon by absorbing and storing carbon dioxide (CO_2) from the atmosphere. These ecosystems encompass a wide range of environments, including forests, grasslands, wetlands, tundras, and deserts, each with its unique characteristics and contributions to carbon sequestration (Figure 2).

Forest Ecosystems: Forests are one of the most significant terrestrial carbon sinks. Trees absorb CO_2 during photosynthesis and store it in their biomass, including leaves, stems, branches, and roots. Additionally, forest soils also store large amounts of carbon. Old-growth forests, with their dense canopies and extensive root

Figure 2. Terrestrial carbon cycle
Courtesy: Foley et al., (2003)

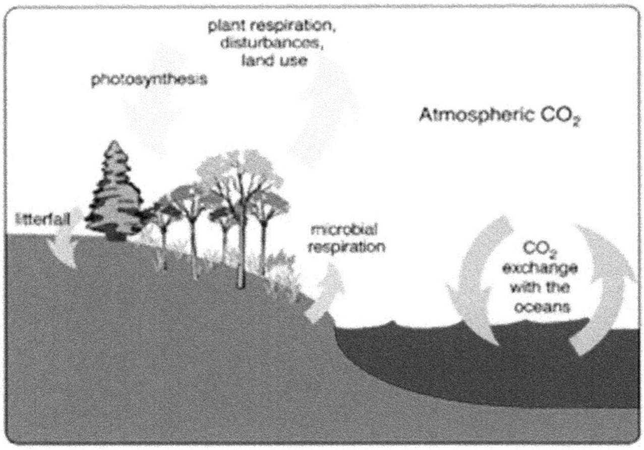

systems, are particularly effective at sequestering carbon. However, disturbances like deforestation, wildfires, and insect outbreaks can release stored carbon back into the atmosphere.

Grassland Ecosystems: Grasslands, including savannas and prairies, are characterized by vast expanses of grasses and herbaceous plants. These ecosystems sequester carbon primarily in their deep root systems, which can penetrate several meters into the soil. Grassland soils contain significant carbon stocks, with some estimates suggesting that they may store more carbon below ground than above ground biomass. However, disturbances such as conversion to agriculture or overgrazing can lead to carbon loss from grassland ecosystems.

Wetland Ecosystems: Wetlands, including marshes, swamps, and bogs, are among the most efficient carbon sinks per unit area. They accumulate organic matter in waterlogged soils, where the anaerobic conditions slow down decomposition rates, allowing carbon to be stored for long periods. Peatlands, a type of wetland characterized by thick layers of partially decomposed plant material (peat), are especially important for carbon sequestration. However, wetland degradation due to drainage, pollution, and development can release stored carbon into the atmosphere.

Tundra Ecosystems: Tundra ecosystems, found in cold regions near the poles and at high altitudes, store large amounts of carbon in their frozen soils, known as permafrost. Permafrost contains vast reserves of organic carbon accumulated over thousands of years, primarily from undecomposed plant material. However, rising temperatures due to climate change are causing permafrost to thaw, releasing stored carbon in the form of CO_2 and methane, which can exacerbate global warming.

Desert Ecosystems: Although often overlooked, desert ecosystems also play a role in carbon sequestration. Desert plants have adapted to arid conditions by developing efficient water and nutrient-use strategies, including extensive root systems that can reach deep into the soil. Desert soils may contain significant carbon stocks, particularly in areas where vegetation cover is more substantial, such as desert oases or shrublands. However, desertification caused by land degradation and climate change can reduce carbon storage capacity in these ecosystems.

Overall, terrestrial ecosystems are vital guardians of carbon, helping to mitigate the impacts of anthropogenic CO_2 emissions on climate change. Protecting and restoring these ecosystems is essential for maintaining their carbon sequestration potential and ensuring a stable climate for future generations. Efforts to conserve forests, restore degraded lands, and promote sustainable land management practices can enhance the role of terrestrial ecosystems as carbon sinks in the fight against climate change.

Plant Diversity and Carbon Capture

Plant diversity plays a significant role in carbon capture and storage, primarily through the process of photosynthesis. Photosynthesis is the biochemical process by which green plants, algae, and some bacteria convert carbon dioxide (CO_2) from the atmosphere into organic compounds, primarily glucose, using sunlight as an energy source. This process not only produces oxygen (O_2) as a byproduct but also captures carbon from the atmosphere, thus aiding in carbon sequestration (Gogoi *et al.,* 2021).

Mechanisms of Carbon Capture by Plants

1. **Photosynthesis:** During photosynthesis, plants absorb atmospheric CO_2 through tiny pores called stomata primarily located on the underside of leaves. This CO_2 is then used, along with water and sunlight, to produce glucose and oxygen. The carbon from CO_2 becomes incorporated into the plant's biomass, including leaves, stems, roots, and fruits.
2. **Biomass Production**: Plants vary in their growth rates and biomass production, with some species growing faster and producing more biomass than others. Species with higher biomass production rates have the potential to sequester more carbon.
3. **Root Systems:** The root systems of plants also play a role in carbon sequestration. Some plants, particularly trees and shrubs, have extensive root systems that can penetrate deep into the soil, storing carbon below ground.

Importance of Plant Diversity in Carbon Capture

1. **Increased Biomass Production:** Plant diversity can lead to increased overall biomass production within ecosystems. Ecosystems with higher plant diversity often contain a variety of species with different growth rates and ecological niches. This diversity can result in more efficient resource use and higher total biomass accumulation, leading to greater carbon capture and storage.
2. **Enhanced Ecosystem Services:** Diverse plant communities provide various ecosystem services that support carbon sequestration. For example, diverse ecosystems are often more resilient to environmental stressors such as drought, pests, and diseases, which can impact plant growth and carbon uptake.
3. **Soil Carbon Storage:** Plant diversity can also enhance soil carbon storage through interactions between plant roots and soil microorganisms. Diverse plant communities contribute a variety of organic matter to the soil through litterfall and root turnover, which can stimulate microbial activity and enhance soil carbon sequestration.
4. **Habitat Heterogeneity:** Diverse plant communities create heterogeneous habitats with varied microenvironments and niches. This heterogeneity can support a wide range of organisms, including decomposers and soil fauna, which contribute to the cycling of organic matter and the formation of stable soil organic carbon.

Conservation and Restoration Implications

1. **Conservation of Natural Habitats:** Protecting and conserving natural habitats with high plant diversity, such as tropical rainforests and grasslands, is essential for maintaining carbon sequestration capacity and biodiversity.
2. **Ecological Restoration:** Restoring degraded ecosystems through reforestation, afforestation, and revegetation efforts can enhance carbon capture and storage while promoting plant diversity and ecosystem resilience.

Below-Ground Biodiversity: Soil Microbes and Carbon Sequestration

Below-ground biodiversity, particularly involving soil microbes, plays a crucial role in carbon sequestration processes. Soil microbes are microscopic organisms such as bacteria, fungi, and archaea that inhabit the soil and are essential for nutrient cycling, organic matter decomposition, and soil health (Figure 3). These microbes interact with plant roots and organic matter, influencing carbon storage dynamics in terrestrial ecosystems (Bhattacharyya *et al.,* 2022).

Role of Soil Microbes in Carbon Sequestration:

1. **Organic Matter Decomposition:** Soil microbes decompose organic matter, including dead plant material and root exudates. During this process, they release carbon dioxide (CO_2) into the atmosphere. However, a portion of the decomposed carbon is converted into stable forms and retained in the soil.
2. **Soil Aggregation:** Certain microbial communities produce extracellular substances like glomalin and polysaccharides that promote soil aggregation. Aggregates provide physical protection to organic carbon, preventing its rapid decomposition and enhancing long-term carbon storage.
3. **Nutrient Cycling:** Soil microbes are involved in nutrient cycling processes such as nitrogen fixation, mineralization, and immobilization. Efficient nutrient cycling supports plant growth and productivity, leading to increased carbon inputs into the soil through litterfall and root turnover.
4. **Mycorrhizal Associations:** Mycorrhizal fungi form symbiotic associations with plant roots, facilitating nutrient uptake in exchange for carbon compounds. These associations enhance plant productivity and contribute to carbon sequestration by increasing root biomass and stimulating soil organic matter formation.

Factors Influencing Below-Ground Biodiversity and Carbon Sequestration

1. **Land Management Practices:** Agricultural practices such as tillage, crop rotation, and fertilizer application influence soil microbial communities and their activities. Sustainable management practices that promote soil health and organic matter accumulation enhance carbon sequestration potential.
2. **Soil Properties:** Soil texture, pH, moisture content, and nutrient availability influence microbial diversity and activity. Soils with favorable conditions for microbial growth tend to have higher carbon storage capacity.
3. **Plant Diversity:** Plant species composition and diversity influence below-ground microbial communities through root exudates and litter quality. Diverse plant communities support a broader range of microbial functions, enhancing carbon sequestration processes.
4. **Climate:** Climatic factors such as temperature and precipitation affect microbial activity and decomposition rates, ultimately influencing carbon turnover and storage in soils.

Implications for Carbon Sequestration Strategies

Figure 3. Soil carbon sequestration
Courtesy: *Bhattacharyya et al., (2022)*

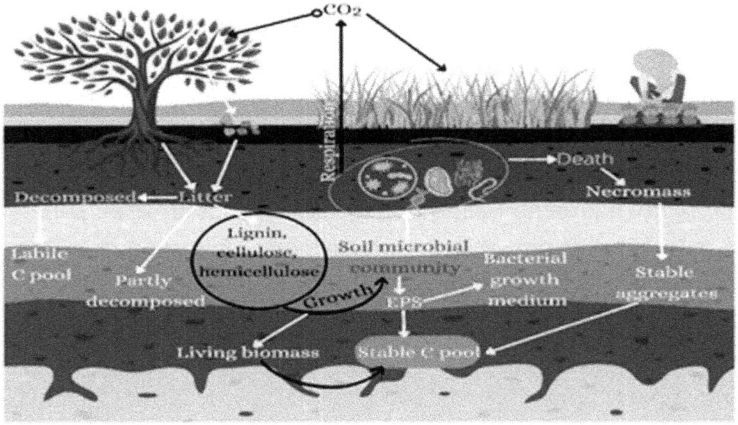

1. **Conservation Agriculture**: Practices such as no-till farming, cover cropping, and agroforestry promote soil microbial diversity and carbon sequestration while reducing soil disturbance and erosion.
2. **Restoration Ecology:** Restoring degraded ecosystems through revegetation and soil rehabilitation efforts can enhance below-ground biodiversity and increase carbon sequestration rates.
3. **Microbial Amendments:** Introducing beneficial microbial inoculants or organic amendments can enhance soil microbial activity and carbon stabilization, contributing to climate change mitigation efforts.

Aquatic Environments: Navigating Carbon Storage

Aquatic carbon sequestration denotes the extended storage of atmospheric carbon over centuries, encompassing sediments, biomass, and water. Aquatic ecosystems spanning from headwaters to deep ocean zones significantly influence the global carbon cycle. The long-term sequestration of carbon is facilitated through diverse mechanisms within lakes, wetlands, estuaries, coastal blue carbon habitats, and deeper marine environments. This process involves not only retaining organic carbon within local biomass and vegetated sediments but also transporting carbon away from its origin to be deposited elsewhere in lakes, rivers, glaciers, groundwater, and the ocean.

While certain ecosystems like lakes, tidal wetlands, and the ocean harbor substantial carbon reserves, others such as rivers and streams may act as persistent sources of carbon dioxide to the atmosphere. The extensive interconnectivity, alternating dynamics between being sources and sinks, and significant spatial and temporal variations frequently observed in aquatic ecosystems pose substantial challenges in accurately quantifying their net carbon sequestration capacity and understanding their potential for climate change mitigation (Williamson and Gattuso, 2022).

Role of Aquatic Flora in Carbon Storage

Aquatic photosynthetic organisms, predominantly phytoplankton, play a crucial role in global carbon sequestration, accounting for approximately 50% of the total. The 1.0 kg of cultivated microalgae can absorb 1.83 kg of CO_2. Algae utilize three distinct processes photoautotrophic, heterotrophic, and mixotrophic metabolisms to sequestrate CO_2. Microalgae acquire inorganic carbon through three primary methods:

(i) Conversion of bicarbonates into CO_2 facilitated by extracellular carbonic anhydrase, which easily permeates cell membranes
(ii) Direct absorption of CO_2 through the plasma membrane
(iii) Direct uptake of bicarbonates via specific carriers in the membrane, also referred to as dissolved inorganic carbon (DIC) pumps.

Carbon dioxide capture primarily takes place through photoautotrophic metabolism, wherein inorganic carbons are converted into carbohydrates aided by light. The fixation of carbon predominantly occurs through the Calvin–Benson cycle (Fig 4).

Factors Affecting Carbon Sequestration

Various environmental factors, including temperature, pH, alkalinity, etc., directly affect the rate of supply of inorganic carbon to phytoplankton.

Carbon Dioxide

Carbon dioxide serves as a fundamental necessity for the growth and development of phytoplankton, facilitating the construction of the carbon skeleton through photosynthesis. Carbon dioxide fixation involves its conversion from the gaseous phase to the liquid phase. The tolerance to carbon dioxide concentration varies among different species of plants. When dissolved gases interact with water and oxygen, new species are generated, each with its unique solubility and potential influence on the growth system of plants.

Figure 4. Carbon sequestration in microalgae species
Courtesy: *Ighalo et al., (2022)*

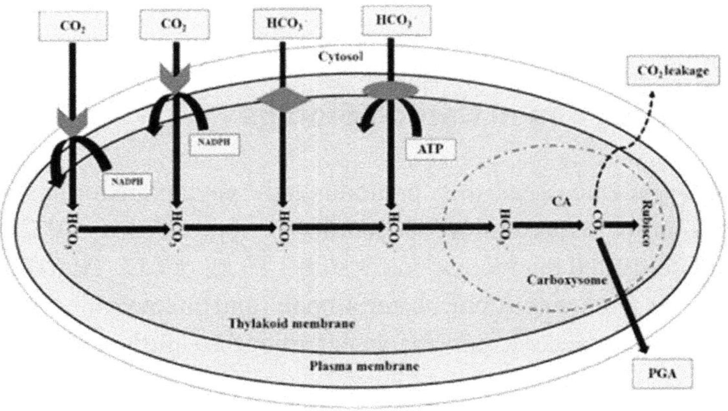

pH

The pH for the growth of phytoplankton plays a crucial role in governing nutrient uptake, photosynthetic activity, and carbon sequestration efficiency. However, an alkaline (high) pH proves advantageous for CO_2 absorption. This is because an alkaline pH enhances the availability of free CO_2 in the culture medium, thereby favoring the growth of algae that are tolerant to high CO_2 levels.

Temperature

Temperature plays a crucial role in the growth and maturation of photosynthetic organisms. A temperature range of 5–40 °C, with the most favorable temperature falling between 0–30 °C. It's important to note that the ideal temperature can differ depending on the species.

Diverse Fauna Distribution in Marine Species for Carbon Sequestration

Marine creatures can capture carbon through various natural processes, including storing it in their bodies, excreting carbon-rich waste products that descend into the depths of the ocean, and aiding in the fertilization or protection of marine plants. The transportation of carbon by mesopelagic fishes to the deep sea is of particular interest, not only due to its direct nature but also because it represents a potentially intact biological mechanism for the cycling of oceanic carbon by marine vertebrates.

Figure 5. A nine mechanisms through which marine vertebrates play roles in the oceanic carbon cycle
Courtesy: *Lutz et al., (2018)*

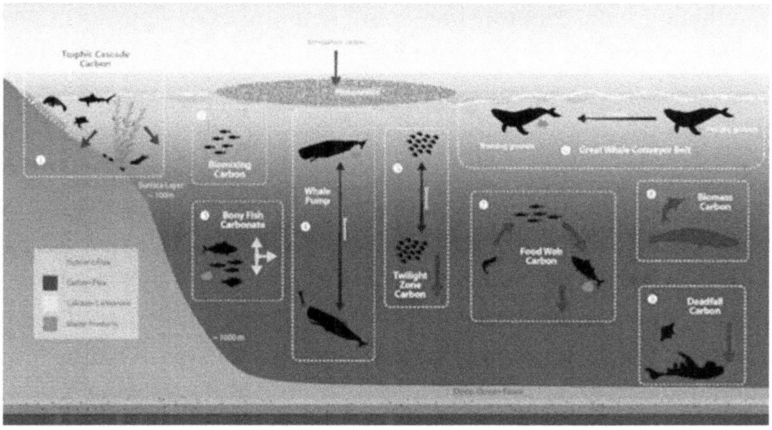

Scientists are increasingly acknowledging that vertebrates such as fish, seabirds, and marine mammals can play a role in sequestering carbon from the atmosphere.

Biomass Carbon refers to the carbon stored in living organisms, a feature present in all marine vertebrates. Massive creatures like whales, weighing up to 50 tons and with lifespans exceeding 200 years, can store significant amounts of carbon over extended periods. Upon their demise, their carcasses descend to the ocean floor, carrying with them a lifetime's worth of trapped carbon. This process, termed Deadfall Carbon, is the carbon potentially buried in sediments on the deep seafloor, where it could remain locked away from the atmosphere for millions of years (Figure 5).

Fish

Fish excrement also contributes to the trapping of carbon. Certain fish engage in daily vertical migrations, ascending to the surface at night to feed and then descending to deeper waters during the day. In these deeper regions, they release fecal pellets rich in carbon that sink rapidly. This phenomenon is referred to as Twilight Zone Carbon. These fish can incline to depths of 1,000 feet or more, and their fecal pellets have the potential to sink even farther. Twilight Zone Carbon can be sequestered for extended periods, ranging from tens to hundreds of years because the water at these depths takes a considerable amount of time to circulate back towards the surface (Sala *et al.,* 2021).

The fish biomass could decline by 10 to 25% due to the high greenhouse gas emissions. These declines, driven by climate change and fishing activities, have significant implications for biodiversity, ecosystem resilience, and the well-being of coastal communities. Restoring fish populations is unlikely to have a substantial direct impact on blue carbon sequestration, as indicated by current assessments. However, reducing industrial trawling could help preserve significant amounts of carbon stored in sediment on the seafloor.

Whales

Whales contribute to the biological pump by fertilizing the surrounding ocean with iron-rich excrement, which promotes the production of fast-sinking particulate organic carbon. Additionally, when a whale dies, its carcass sinks to the seafloor in an event known as a whale fall, with a portion of it potentially becoming buried in ocean sediments. The restoration of whale populations to pre-whaling levels could lead to additional carbon sequestration of 0.16 - 6.7 million metric tons of carbon per year, depending on the population size.

However, whales face threats from climate impacts, increased shipping traffic, and pollution. Despite their potential significance in the biological carbon pump, whale's role in marine carbon sequestration is minimal. Moreover, the time scale and uncertainties associated with population recovery, compounded by the threats posed by climate change, further complicate the calculation of additional carbon sequestration that could be attributed to these species (Maldonado *et al.,* 2016).

Calcifying Organisms

Calcifying organisms such as oysters are filter feeders that deposit organic carbon onto the seabed. However, they also release CO_2 through respiration and the calcification process of shell formation. Calcification releases CO_2 into the atmosphere, while dissolution represents an oceanic carbon sink. The rates of calcification and dissolution vary based on ocean acidity and the concentration of dissolved inorganic carbon in the water. Currently, more CO_2 is released into the atmosphere by reefs than is captured through dissolution. However, this balance may shift in the future due to ocean acidification favoring dissolution (Howard *et al.,* 2017). Calcification is prevalent in other coastal ecosystems, such as seagrasses, which can support the growth of shelled molluscs. Accurate carbon accounting in these ecosystems requires site-specific assessments to evaluate the balance between organic and inorganic carbon production and sequestration.

Marine Biodiversity: Blue Carbon Frontiers

Blue Carbon Ecosystems encompass coastal, estuarine, and marine photosynthetic environments capable of absorbing significant amounts of carbon dioxide from the atmosphere or surroundings and storing it for extended periods—a process known as carbon sequestration or biosequestration. These ecosystems include salt marshes, seagrass meadows, mangrove forests, macroalgae communities like kelp forests, and freshwater tidal ecosystems such as coastal bald cypress floodplain forests. Within these blue carbon systems, the numerous grass blades of seagrass meadows, salt marshes, and the prop roots in mangrove forests act to slow down water currents, capturing particles that settle out of these decelerated currents. Over time, this process leads to the accumulation of soils and the burial of carbon-rich organic materials, such as dead plant matter. As these soils are submerged, they become anaerobic (lacking oxygen) and are not subject to periodic fires like terrestrial ecosystems. Consequently, there is minimal decomposition (or loss of carbon to the atmosphere through burning), facilitating the buildup and long-term sequestration of blue carbon.

Coral Reefs

The coral reef is the most biologically diverse marine ecosystem, with an estimated annual sequestration of 900 million tons of carbon on a global scale. The primary productivity of coral reefs is up to 300–5 000 g C m^{-2} a^{-1}, while that of non-coral reef systems only accounts for 50–600 g C m^{-2} a^{-1}. Although the role of a potential carbon sink has long been recognized, coral reefs have been defined as a carbon source due to CO_2 release during calcification. The high productivity of coral reef ecosystems relies mainly on photosynthetic dinoflagellates in the family Symbiodiniaceae (collectively called zooxanthellae) which are symbiotic with corals. Zooxanthellae provide 95% of photosynthates (sugars, amino acids, oxygen) to corals for their growth and calcification, while corals provide zooxanthellae with metabolic wastes (CO_2, nitrogen, and phosphorus) as nutrients (Fig 6).

During the formation of reefs, a significant quantity of carbonate is deposited, and the annual deposition of $CaCO_3$ in coral reef areas is estimated at 0.084 Pg C (1 Pg = 10^15 g), representing approximately 23%–26% of the global annual $CaCO_3$ deposition. Despite this, coral reefs can continuously accumulate and store carbon in the form of $CaCO_3$, which constitutes their calcareous skeleton. Coral reefs serve as a carbon sink, sequestering nearly 70 to 90 Megatons of carbon annually (Fig 7). While coral reefs release about 1.86 Megatons of carbon per year through calcification, they can also store between 70 to 90 Megatons of carbon annually through the accumulation of $CaCO_3$ in their skeleton. Considering both the released CO_2 and the accumulated $CaCO_3$ involved in the formation and accumulation of

Figure 6. Modes and pathways of enhancing carbon sequestration in coral reef ecosystems
Courtesy: *Shi et al., (2021)*

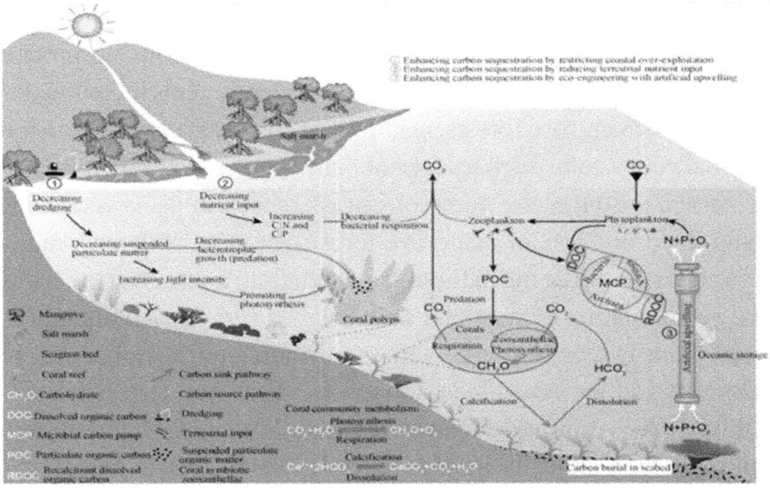

coral skeleton, the net carbon storage by coral reefs ranges between 68.14 and 88.14 Megatons of carbon per year.

A significant portion of the carbon will contribute to the structure of coral reefs through the ongoing accumulation of the calcareous skeleton. A portion of the carbon will be integrated into the framework of coral reefs, while another portion will re-enter the trophic chain through processes like bioerosion, such as the scraping

Figure 7. Carbon exchanges within coral reefs in the baseline scenario (A) and an ocean acidification scenario (B)
Courtesy: *Coral guardian, https://www.coralguardian.org/en/carbon-balance-in-corals*

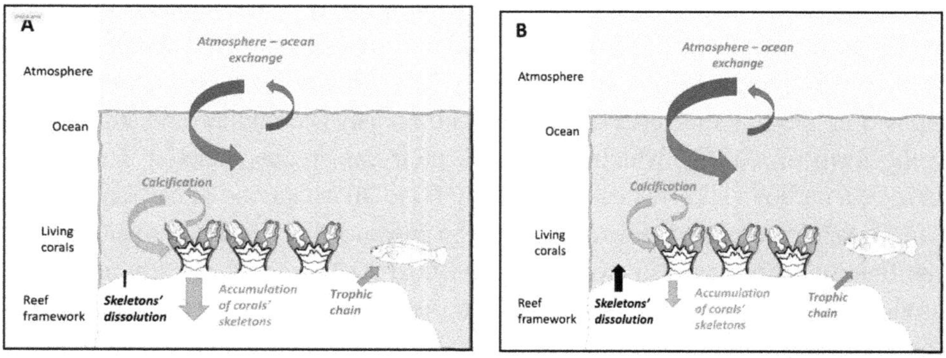

and boring actions of reef organisms. Ultimately, some amount of carbon will be released into the water through natural or chemical dissolution.

Ocean acidification, which is linked to the human-driven rise in atmospheric CO_2 levels, poses a threat to the balance among these processes. With higher levels of CO_2 in the atmosphere, the oceans absorb more of it, leading to an increase in the concentration of hydrogen ions in the water. This rise in hydrogen ions results in acidification of the environment, disrupting the mechanisms involved in calcification. Consequently, there is a decrease in calcification, an increase in coral dissolution, and an overall erosion of biodiversity.

However, ocean acidification driven by climate change undermines the formation and accumulation of coral reef skeletons. This phenomenon accelerates the dissolution of skeletons while slowing down the calcification process, posing a significant threat to the structural integrity of an entire ecosystem.

Kelp Forest

Kelp forests be prominent as one of the most captivating and vital ecosystems within oceans. These underwater forests, predominantly characterized by towering giant kelp, create dense canopies that offer refuge and sustenance for a diverse array of marine species. They thrive in cold, nutrient-rich waters along coastlines worldwide, often observed along rocky shores in areas with high nutrient levels and cool ocean temperatures, such as upwelling zones (Figure 8). Kelp forests hold immense value, estimated at around $500 billion globally, and play a crucial role in sequestering carbon dioxide, capturing approximately 4.5 million tonnes of CO_2 from seawater annually.

The lifespan of kelp varies depending on the species, ranging from up to a year to longer durations. Throughout this time, they can reach impressive heights, spanning from two to 30 meters (98 ft), with growth rates of up to 61 centimeters (2 ft) per day. As they grow, kelp absorbs carbon dioxide from the surrounding environment, sequestering up to 600 million tonnes annually. Through the process of photosynthesis, carbon dioxide is converted into biomass. Once carbon dioxide is assimilated into the kelp, it sinks to the ocean floor, where it can be stored underground or harvested for various purposes. The rapid growth rate of kelp compared to terrestrial forests gives it an advantage in sequestering carbon at a faster pace (Figure 9).

The decline of kelp forests is expected to intensify with the rise in global sea and atmospheric temperatures. Human-induced environmental pressures, including coastal development, poor water quality, coastal pollution, climate-induced marine heatwaves, and drought, further exacerbate the health of kelp forests. As a kelp forest deteriorates, it releases previously sequestered carbon dioxide back into the atmosphere, transitioning from a carbon sink to a carbon source. This global decline

Figure 8. Carbon sequestration and storage in kelp forests
Courtesy: *Krause-Jensen & Duarte, (2016)*

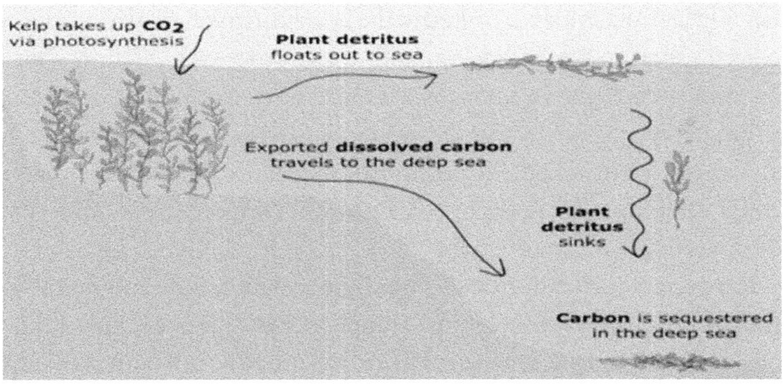

in kelp forests results in the loss of critical carbon sequestration potential, alongside the erosion of valuable ecosystem services such as habitat provision for various species, primary production, and nutrient cycling.

Mangroves

Mangrove forests cover just under 1% of the world's forested land, occupying a total area of 15.2 million hectares (152,000 square kilometers) globally. Asia holds the largest share of mangrove forests worldwide, accounting for 42%, followed by Africa

Figure 9. Mean annual total carbon sequestration and global distribution in teragrams of carbon and km²
Courtesy: *Aarnio, (2022)*

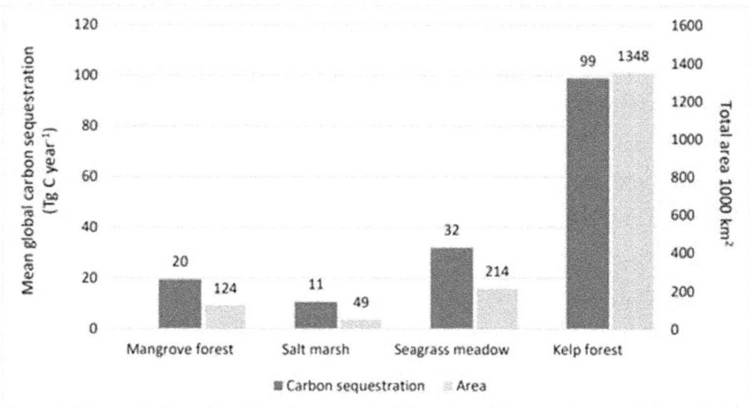

Figure 10. Differences in whole-ecosystem carbon stocks among boreal, temperate and tropical terrestrial forests, and subtropical and tropical mangrove forests
Courtesy: *Alongi, (2012)*

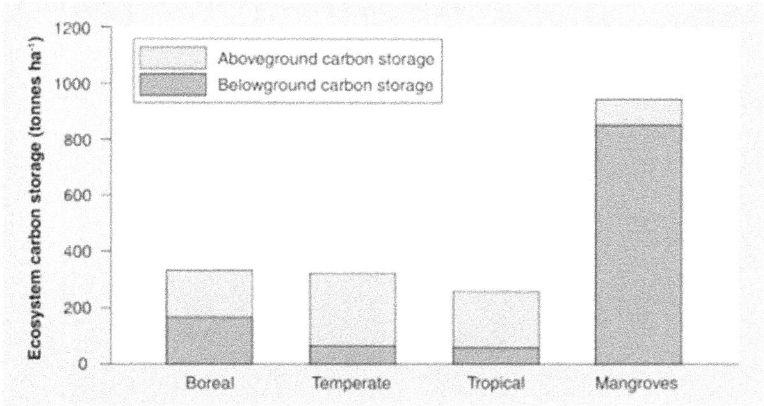

(21%), North/Central America (15%), Oceania (12%), and South America (11%). In South Asia, India possesses approximately 3% of the total mangrove cover. It serve as highly efficient carbon sinks, storing three to five times more carbon per unit area than tropical forests and sequestering carbon at a rate two to four times faster than forests like the Amazon. However, when mangrove forests are cut down, they release more greenhouse gases compared to other forested areas.

Carbon sequestration in mangrove systems occurs through the entrapment and burial of sediments by roots and pneumatophores, as well as the net growth of forest biomass. While mangroves accumulate carbon in tree biomass, a significant portion of this carbon is lost in the short and medium term due to clear-cutting, human activities, decomposition, and export to adjacent ecosystems. In the long term, carbon is primarily stored underground as soil carbon and, under suitable conditions, as peat (organic matter). The accumulation of carbon in mangroves involves direct inputs of mangrove carbon into the soil pool and increased rates of sediment accumulation. The amount of carbon stored in mangrove soils varies widely, ranging from less than 0.1% to over 40% of soil dry weight, with a median value of 2.2% (Fig 10).

Mangroves play a crucial role in carbon sequestration, with the capacity to store and stockpile carbon from the atmosphere during their growth period ranging from 50 to 220 metric tons per acre. Globally, mangroves can sequester more than 24 million metric tons of carbon annually. Despite their relatively small contribution to global forest carbon sequestration, mangroves significantly contribute to carbon burial in the global coastal ocean, accounting for an average of 14% of carbon

Figure 11. Annual rates of carbon burial estimated in various mangrove forests worldwide
Courtesy: *Alongi, (2012)*

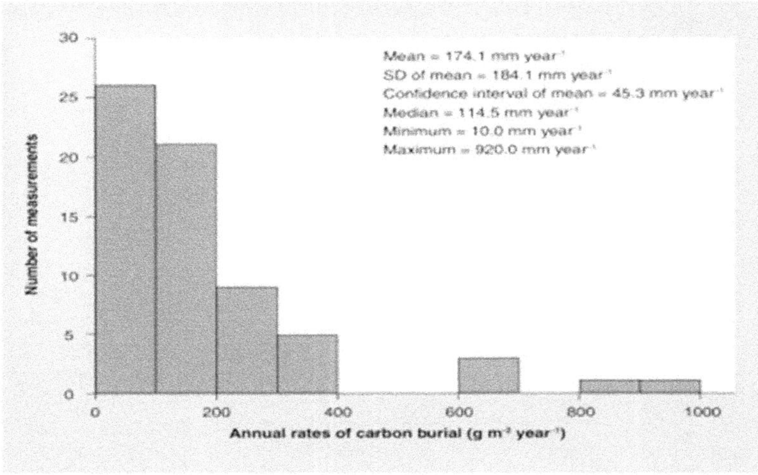

sequestration in the world's oceans, despite occupying only 0.5% of total coastal ocean area (Fig 11).

The Sundarbans, the world's largest mangrove forest, stands out as particularly efficient in carbon sequestration. The carbon sequestration potential of the Sundarbans and comparing carbon uptake rates with carbon emission rates from a coal-based power plant in Kolaghat, India. Through continuous monthly measurements and analysis against carbon pollution levels in Kolkata, they established a control group. The Sundarbans managed to sequester 98% of the carbon emitted by the power plant in one year. This underscores the critical importance of global mangrove protection efforts to sustain these impressive sequestration rates (Ray and Jana 2017).

Unique Mechanisms in Marine Species for Carbon Sequestration

The marine microbial community plays a pivotal role in marine ecosystems, influencing diversity, functions, and biotic interactions. These interactions include relationships between macrophytes and phytoplankton, algae and bacteria, as well as bacteria and fungi, which contribute to the cycling of dissolved organic matter (DOM) and potential carbon (C) storage mechanisms.

Biotic interactions on DOM involve the microbial loop and microbially-driven DOM degradation, facilitated by the microbial carbon pump (MCP) mechanism. Initially, marine microorganisms engage in intricate food webs, including the

Figure 12. Carbon sequestration in microbial marine ecosystem
Courtesy: *He et al., (2022)*

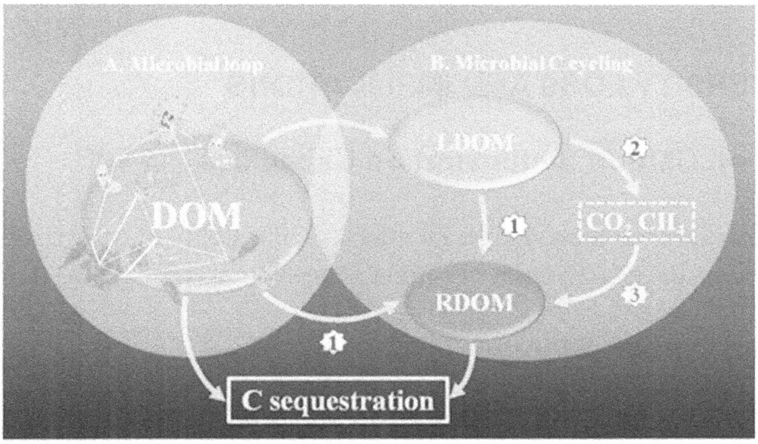

degradation of DOM derived from macroorganisms. Necromass from both macroorganisms and microorganisms may sink to the seabed, sequestering carbon from upper waters.

Microbially-driven DOM degradation produces various products, such as labile dissolved organic carbon (LDOM), primarily utilized by microorganisms, leading to the production of CO_2 and/or CH_4. CO_2 can be re-fixed by autotrophic or chemo-autotrophic microorganisms. Some DOM is further transformed into refractory dissolved organic carbon (RDOC) or refractory DOM (RDOM) via the MCP mechanism for carbon sequestration (Figure 12).

Future Strategies to Enhance Carbon Sequestration May Involve

1. Selecting microbial taxa/genes to promote efficient transformation of DOM and LDOM into RDOM.
2. Mitigating CH_4 emissions by inhibiting CH_4 production or enhancing CH_4 oxidation.
3. Re-transforming CO_2 and CH_4 into organic carbon.

These strategies have the potential to augment carbon sequestration in marine ecosystems, contributing to climate change mitigation efforts.

The adoption of these approaches to enhance carbon sequestration in marine environments necessitates the prompt advancement of novel scientific and technological solutions in microbial ecology, microbiome manipulation, and geoengineering.

BIODIVERSITY LOSS AND IMPLICATIONS

Impact on Carbon Sequestration Capacities

Biodiversity loss can have significant effects on the ability of ecosystems to store carbon and on the effectiveness of carbon sequestration capacities.

Reduced Carbon Storage

Biodiverse ecosystems, like old forests, wetlands, and coral reefs, typically possess substantial carbon storage capabilities because of the variety of plant and animal species. However, when biodiversity diminishes due to causes like habitat destruction or invasive species, these ecosystem's capacity to sequester carbon can be undermined. For instance, the loss of biodiversity through deforestation or forest degradation diminishes the ability of the forest to absorb and retain carbon dioxide from the atmosphere.

Altered Ecosystem Functions

Diminishing biodiversity can disturb vital ecosystem functions, including nutrient cycling, water regulation, and soil formation, all of which are crucial for sustaining carbon sequestration processes. For example, alterations in the composition of plant species can impact rates of photosynthesis and decomposition, thereby affecting the absorption and release of carbon in terrestrial ecosystems.

Loss of Carbon-Rich Habitats

Biodiverse ecosystems frequently encompass carbon-rich habitats like mangrove forests, peatlands, and seagrass meadows, which have notably high rates of carbon sequestration compared to other ecosystems. However, the destruction or degradation of these habitats not only results in biodiversity loss but also releases substantial amounts of stored carbon into the atmosphere, further intensifying climate change.

Impacts on Restoration Efforts

The decline in biodiversity can impede the effectiveness of restoration endeavors intended to sustain carbon sequestration capacities. The restoration of degraded ecosystems typically involves reintroducing a variety of native species to reconstruct ecological functions and resilience. However, in regions where biodiversity loss is severe, revitalizing ecosystem complexity and functionality may prove challenging, thus restricting the efficacy of restoration endeavors in sequestering carbon.

Ocean Acidification

Rising levels of carbon dioxide in the atmosphere contribute to the acidification of oceans, a process that poses challenges for marine organisms such as corals, plankton, and shellfish in maintaining their protective coatings. Consequently, populations of these species decline, along with those of species dependent on them for food and habitat.

CONSEQUENCES FOR CLIMATE CHANGE MITIGATION

Reduced Carbon Sequestration

Diverse ecosystems like forests, wetlands, and grasslands play a crucial role in absorbing carbon dioxide from the atmosphere through photosynthesis and storing it in biomass and soils. However, when biodiversity declines due to factors like deforestation, habitat degradation, or species loss, the ability of these ecosystems to capture and retain carbon decreases. This decline in carbon sequestration worsens climate change by enabling more greenhouse gases to build up in the atmosphere.

Impacts on Carbon Markets and Offsets

Carbon markets and offset programs depend on actions that boost carbon sequestration, like reforestation, afforestation, and ecosystem restoration, to mitigate greenhouse gas emissions. However, biodiversity loss can compromise the efficacy of these efforts by diminishing the resilience and carbon storage capability of ecosystems. Consequently, this may pose obstacles to meeting emission reduction goals and upholding the credibility of carbon offset projects.

Vulnerability to Climate Impacts

The loss of biodiversity improves the susceptibility of ecosystems to the consequences of climate change, such as severe weather occurrences, rising sea levels, and habitat deterioration. This increased vulnerability can exacerbate biodiversity loss, forming a detrimental cycle that impedes ecosystems capacity to provide crucial climate regulation services, including carbon sequestration.

From the perspective of climate mitigation, only interventions in management that result in additional carbon sequestration and long-term storage, compared to scenarios without intervention, are relevant. Carbon storage is evaluated across various timeframes in the literature, ranging from short-term (centennial) to long-term (millennia). Only long-term storage offers solutions deemed permanent for mitigation. However, employing shorter time scales for climate mitigation strategies can yield benefits such as averting climatic tipping points. It is imperative to be able to consistently measure and monitor carbon and other greenhouse gas (GHG) fluxes and stores over extended periods when determining whether a sequestration process is indeed additional and contributes significantly to climate mitigation (Macreadie *et al.,* 2017).

The conservation and restoration of coastal ecosystems might not yield a net "carbon positive" outcome when all ecosystem components of management practices are taken into account, such as in coastal regions where calcification surpasses carbonate dissolution (e.g., certain seagrass meadows). Accounting for all ecosystem synergies and fluxes becomes even more complex in the open ocean. Continuous monitoring and quantification of these processes and rates over time are necessary but can be prohibitively expensive.

Strategies for Enhanced Carbon Sequestration Restoring Terrestrial Biodiversity

Restoring terrestrial biodiversity plays a crucial role in combating climate change by enhancing carbon sequestration. As ecosystems face unprecedented threats from human activities, implementing strategies to restore biodiversity can mitigate carbon emissions while promoting ecosystem resilience and sustainability. Biodiversity encompasses the variety of life forms within ecosystems, including plants, animals, and microorganisms. Diverse ecosystems tend to sequester more carbon due to increased productivity, nutrient cycling, and ecosystem stability (Cardinale et al., 2012). Different species contribute to carbon sequestration through various mechanisms such as photosynthesis, decomposition, and soil formation (Hooper et al., 2012).

Reforestation and Afforestation Efforts

Planting trees and restoring forest ecosystems can significantly enhance carbon sequestration (Houghton *et al.,* 2015). Selecting native tree species and diverse vegetation types promotes ecosystem resilience and supports local biodiversity (Chazdon *et al.,* 2016). Agroforestry practices integrate trees into agricultural landscapes, providing additional carbon storage while enhancing soil fertility and biodiversity.

Ecological Restoration of Degraded Lands

Rehabilitating degraded ecosystems, such as grasslands, wetlands, and mangroves, can restore carbon sequestration capacity (Bastin *et al.,* 2019). Employing techniques like re-vegetation, soil amendment, and erosion control helps rebuild ecosystem functions and biodiversity (Clewell and Aronson, 2006). Restoration efforts should prioritize native species and consider the ecological needs of target ecosystems to maximize carbon storage potential (CBD, 2020).

Conservation and Protection of High-Biodiversity Areas

Preserving intact ecosystems and biodiversity-rich areas prevents carbon loss from deforestation and habitat destruction (Watson *et al.,* 2018). Establishing protected areas, wildlife corridors, and marine reserves safeguards critical habitats and promotes carbon sequestration (Butchart *et al.,* 2015). Integrating traditional ecological knowledge and engaging local communities in conservation efforts enhances the effectiveness and sustainability of protected areas (Berkes, 2018).

Sustainable Land Management Practices

Adopting sustainable land management practices, such as organic farming and agroecology, fosters biodiversity while sequestering carbon (Pretty *et al.,* 2018). Practices like no-till agriculture, cover cropping, and crop rotation improve soil health, increase carbon storage, and reduce greenhouse gas emissions (Giller *et al.,* 2015). Restoring riparian buffers, hedgerows, and native vegetation in agricultural landscapes enhances biodiversity and ecosystem services while mitigating climate change impacts (Gonthier *et al.,* 2014).

Enhancing Soil Carbon Storage

Promoting soil carbon sequestration through practices like biochar application, composting, and soil conservation techniques (Lehmann *et al.,* 2015). Encouraging regenerative agriculture practices that build soil organic matter and microbial diversity enhances carbon storage capacity (Paustian *et al.,* 2016). Restoring degraded soils and implementing sustainable land use practices improve soil structure, fertility, and resilience to climate change (FAO, 2015).

Aquatic and marine ecosystems are vital components of Earth's biosphere, providing numerous ecological, economic, and social benefits. However, these ecosystems face various threats, including pollution, habitat degradation, overfishing, and climate change. Protecting and rehabilitating aquatic and marine ecosystems is essential to preserve biodiversity, maintain ecosystem services, and ensure sustainable resource management.

Establishing Marine Protected Areas (MPAs)

MPAs are designated areas where human activities are regulated to conserve biodiversity and ecosystem integrity. Implementing MPAs helps protect marine habitats, species, and important ecological processes (Halpern *et al.,* 2019). Well-designed and effectively managed MPAs can enhance fish stocks, preserve biodiversity hotspots, and mitigate climate change impacts.

Sustainable Fisheries Management

Adopting sustainable fishing practices, such as quotas, gear restrictions, and seasonal closures, is crucial for maintaining fish populations and ecosystem health. Implementing ecosystem-based fisheries management approaches considers the interconnections between species, habitats, and human activities. Collaboration between governments, fishing communities, and conservation organizations is essential for effective fisheries management and marine conservation (Sumaila *et al.,* 2012).

Restoring Coastal Habitats

Rehabilitating degraded coastal habitats, such as mangroves, salt marshes, and seagrass beds, enhances biodiversity and provides valuable ecosystem services (Barbier *et al.,* 2011). Restored coastal habitats act as nurseries for fish, protect shorelines from erosion, and sequester carbon, contributing to climate change mitigation. Engaging local communities in coastal restoration projects fosters stewardship and promotes sustainable coastal management practices.

Reducing Pollution and Marine Debris

Implementing pollution control measures, such as wastewater treatment, stormwater management, and plastic waste reduction, minimizes contamination of aquatic environments (Galgani *et al.,* 2015). Public awareness campaigns and community initiatives raise awareness about the impacts of marine debris and promote behavior change to reduce littering and plastic consumption (Jambeck *et al.,* 2015).

Climate Change Adaptation and Resilience

Enhancing the resilience of aquatic and marine ecosystems to climate change involves implementing adaptation strategies such as habitat restoration, assisted migration, and ecosystem-based approaches (CBD, 2014). Protecting natural carbon sinks like seagrass meadows and mangroves helps mitigate climate change by sequestering carbon and reducing greenhouse gas emissions (Howard et al., 2017). Integrating climate change considerations into marine spatial planning and conservation strategies ensures the long-term sustainability of aquatic and marine ecosystems (Douvere *et al.,* 2019).

CONCLUSION

The complex interactions among terrestrial, aquatic, and marine ecosystems are crucial for regulating the global carbon cycle and maintaining climate stability. This chapter has delved into the varied processes by which these ecosystems capture and store carbon, spanning from terrestrial plants absorbing carbon dioxide through photosynthesis to the deposition of organic material in marine sediment layers. On land, forests, grasslands, and wetlands act as significant carbon sinks by absorbing carbon dioxide and storing it in plants and soil. Protecting and restoring these ecosystems is crucial for maximizing their carbon storage potential and preserving biodiversity. In freshwater and marine environments, carbon sequestration occurs through processes such as the production of organic matter and its deposition in sediments. However, these ecosystems are under threat from human activities like pollution and habitat destruction, which could hinder their ability to sequester carbon effectively.

However, these ecosystems are under threat from various factors such as deforestation, pollution, habitat degradation, and the impacts of climate change. These threats pose significant risks to their ability to effectively sequester carbon. Tackling these challenges demands unified action at the local, regional, and global levels. It's imperative to implement measures aimed at conserving and restoring ecosystems, encouraging sustainable practices in land and ocean management, and reducing greenhouse gas emissions. Only through collaborative efforts can safeguard these vital ecosystems and their crucial role in carbon sequestration. By harnessing their carbon sequestration capabilities, we can not only mitigate climate change impacts but also protect biodiversity and ensure a sustainable future for generations to come.

REFERENCES

Aarnio, S. (2022). *Blue carbon: The value of carbon sequestration in kelp forests compared to traditional blue carbon ecosystems.*

Alongi, D. M. (2012). Carbon sequestration in mangrove forests. *Carbon Management, 3*(3), 313–322. doi:10.4155/cmt.12.20

Barbier, E. B., Koch, E. W., Silliman, B. R., Hacker, S. D., Wolanski, E., Primavera, J., & Reed, D. J. (2011). Coastal ecosystem-based management with nonlinear ecological functions and values. *Science, 319*(5861), 321–323. doi:10.1126/science.1150349 PMID:18202288

Bastin, J. F., Finegold, Y., Garcia, C., Mollicone, D., Rezende, M., Routh, D., & Crowther, T. W. (2019). The global tree restoration potential. *Science, 365*(6448), 76–79. doi:10.1126/science.aax0848 PMID:31273120

Berkes, F. (2017). *Sacred ecology*. Routledge. doi:10.4324/9781315114644

Bhattacharyya, S. S., Ros, G. H., Furtak, K., Iqbal, H. M., & Parra-Saldívar, R. (2022). Soil carbon sequestration–An interplay between soil microbial community and soil organic matter dynamics. *The Science of the Total Environment, 815*, 152928. doi:10.1016/j.scitotenv.2022.152928 PMID:34999062

Butchart, S. H., Clarke, M., Smith, R. J., Sykes, R. E., Scharlemann, J. P., Harfoot, M., & Burgess, N. D. (2015). Shortfalls and solutions for meeting national and global conservation area targets. *Conservation Letters, 8*(5), 329–337. doi:10.1111/conl.12158

CBD (Convention on Biological Diversity). (2014). *CBD Technical Series No. 75 - Ecosystem-based Adaptation in Marine and Coastal Ecosystems: Opportunities and Challenges*. Secretariat of the Convention on Biological Diversity.

CBD (Convention on Biological Diversity). (2020). *Global Biodiversity Outlook 5*. Secretariat of the Convention on Biological Diversity.

Chazdon, R. L., Broadbent, E. N., Rozendaal, D. M., Bongers, F., Zambrano, A. M. A., Aide, T. M., & Poorter, L. (2016). Carbon sequestration potential of second-growth forest regeneration in the Latin American tropics. *Science Advances, 2*(5), e1501639. doi:10.1126/sciadv.1501639 PMID:27386528

Clewell, A. F., & Aronson, J. (2006). Motivations for the restoration of ecosystems. *Conservation Biology, 20*(2), 420–428. doi:10.1111/j.1523-1739.2006.00340.x PMID:16903103

Coral Guardian. 2022, https://www.coralguardian.org/en/carbon-balance-in-corals/

Douvere, F., et al. (2019). Marine Spatial Planning: A Step-by-Step Approach Toward Ecosystem-Based Management. *Intergovernmental Oceanographic Commission Manuals and Guides, 83*.

FAO (Food and Agriculture Organization of the United Nations). (2015). *Status of the World's Soil Resources: Main report*. Rome.

Farrelly, D. J., Everard, C. D., Fagan, C. C., & McDonnell, K. P. (2013). Carbon sequestration and the role of biological carbon mitigation: A review. *Renewable & Sustainable Energy Reviews, 21*, 712–727. doi:10.1016/j.rser.2012.12.038

Galgani, F., Hanke, G., Werner, S. D. V. L., & De Vrees, L. (2015). Marine litter within the European marine strategy framework directive. *ICES Journal of Marine Science, 70*(6), 1055–1064. doi:10.1093/icesjms/fst122

Giller, K. E., Witter, E., Corbeels, M., & Tittonell, P. (2015). Conservation agriculture and smallholder farming in Africa: The heretics' view. *Field Crops Research, 114*(1), 23–34. doi:10.1016/j.fcr.2009.06.017

Gogoi, A., Ahirwal, J., & Sahoo, U. K. (2021). Plant biodiversity and carbon sequestration potential of the planted forest in Brahmaputra flood plains. *Journal of Environmental Management, 280*, 111671. doi:10.1016/j.jenvman.2020.111671 PMID:33218834

Gonthier, D. J., Ennis, K. K., Farinas, S., Hsieh, H. Y., Iverson, A. L., Batáry, P., & Perfecto, I. (2014). Biodiversity conservation in agriculture requires a multi-scale approach. *Proceedings of the Royal Society B: Biological Sciences, 281*(1791), 20141358. 10.1098/rspb.2014.1358

Halpern, B. S., Frazier, M., Afflerbach, J., Lowndes, J. S., Micheli, F., O'Hara, C., & Selkoe, K. A. (2019). Recent pace of change in human impact on the world's ocean. *Scientific Reports, 9*(1), 11609. doi:10.1038/s41598-019-47201-9 PMID:31406130

He, Z., Lin, L., Wang, X., Qin, W., & Zhang, C. (2022). Carbon storage by marine microorganisms for carbon neutrality. *Frontiers in Marine Science, 9*, 1018397. doi:10.3389/fmars.2022.1018397

Hooper, D. U., Adair, E. C., Cardinale, B. J., Byrnes, J. E., Hungate, B. A., Matulich, K. L., & O'Connor, M. I. (2012). A global synthesis reveals biodiversity loss as a major driver of ecosystem change. *Nature, 486*(7401), 105–108. doi:10.1038/nature11118 PMID:22678289

Houghton, R. A. (2015). Revised estimates of the annual net flux of carbon to the atmosphere from changes in land use and land management 1850–2000. *Tellus. Series B, Chemical and Physical Meteorology, 55*(2), 378–390.

Howard, J., Sutton-Grier, A., Herr, D., Kleypas, J., Landis, E., Mcleod, E., Pidgeon, E., & Simpson, S. (2017). Clarifying the Role of Coastal and Marine Systems in Climate Mitigation. *Frontiers in Ecology and the Environment, 15*(1), 42–50. doi:10.1002/fee.1451

Howard, J., Sutton-Grier, A., Herr, D., Kleypas, J., Landis, E., Mcleod, E., & Simpson, S. (2017). Clarifying the role of coastal and marine systems in climate mitigation. *Frontiers in Ecology and the Environment, 15*(1), 42–50. doi:10.1002/fee.1451

Ighalo, J. O., Dulta, K., Kurniawan, S. B., Omoarukhe, F. O., Ewuzie, U., Eshiemogie, S. O., & Abdullah, S. R. S. (2022). Progress in microalgae application for CO_2 sequestration. *Clean Chemical Engineering, 3*, 100044. doi:10.1016/j.clce.2022.100044

Jambeck, J. R., Geyer, R., Wilcox, C., Siegler, T. R., Perryman, M., Andrady, A., & Law, K. L. (2015). Plastic waste inputs from land into the ocean. *Science, 347*(6223), 768–771. doi:10.1126/science.1260352 PMID:25678662

Krause-Jensen, D., & Duarte, C. M. (2016). Substantial role of macroalgae in marine carbon sequestration. *Nature Geoscience, 9*(10), 737–742. doi:10.1038/ngeo2790

Lal, R. (2008). Carbon sequestration. *Philosophical Transactions of the Royal Society of London. Series B, Biological Sciences, 363*(1492), 815–830. doi:10.1098/rstb.2007.2185 PMID:17761468

Lehmann, J., Rillig, M. C., Thies, J., Masiello, C. A., Hockaday, W. C., & Crowley, D. (2015). Biochar effects on soil biota–a review. *Soil Biology & Biochemistry*, *43*(9), 1812–1836. doi:10.1016/j.soilbio.2011.04.022

Link, J. (2010). *Ecosystem-based fisheries management: confronting tradeoffs*. Cambridge University Press. doi:10.1017/CBO9780511667091

Lutz, S., Bhakta, D., Pearson, H., Vatter, J., & Barnes, R. (2018). *Oceanic Blue Carbon*. Flyer.

Macreadie, P. I., Serrano, O., Maher, D. T., Duarte, C. M., & Beardall, J. (2017). Addressing Calcium Carbonate Cycling in Blue Carbon Accounting. *Limnology and Oceanography Letters*, *2*(6), 195–201. doi:10.1002/lol2.10052

Maldonado, M. T., Surma, S., & Pakhomov, E. A. (2016). Southern Ocean Biological Iron Cycling in the Pre-Whaling and Present Ecosystems. *Philosophical Transactions. Series A, Mathematical, Physical, and Engineering Sciences*, *374*(2081), 20150292. doi:10.1098/rsta.2015.0292 PMID:29035257

Paustian, K., Lehmann, J., Ogle, S., Reay, D., Robertson, G. P., & Smith, P. (2016). Climate-smart soils. *Nature*, *532*(7597), 49–57. doi:10.1038/nature17174 PMID:27078564

Pretty, J., Benton, T. G., Bharucha, Z. P., Dicks, L. V., Flora, C. B., Godfray, H. C. J., & Wratten, S. (2018). Global assessment of agricultural system redesign for sustainable intensification. *Nature Sustainability*, *1*(8), 441–446. doi:10.1038/s41893-018-0114-0

Ray, R., & Jana, T. K. (2017). Carbon sequestration by mangrove forest: One approach for managing carbon dioxide emission from coal-based power plant. *Atmospheric Environment*, *171*, 149–154. doi:10.1016/j.atmosenv.2017.10.019

Sala, E., Mayorga, J., Bradley, D., Cabral, R. B., Atwood, T. B., Auber, A., Cheung, W., Costello, C., Ferretti, F., Friedlander, A. M., Gaines, S. D., Garilao, C., Goodell, W., Halpern, B. S., Hinson, A., Kaschner, K., Kesner-Reyes, K., Leprieur, F., McGowan, J., & Lubchenco, J. (2021). Protecting the Global Ocean for Biodiversity, Food and Climate. *Nature*, *592*(7854), 397–402. doi:10.1038/s41586-021-03371-z PMID:33731930

Shi, T., Zheng, X., Zhang, H., Wang, Q., & Zhong, X. (2021). Coral reefs: Potential blue carbon sinks for climate change mitigation. [Chinese Version]. *Bulletin of Chinese Academy of Sciences*, *36*(3), 270–278.

Sumaila, U. R., Cheung, W., Dyck, A., Gueye, K., Huang, L., Lam, V., & Zeller, D. (2012). Benefits of rebuilding global marine fisheries outweigh costs. *PLoS One*, *7*(7), e40542. doi:10.1371/journal.pone.0040542 PMID:22808187

UNEP (United Nations Environment Programme). (2016). *Marine Plastic Debris and Microplastics – Global lessons and research to inspire action and guide policy change*. UNEP.

Watson, J. E., Evans, T., Venter, O., Williams, B., Tulloch, A., Stewart, C., & Lindenmayer, D. (2018). The exceptional value of intact forest ecosystems. *Nature Ecology & Evolution*, *2*(4), 599–610. doi:10.1038/s41559-018-0490-x PMID:29483681

Williamson, P., & Gattuso, J. P. (2022). Carbon removal using coastal blue carbon ecosystems is uncertain and unreliable, with questionable climatic cost-effectiveness. *Frontiers in Climate*, *4*, 4. doi:10.3389/fclim.2022.853666

Zeng, N. (2008). Carbon sequestration via wood burial. *Carbon Balance and Management*, *3*(1), 1–12. doi:10.1186/1750-0680-3-1 PMID:18173850

Chapter 14
Urbanization's Ecological Toll: Exploring Its Role in Biodiversity Decline

Vishal M. Makwana
https://orcid.org/0000-0001-6343-9449
Zoology Department, Sir P. P. Institute of Science, M. K. Bhavnagar University, India

Foram P. Patel
D. K. V. Arts and Science College, Saurashtra University, Rajkot, India

Pravinsang P. Dodia
Zoology Department, Sir P. P. Institute of Science, M. K. Bhavnagar University, India

ABSTRACT

In this book chapter, the authors explore the intricate relationship between urbanization and biodiversity conservation. Despite urban environments being traditionally associated with habitat loss and fragmentation, they hold immense potential as critical arenas for safeguarding biodiversity. The authors delve into innovative conservation strategies that capitalize on the unique opportunities offered by urban settings, including the utilization of unconventional spaces and the integration of Indigenous perspectives. Through case studies and examples, they highlight the importance of community involvement and citizen science initiatives in fostering urban biodiversity. By reconceptualizing urban areas as vital components of conservation efforts, the authors aim to ensure the survival and recovery of threatened species within their natural habitats. This chapter offers insights into how urban landscapes can be transformed into resilient ecosystems that coexist harmoniously with human communities.

DOI: 10.4018/979-8-3693-6950-0.ch014

INTRODUCTION

Biodiversity, a term that encapsulates the variety and variability of life on Earth, plays a pivotal role in maintaining ecological balance and supporting human well-being. This concept encompasses the diversity within species, between species, and of ecosystems, underscoring the interconnectedness of all living things. The importance of biodiversity lies not only in its contribution to the natural processes that sustain life, such as water purification, pollination of crops, and regulation of the climate, but also in its cultural, economic, and scientific value to humans (Díaz et al., 2019). Healthy, biodiverse ecosystems offer critical services that support agricultural productivity, provide medicinal resources, and help mitigate the impacts of natural disasters. As such, the preservation of biodiversity is fundamental to the resilience of our natural environment and the prosperity of human societies. Urban areas can harbour significant biodiversity and provide crucial ecosystem services to a large portion of the world's population (Spotswood et al., 2021). However, human activities pose significant threats to biodiversity, including habitat destruction, pollution, climate change, overexploitation, and invasive species, necessitating effective conservation strategies (Duraiappah et. al, 2005).

The current chapter addresses the following objectives in a detailed and structured manner: It assesses how urbanization influences biodiversity, specifically examining the changes in species diversity within urban settings. It delves into the problem of habitat fragmentation resulting from urban expansion and discusses its effects on the conservation of various species. It outlines the challenges urbanization poses to preserving biodiversity in cities, including the impact on native species and the proliferation of non-native species. It investigates potential avenues for promoting biodiversity conservation in urban areas, showcasing effective methods and practices that can counteract the adverse effects of urbanization.

URBANIZATION PROCESSES AND PATTERNS

Urban Growth Dynamics

Urban growth refers to the expansion of cities and towns, while urbanization denotes the shift of a population to urban areas (McDonald et al., 2019). In 2000, the United Nations Population Division (UNPD) reported that 2.9 billion people lived in cities. This number increased to 4 billion by 2015. Predictions indicate that by 2030, cities will grow by an additional 1.2 billion people, mainly in China, India, Nigeria, and Indonesia. Alongside population growth, urban areas are expected to expand by 1.2 to 1.8 million km2 from 2000 to 2030. From 1990 to 2015, the urban population jumped from 2.3 billion to 4 billion. It's estimated that 55% of the world's population lived in urban areas in 2018, a number expected to reach 70% by 2050 as per United Nations Population Division. The future of urbanization could vary widely, with urban populations potentially making up 60% to 92% of the global population by the century's end, depending on government policies and societal trends (Jiang & O'Neill, 2017).

However, estimating the size of the urban population and the area covered by urban regions can be challenging due to differences in how countries define and measure urban areas (Montgomery, 2003). Definitions of 'urban' vary, often based on population density, built-up area, or a composite of factors. Besides, despite these challenges, there is agreement that urban populations have been growing significantly and are expected to reach 5.2 billion by 2030. The total urban area is also predicted to expand, with estimates of the future urban land area ranging widely. Seto and colleagues predict that urban areas will cover 1.9 million km^2 by 2030, but there are other forecasts too. Estimates for urban growth vary, ranging from 1.3 million km^2 to 1.7 million km2 between 2020 and 2050. Urbanization has also shown to concentrate populations in denser areas, reducing the overall land needed for human settlement compared to if the same population growth occurred in rural areas (Doxsey-Whitfield et al., 2015).

The rapid expansion and pace of urbanization exert complex effects on the environment worldwide. Urban development is increasingly intersecting with areas rich in biodiversity, leading to significant ecological consequences. This encroachment often results in habitat destruction, fragmentation, and alteration, with adverse effects on local flora and fauna (Hahs et al., 2023). The spread and organization of urban spaces can drastically alter community biodiversity, pushing out native species and diminishing the variety of roles played within ecosystems (Hahs et al., 2023). Several factors influence the extent of habitat impacted by urban population growth, including urban population density, urban form, and the location of urban growth. Notably, urban areas are often situated in productive and biodiverse regions, such as coastlines and floodplains, exacerbating the global impact of urban growth on biodiversity (Polaina, 2015).

The design of cities, including their size, density, and layout, influences their impact on biodiversity (Frey, 2003; Alberti et al., 2007). The way cities consume and dispose of resources, known as urban metabolism, also has a significant effect (Kennedy et al., 2011). These factors are influenced by income levels, technology, infrastructure, societal behaviours, and cultural norms. Decisions made by people shape the urban environment and its effects on nature (Montgomery et al., 2003. Urbanization's impact on biodiversity can be categorized as either direct or indirect. Direct impacts involve the conversion of natural habitats to urban land uses, leading to changes in land cover and habitat loss. Fragmentation of natural habitats through urban expansion also results in altered abiotic and biotic conditions, such as increased edge effects and reduced connectivity among remaining habitat patches. On the other hand, indirect impacts are mediated by intermediate processes, encompassing the consumption of resources (e.g., energy and food) and the release of solid, liquid, and gaseous wastes from urban areas (McDonald et al., 2019).

The combined effects of urban structure and energy and material flows influence local land use and human activities, which then have direct effects on biodiversity (Sanderson, 2018). The nature and severity of these direct effects are shaped by the design of urban areas. More compact and densely built cities tend to have a lesser environmental footprint. The structure and functional processes of urban environments influence the consumption of resources, levels of pollution, and trade patterns within these areas, subsequently resulting in indirect effects on biodiversity. These indirect effects extend across different spatial scales, ranging from local instances, such as the effect of sewage on downstream coral reefs, through regional impacts, exemplified by the diversion of water for urban purposes, to global consequences, including the emissions of greenhouse gases (GHGs). The majority of the world's population and economic activities are concentrated in urban areas, making these areas significant contributors to pollution and high demand for resources (McDonald et al., 2019). Research has indicated that the increase in urban populations not only impacts but will persist in influencing land utilization significantly (Angel et al., 2012), as well as energy use and its consequent effects on climate change (Güneralp et al., 2017).

Urban Fragmentation and Habitat Loss

For a more ecologically meaningful assessment, examining the impact on specific biomes is crucial. The temperate broadleaf forest biome is forecasted to experience the highest natural habitat conversion to urban land uses, followed by the tropical moist forest biome. Notably, the latter biome, predominantly in middle- and low-income countries, faces rapid urbanization. Proportionally, the mangrove biome is anticipated to be the most affected by urban growth, covering approximately 2.9% of its total area. The Mediterranean biome is also forecasted to be significantly impacted (0.6%), while the tundra and boreal forest/taiga biomes are expected to be minimally affected due to the scarcity of urban areas in these regions (McDonald et al., 2019). The scarcity of research on the indirect effects on biodiversity is alarming, especially since global data indicate these effects might be more significant than direct ones. A prime example is the impact of food consumption in cities, identified as a major factor in urban environmental footprints. Research (Moore, 2013) has highlighted food consumption as a key contributor to a city's environmental impact. According to the Food and Agriculture Organization, the world had 48.7 million km2 of agricultural land in 2016, and with 54.4% of the population living in urban areas

Figure 1. Mechanisms contributing to biodiversity decline

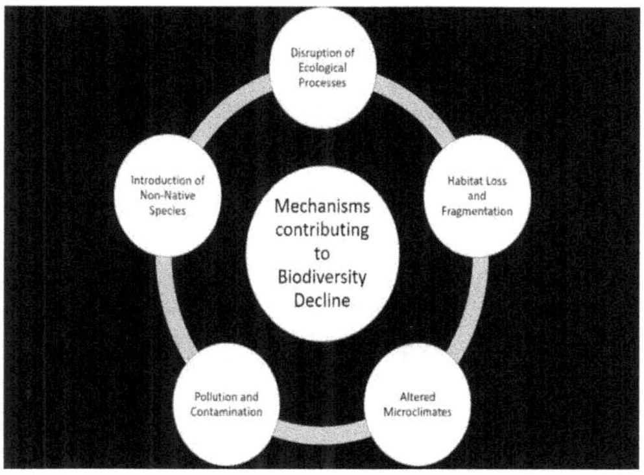

that year, urban food consumption requires over half of this land, about 26.5 million km2. This area is 36 times larger than the urban areas themselves, underscoring the vast indirect environmental impact of urban food demand.

However, equating urban and rural food consumption is overly simplistic, as urbanization often leads to increased caloric intake due to changes in lifestyle and economic status. This increase is typically characterized by a shift towards more animal products, fats, and vegetables. The growth in urban caloric demand does not always translate into agricultural expansion; some of the demand is met through intensification on existing agricultural land, which can still affect biodiversity. Nonetheless, a portion of the increased demand will lead to new agricultural land, potentially causing significant biodiversity loss (Chaplin-Kramer et al., 2015). However, the complex food supply chains serving urban areas, which vary by location, it's clear that urban dietary needs have a substantial impact on agricultural land use in certain regions (Seto & Ramankutty, 2016). Urbanization induces habitat loss and fragmentation, disrupts ecological processes, introduces non-native species, pollutes habitats, and alters microclimates, all contributing to biodiversity decline in urban areas (Figure 1).

MECHANISMS OF BIODIVERSITY LOSS IN URBAN ENVIRONMENTS

The health and longevity of populations depend significantly on the availability, quality, and spatial arrangement of their natural habitats (Flesch, 2017). The primary direct impact of urban growth on biodiversity is the loss of natural habitat due to the conversion to urban land uses. Between 1992 and 2000, 16% of total natural habitat loss, equivalent to 190,000 km^2, occurred due to urban expansion. An additional

290,000 km^2 is estimated to be converted between 2000 and 2030 (McDonald et al., 2019). Examining the forecasted direct urban impacts on natural habitat loss (2000–2030), the United States, Brazil, Nigeria, and China are projected to lose over 10,000 km^2 each. However, other countries on every continent are also expected to experience substantial urban-caused habitat loss. Analyzing the relative impact of urban growth on a country's area, islands like Hong Kong S.A.R., Mauritius, and Puerto Rico stand out with the greatest impact.

Habitat Loss, Alteration, Fragmentation, and Its Ecological Consequences

On a global scale, the conversion of natural habitat to urban areas reduces biodiversity, endangering sensitive, endemic species (Elmqvist et al., 2013). McDonald et al. (2018) estimated that 13% of terrestrial vertebrate ecoregional endemics face high threats from global urban expansion. Newbold and colleagues (2015) studied species richness and abundance in urban areas influenced by human activities. They found that sites with minimal use had slightly lower species richness and abundance than primary vegetation, while sites with light and intense use showed significantly lower levels. In their seminal work, McKinney (2008) conducted a thorough investigation into the ramifications of urban development on the richness of plant and animal species. The findings indicate a discernible decline in species richness in areas subjected to intense urbanization. Conversely, urban environments characterized by moderate development exhibit variable impacts on the diversity of species across different taxonomic categories. On a global scale, urban centres are observed to harbour a selective assortment of species originating from regional species pools, while simultaneously supporting diverse arrays of non-native species across various taxonomic groups (Aronson et al., 2014).

Urban areas can introduce new types of habitats and interactions among species, altering the natural balance and favouring certain species over others. This often leads to a dominance of species that can thrive in altered environments, at the expense of specialist species that require specific conditions to survive. Besides, Urbanization also impacts animal traits, particularly those related to reproduction. For example, in urban settings, some amphibians and reptiles exhibit decreased mobility, more specialized diets, and smaller clutch sizes. These changes suggest a deep-seated influence of urbanization on the evolutionary and ecological dynamics of these species. Furthermore, urbanization can lead to a shift in the traits within communities, such as body size and dietary habits, with varied impacts across different taxa. Some species may adapt by becoming more specialized or generalist in their feeding strategies, depending on the availability and distribution of food resources (Hahs et al., 2023). Urbanization has a tremendous influence on butterfly diversity in urban settings.

While certain adapted species survive in urban areas, total diversity is frequently reduced owing to habitat loss, fragmentation, and pollution. Green spaces, such as parks and gardens, can counteract these impacts by providing vital supplies and habitat connections, resulting in a more diversified butterfly population even inside urban areas. However, protecting and improving these natural spaces is critical for sustaining butterfly variety despite growing urban settings (Trivedi et al., 2022).

Habitat fragmentation involves breaking down larger habitats into smaller, more isolated sections. Increased habitat fragmentation significantly impacts biodiversity worldwide. It leads to less available habitat and greater isolation of habitat patches. In response, the dispersal abilities of species may change (Makwana et al., 2023). The fragmentation of habitats due to urban expansion isolates wildlife populations, making them more vulnerable to extinction. This process leads to groups of populations with restricted connections, affecting species' evolution and survival by limiting gene exchange and reducing the number of individuals (Travis et al., 2013). Dispersal, the movement of individuals that facilitates gene flow across areas, is a critical characteristic of species. It can develop as a way to counteract habitat fragmentation, enabling organisms to move between fragmented patches and settle in new ones, which helps in preserving these population groups. Nonetheless, under certain conditions, dispersal might not be advantageous. Fragmentation creates a more varied landscape, increasing the costs and risks associated with moving, such as the chance of entering an unfriendly environment, which could lead to a preference for lesser dispersal. This phenomenon has been supported by both theoretical and empirical research across different species groups (Duputié & Massol, 2013).

Besides, habitat fragmentation often results in a preference for species with lower dispersal capabilities due to the costs and spatial challenges associated with dispersal. Tropical forests, crucial for global biodiversity, are being cleared at an increasing pace, leading to severe degradation and fragmentation. This fragmentation poses significant threats to the survival of sensitive and endangered species relying on these habitats. The reduction, fragmentation, and deterioration of habitats often lead to poorer habitat quality, an increase in isolated habitat patches, and a reduction in the size of these patches (Arroyo-Rodríguez and Dias 2010). Such habitat alterations can intensify edge effects and the distances between habitat patches, adversely affecting ecological functions including reproduction, dispersal, mortality, and species interactions (Smith et al. 2017). Consequently, conserving and restoring natural habitats is crucial for maintaining species populations in fragmented landscapes.

Recent advancements in macroecology have highlighted the variability in how populations across a species' geographic range react to habitat changes. Contrary to the traditional view that all members of a species uniformly respond to environmental changes, it's now understood that the impact of habitat alterations can differ across a species' range. Factors such as proximity to the geographic range edge and environmental suitability play roles in this variation. According to the centre-periphery hypothesis, populations tend to be more abundant in the central areas of their range, decreasing towards the edges where conditions are less optimal (Sagarin and Gaines, 2006). Habitat changes may exacerbate the decline in species abundance at the range's edges. Furthermore, the physiological tolerance hypothesis posits that a species' resistance to environmental stressors, like temperature and rainfall, may differ throughout its range, affecting its reaction to habitat disturbances. Populations near their physiological limits may be more affected by habitat changes than those in more favourable conditions (Williams and Newbold, 2021).

However, the relationship between geographic distance and environmental suitability is complex. Species abundance doesn't always uniformly decrease from a range's centre towards its edges, and the interplay between environmental gradients and geographic distance can significantly influence species' responses to habitat changes (Williams and Newbold, 2021). Understanding the intricate dynamics between environmental gradients and the geographic edges of ranges is key to grasping how populations react to habitat alterations. Despite the importance of these dynamics, there has been insufficient focus on understanding the macroecological factors that influence species' reactions to habitat changes. Such knowledge is vital for formulating effective conservation strategies (Oliver et al., 2016). Without understanding how populations across different ranges respond to habitat changes, conservation efforts may not be successful, potentially leading to severe population declines and negative effects on ecosystem services (Betts et al., 2017).

Introduction of Invasive Species and Its Ecological Consequences

The process of urbanization often results in the substitution of natural green spaces with impervious materials and constructions, elevating pollution levels. This environmental alteration generally leads to biodiversity loss and fosters an ecosystem predominantly inhabited by a simplified community of species. Such communities are characterized by a decline in specialized, sensitive taxa and a proliferation of generalist, tolerant taxa (Faeth et al., 2011). Furthermore, the increased mobility of people and goods enhances the capacity for species to be transported across different urban areas, thus urban ecosystems are known to harbour a significant number of alien species. Alien species are those non-native species introduced into new areas primarily through human activities.

An alien species is considered invasive when it is capable of establishing large, self-sustaining populations across multiple life cycles, far from its original point of introduction (Richardson et al., 2011). Many invasive species find urban environments conducive to thriving, often reaching high population densities, which can have various adverse effects on human well-being and ecological systems (Shochat et al., 2010). Although a diverse mix of competitors and predators could theoretically offer biotic resistance to such invasions, evidence of this effect in urban contexts remains elusive. Despite longstanding recognition of the presence of invasive species within urban environments, the underlying mechanisms that facilitate their successful invasion remain poorly understood (Cadotte et. al., 2017). It has been hypothesized that the successful invasion of species in urban areas may be due to the possession of certain traits that confer advantages within these environments. For instance, traits such as the production of fleshy fruits or a diminished fear of humans have been linked to successful urban invasions (Carrete & Tella, 2011). Moreover, urbanization itself may act as a potent selective pressure, leading to modifications in traits that augment a species' invasive potential (Alberti et al., 2017).

The propensity of a species to invade is intimately tied to its life-history traits. Species exhibiting 'rapid' life-history traits, including swift growth and high fecundity, are believed to possess a greater potential for invasion due to these characteristics enhancing their survival and reproduction rates. Life-history theory suggests that these traits are largely dictated by nutritional trade-offs, as organisms have finite nutritional resources to allocate towards growth, reproduction, and coping with biotic challenges. Urbanization, by potentially increasing the availability of food resources, could alleviate these nutritional trade-offs, enabling organisms to invest in traits conducive to faster life histories and thereby augmenting their invasive potential (Shik & Dussutour, 2020). However, the degree to which changes in traits linked to invasive success and the ecological dynamics that facilitate invasions in urban settings are understood remains limited.

Boscutti et al. (2022) highlighted the considerable threat posed by the proliferation of exotic plant species to biodiversity, affecting population, community, and ecosystem levels, necessitating a landscape-oriented approach to understanding plant invasions across spatial and temporal scales. The study emphasized that human-altered landscapes, particularly due to agriculture and urbanization, are pivotal in fostering a greater diversity of exotic species by creating vacant niches that facilitate their spread. Urban landscapes, characterized by disturbed habitats like roadsides and gardens, emerge as significant sources for the spread of exotic plant species into surrounding areas, though the impacts of landscape configuration in such environments have been less frequently explored. Boscutti et al. (2022) uncovered that in fragmented natural habitats within intensive landscapes, smaller fragments and edges between natural and disturbed habitats often exhibit a higher density and richness of exotic species, suggesting a correlation between patch shape complexity and the facilitation of exotic plant spread across human-impacted landscapes.

The research underscored the importance of scaling up the study of spatial configuration effects from habitat to landscape levels to predict the impact of land-use changes on plant invasions accurately, taking into account variations across spatial scales. Despite a focus on the relationship between native and alien species richness in multi-scale approaches, the study by Boscutti et al. (2022) delved into the scale-dependence of abiotic drivers of plant invasions, highlighting the differential role of factors like climate, soil, and habitat types at broader spatial scales versus disturbance and biotic interactions at finer scales. In the European context, the distinction between archaeophytes and neophytes is crucial, with each group exhibiting specific distribution patterns and associations with different land uses archaeophytes with agriculture and neophytes with urban areas. Boscutti et al. (2022) found that urban expansion and shape complexity significantly influence neophyte invasions, with shape complexity being a primary abiotic driver at larger spatial scales, and the type of recipient habitat and urban coverage influencing exotic species success at habitat patch scales.

Furthermore, Boscutti et al. (2022) observed that urban areas often act as primary sources for exotic plant invasions, with roads serving as effective dispersal corridors. The study also indicated that semi-natural habitats are more prone to invasion within urban-dominated matrices, highlighting the shared response of neophyte and archaeophyte groups to urban effects and the role of agricultural cover in facilitating archaeophyte spread. This comprehensive analysis underscores the complex interplay between landscape composition and configuration in shaping exotic plant invasions, emphasizing the need for multi-scale approaches to understand and manage biodiversity threats effectively. Marques et al. (2020) conduct a comprehensive study on the influence of urban disturbances on the invasive potential and success of species, focusing on the guppy, *Poecilia reticulata*, as a case study. They propose that urban environments provide increased food resources, thereby enabling invasive species like guppies to develop traits that enhance their invasive capabilities and success rates (Richardson et al., 2011; Marques et al., 2020). By comparing environmental conditions and life-history traits of guppies in urban versus non-urban stream settings, they aim to understand how urbanization affects these species' invasiveness, with an anticipation that urban guppy populations would exhibit increased densities and body sizes due to better food availability.

The study specifically examines the availability and consumption of chironomids in urban streams as a key factor in supporting guppy proliferation through increased somatic growth and reproduction, suggesting a potential easing of life-history trade-offs in urban environments. Additionally, Marques et al. (2020) consider the role of fish biodiversity in urban streams and its effect on guppy invasion success, hypothesizing that higher biodiversity could reduce guppy densities and reproductive investments, although their findings indicate a minimal effect of biodiversity on mitigating guppy invasion in urban settings.

Their research underscores the impact of urbanization on invasive species by demonstrating that guppies in urban streams of Rio de Janeiro not only thrive in terms of population density and body size but also possibly benefit from the altered environmental conditions characteristic of the 'urban stream syndrome' (Borges et al., 2015). The increased availability of food resources, particularly chironomids, is highlighted as a critical factor enabling the observed enhanced traits in urban guppies, which may facilitate their invasive success by allowing for a relaxation of life-history trade-offs associated with growth, reproduction, and biotic interactions. However, Marques et al. (2020) also note the complexity of biotic interactions in urban ecosystems and the limited role of biodiversity in providing resistance against invasive species, pointing to the need for further investigation into the factors that contribute to the success of invasive species in urban environments and the conservation potential of urban biodiversity. This study contributes significantly to the understanding of how urbanization influences invasive species, offering insights into the mechanisms driving their success and the potential challenges for biodiversity conservation in urban ecosystems.

Pollution and Its Impact on Urban Biodiversity

Pollution, increased noise, and altered climatic conditions in urban environments pose additional threats, particularly to specialist species that rely on specific environmental conditions. These changes can disrupt the ecological processes these species depend on, leading to a decrease in food resources and overall suitability of the urban landscape for their survival (Hahs et al., 2023). Urbanization impacts many species negatively through changes such as increased temperatures (known as the urban heat island effect), pollution in the air, soil and water (Makwana et al., 2024), and increased noise and light pollution. However, urban environments can also support biodiversity conservation and serve as vital habitats for certain species like birds of prey, insect pollinators, and arthropods (Buchholz et al., 2018). Environmental stress from urbanization leads to various consequences for organisms, including changes in predation, new species interactions, and food availability. It can also cause an increase in physical and genetic diversity within species. A common outcome of such stress is fluctuating asymmetry in bilateral features, a sign of developmental instability that serves as a stress indicator. This measure has been applied to study the effects of urbanization on different animal groups, including mammals and insects.

Taxonomic Bias and Understudied Areas

The scientific literature concerning the direct impacts of urban growth on biodiversity exhibits significant geographic biases, with studies primarily concentrated in certain regions while others remain understudied, potentially due to language barriers or a need for further investigation. Studies on the direct impact of urban growth on biodiversity are predominantly conducted in high-income countries, despite these nations having a smaller global population. There is a notable absence of scientific research from middle- and low-income countries, which are projected to experience the most significant habitat loss due to urbanization. High-income countries primarily concentrate their research efforts on studying the impact of urban growth on temperate broadleaf forests, while tropical moist forests, prevalent in middle- and low-income countries, receive less attention. Mediterranean habitats are more frequently studied, largely due to research conducted in high-income European countries. There is a notable taxonomic bias in the literature, with a predominant focus on terrestrial taxa, particularly plants and birds. Indirect impacts on biodiversity, especially in marine taxa, are under-researched despite potential increases (McDonald et al., 2019).

Studies examining the indirect impacts of urban growth on biodiversity, concerning population or area, are relatively scarce compared to direct impacts. Of the studies that do investigate indirect impacts, a significant majority originate from high-income countries, indicating a bias. Besides, the majority of indirect impact studies focus on freshwater systems, particularly exploring the effects on macroinvertebrates. Some studies delve into long-distance teleconnections, where the resource demands of urban areas affect landscapes far away. Urban environments play a crucial role in biodiversity conservation, research efforts need to address geographic biases and expand to understudied regions, taxa, and indirect impacts to achieve comprehensive understanding and effective conservation strategies.

UNDERSTANDING URBANIZATION'S ECOLOGICAL TOLL

When natural habitats transform into urban areas, certain species are affected locally. Sensitive species may be lost, while tolerant ones like Eastern grey squirrels (*Sciurus carolinensis*) in eastern North America can adapt, and synanthropic species, such as Norway rats (*Rattus norvegicus*) or house sparrows (*Passer domesticus*), follow human presence (Shochat et al., 2010). The impact on species richness depends on the balance of these subsets (Faeth, 2011). Nonetheless, the paucity of comprehensive data concerning taxa beyond flora and avifauna poses a significant barrier to elucidating generalized patterns of species diversity in urban locales. The overlapping presence of humans and productive regions presents a challenge for biodiversity, highlighting the importance of understanding how urban expansion affects varied urban ecosystems, where research is currently limited. Despite significant environmental changes such as vegetation removal, increased housing density, and the urban heat island effect, however, it's unclear if productivity still plays a crucial role in species abundance and distribution within urban areas (Threlfall et al., 2012). These changes may alter food chain dynamics, especially affecting primary consumers like insects, which react to changes in nutrients, temperature, and productivity. This could, in turn, impact secondary consumers. Studies have shown that urban areas with more vegetation tend to have higher insect densities, which can influence the diversity and abundance of other species, including microbats (Kalcounis-Rueppell, 2007). Key findings from a study conducted by Threlfall et al. (2012) indicate that ambient temperature, housing density, and the presence of fertile shale geologies are significant factors affecting insect biomass and bat foraging behavior. Specifically, suburban areas with fertile shale geologies have higher insect biomass than those with poor-quality sandstone, and bat foraging is most active in suburban bushland and riparian zones located in fertile areas. These insights emphasize the importance of conserving and restoring natural habitats, especially in fertile regions, to support bat foraging habits and biodiversity.

Natural climate shifts and human activities have significantly contributed to the reduction and division of habitats worldwide, resulting in 70% of the world's remaining forests being located within 1 km of a forest edge (Haddad et al., 2015). Ecological research has indicated that the edge effect, the impact of these forest edges on the environment, affects the composition, richness, and number of species across various groups, leading to a global decrease in biodiversity. This edge effect not only diminishes species numbers and the connections between them but is also believed to reduce their genetic diversity. Earlier theoretical research suggested that populations near the edges of habitats have shorter times to the most recent common ancestor, implying they have less genetic diversity than those in more central locations. Nonetheless, it remains a complex task to accurately predict the edge effect's impact on genetic diversity within the varied and complex nature of real-world habitat fragments, where multiple edge effects might interact. Fragmentation and edge effects, while individually impactful, synergistically exacerbate the degradation of natural forests. Hasui et al. (2023) identified four distinct patterns of sensitivity that did not correlate with characteristics specific to species. They discovered that the combination of biogeographic and landscape factors influences population sizes, leading to effects that can either counteract or enhance each other. According to their findings, these interactions can alter how populations react to changes in their habitats, with variations observed based on the geographic location of their range or the suitability of their environment (Hasui et al., 2023). The impact of habitat changes on species can vary greatly, influenced by whether these changes work together or against each other, as well as the specific habitat needs of different species. Species that specialize in forest habitats and have limited ability to move, like *Drymophila ochropyga* and *Antilophia galeata*, have experienced a combined decrease in numbers due to habitat changes. Conversely, species such as *Myiothlypis flaveola*, *Myarchus ferox*, and *Phaethornis squalidus* have shown varied responses based on their geographic location or how well the environment meets their needs, which might appear to challenge the assumption that species always choose the best habitats for their survival (Jacob et al. 2018; Banks-Leite et al. 2022). This suggests that sometimes, choosing fewer ideal habitats might be beneficial for avoiding competition with other species. Therefore, the relationship between habitat changes and their effects on species' numbers and where they live is intricate, shaped by the species' preferences, interactions with other species, and how environmental pressures interact. Recognizing these nuanced dynamics is essential for the successful conservation and management of biodiversity amid ongoing changes to habitats.

Urban stress can also lead to variations within species in physiological, ecological, and behavioural traits (Gunn et al., 2021). Urban-dwelling individuals might exhibit distinct physiological or behavioural traits compared to those in rural areas. For instance, urban animals often show greater behavioural flexibility and are more likely to take risks in the presence of predators (Lapiedra et al., 2016). Such stress-related behaviours are useful for understanding individual behavioural variations, which can significantly influence survival. Furthermore, urbanization can affect males and females differently, leading to sex-specific trait variations. While some studies find that personality traits do not vary significantly between sexes across many animals, others highlight sex-specific differences in certain behaviours, which may influence the evolution of these traits (Mezőfi et al., 2022). Additionally, male and female responses to environmental stress can differ, as seen in some species' reactions to food scarcity and temperature changes (Roodt et al., 2022).

The impact of urbanization on animal behavior is evident in both bird species and insects, such as grasshoppers, with urban individuals displaying different traits compared to their rural counterparts. The research conducted by Rech et al. (2022) reveals that urbanization imposes significant environmental stress on individuals, as evidenced by increased developmental asymmetry in *Chorthippus biguttulus*, particularly in the tegmina and femora. This fluctuating asymmetry, correlating strongly with urbanization levels, is a reliable indicator of environmental stress. The study suggests that urban stressors such as artificial lighting, noise pollution, and chemical pollutants, among others, could contribute to these observed asymmetries, although pinpointing the primary factors remains a challenge.

Furthermore, Rech et al. (2022) observed that urbanization impacts the physical performance aspects of *C. biguttulus*, with body parts essential for specific functions (e.g., femora for movement) displaying lower levels of asymmetry compared to others (e.g., tegmina). This implies that critical functional body parts are more resistant to the effects of urban stressors. The study also noted that urbanized individuals exhibit bolder behaviours in response to simulated attacks and show increased activity levels, potentially as an adaptation to the heightened risk of predation in urban environments. These behavioural adaptations, including lower tonic immobility and greater mobility, could influence survival and escape capabilities from predators, though their impact on overall fitness in urban settings needs further examination. Additionally, the research highlighted the absence of sex differences in the expression of boldness, activity levels, and asymmetry, indicating a uniform response to urban stressors across sexes in *C. biguttulus*.

The interconnection between human activities, particularly urbanization, and river water pollution is a global concern that has been extensively documented. In China, a 2001 study highlighted the alarming state of water quality, with nearly half of the tested water systems failing to meet basic standards, and significant pollution in both surface and groundwater due to industrial discharges (Tang, 2022). The Chinese government's response has included substantial investments in river channel governance and the appointment of river managers to combat pollution, yet challenges persist in effectively managing and purifying polluted water bodies. Similarly, the rapid urbanization in Brazil, particularly in major cities, has exacerbated water pollution issues, notably highlighted during the Olympic Games in Rio de Janeiro. The lack of effective waste management systems and regulatory oversight has led to severe contamination of water channels, posing health risks and environmental concerns (Tang, 2022). In the United States, historical and ongoing struggles with river pollution have been observed alongside urbanization. Chicago, for example, implemented significant infrastructural improvements in the 19th century to address cholera and dysentery outbreaks linked to water contamination. The Water Pollution Control Act of 1972 marked a significant legislative effort to improve water quality nationwide. Despite these efforts, challenges remain in managing pollution effectively in the face of continuing urban and industrial development (Tang, 2022). China's urbanization rate, although lower than in some developed and less developed nations, has prompted significant government action in river channel management, as seen in the Suzhou River and Shenzhen-Hong Kong Boundary River improvement projects. These initiatives have aimed at enhancing water quality and promoting sustainable development but underscore the complex, ongoing battle against urbanization-induced environmental degradation (Tang, 2022). The examination of urbanization's impact on river ecosystems across different regions and development stages reveals a common narrative of environmental strain. While government interventions and policies have evolved to address the issue, the persistent challenges of implementation, alongside the need for greater environmental consciousness and sustainable practices, remain critical in mitigating the adverse effects of urbanization on river water quality.

STRATEGIES FOR BALANCING URBANIZATION AND BIODIVERSITY CONSERVATION

Identification of Priority Areas at Risk

Global analyses have identified key areas where urban growth poses a significant risk to biodiversity. Thirty priority ecoregions containing 78% of vertebrate endemics threatened by urban growth are identified. However, published studies on urban biodiversity often overlook these priority ecoregions, indicating a need for future research in these critical areas (McDonald et al., 2019).

Strategies for Balancing Urbanization and Biodiversity Conservation

The challenge of urbanization presents a unique opportunity for biodiversity conservation, requiring innovative strategies to mitigate habitat loss and the alteration of landscapes. As urban areas expand, the importance of preserving ecological integrity to ensure the survival of diverse species cannot be overstated. Hahs et al. (2023) emphasizes the necessity of understanding how different taxa respond to urban environments to effectively conserve and promote biodiversity within and around urban areas. This understanding is critical to minimizing the negative impacts of urbanization and supporting a wide range of species amidst growing urban landscapes.

A study by Hending et al. (2023) highlights the urgent need for conservation measures in tropical regions to prevent further fragmentation. Such measures include reforestation and restoration initiatives aimed at connecting isolated forest fragments and reducing edge areas, thereby enhancing habitat quality for the survival of threatened species. Moreover, the research underscores the importance of not generalizing species sensitivity without considering range position and environmental suitability, advocating for the protection of vulnerable areas, especially those at the range edge, which are critical for species sensitive to habitat transformations (Hasui et al., 2023).

Opportunities and Innovations in Urban Biodiversity Conservation

Urban areas, often viewed as the antithesis of natural environments, present untapped opportunities for biodiversity conservation. The prevailing attitude that urban environments are beyond redemption for conservation purposes (Kowarik, 2018) has led to a significant oversight of the conservation potential within cities.

This perspective has resulted in conservation efforts that typically focus on large, intact, or undisturbed areas, excluding urban areas from consideration (Kendal et al., 2017). However, urban environments are arguably prime targets for conservation science, offering a golden opportunity to conserve species and ecosystems under threat while allowing people to engage with nature (Kareiva and Marvier, 2012). This approach not only seeks to conserve urban-restricted threatened species but also to re-engage communities with nature, potentially improving human health and well-being. The conservation of 39 nationally-threatened species, whose distribution is restricted to Australian urban areas, exemplifies the critical role cities play in the last-chance conservation of species within their natural range. The recovery of these species in urban environments challenges traditional conservation practices by necessitating innovative approaches to secure critical habitat in the limited space of cities. This includes embracing biodiversity-sensitive urban design and conservation developments and fostering partnerships among stakeholders to balance competing land-use needs.

Community Engagement Initiatives

The engagement of the community in conservation actions is paramount. Proximity to large human populations, while posing risks, can be advantageous if the community is aware and actively involved in conservation efforts. Successful community conservation efforts, such as those for the Frankston spider-orchid, demonstrate the potential of community engagement in improving habitat and supporting species recovery. Moreover, the inclusion of Indigenous perspectives in conservation efforts is crucial. Approximately half of the species' recovery documents acknowledge the intent to consult with Indigenous communities, highlighting the need to recognize and incorporate Indigenous knowledge and cultural relationships with the land in urban conservation practices (Leiper et al., 2018).

FUTURE RESEARCH DIRECTION

Urban environments, despite their reputation for habitat fragmentation and degradation, possess inherent qualities that can be harnessed to bolster conservation endeavours. They often harbour pockets of remnant vegetation, green spaces, and urban parks, which can serve as critical refuges for biodiversity. Moreover, vacant lots, brownfields, and green roofs present untapped opportunities for habitat restoration and creation. By recognizing and exploiting these unconventional urban habitats, conservationists can enhance biodiversity and establish vital corridors for species movement and gene flow. Furthermore, active engagement with local communities is indispensable for successful urban biodiversity conservation. Involving residents in citizen science initiatives, habitat restoration projects, and environmental education programs fosters a sense of ownership and stewardship over urban green spaces. Empowering communities to participate in conservation activities not only strengthens social cohesion but also amplifies the efficacy and sustainability of conservation initiatives.

In addition, the integration of Indigenous perspectives and traditional ecological knowledge into urban conservation strategies enriches our understanding of ecosystems and augments conservation outcomes. Indigenous communities possess deep-rooted wisdom and intimate connections to the land, offering valuable insights into sustainable land management practices, species interactions, and ecosystem dynamics. By honouring Indigenous worldviews and incorporating traditional ecological knowledge, conservation efforts can become more holistic, culturally relevant, and ecologically resilient. Ultimately, by reconceptualizing urban areas as indispensable components of biodiversity conservation, we can forge a path toward the preservation and recovery of threatened species within their natural habitats. Embracing the multifaceted nature of urban landscapes and leveraging their potential for conservation not only benefits wildlife but also enhances the well-being of human inhabitants. Through collaborative endeavours that prioritize ecological integrity, community engagement, and cultural diversity, we can cultivate urban environments that thrive in harmony with nature, ensuring the long-term survival of species and ecosystems for generations to come.

CONCLUSION

In summary, urban environments play a pivotal role in biodiversity conservation, particularly for species whose existence is intricately linked to these landscapes. Achieving a harmonious balance between urbanization and biodiversity preservation demands the implementation of innovative conservation approaches that capitalize on the distinctive characteristics of urban settings. This entails acknowledging the capacity of unconventional urban spaces to contribute to conservation efforts, nurturing community engagement, and incorporating Indigenous knowledge systems into conservation practices. By reconceptualizing urban areas as indispensable facets of biodiversity conservation strategies, we can safeguard the persistence and resurgence of imperilled species within their native ecosystems.

REFERENCES

Alberti, M., Correa, C., Marzluff, J. M., Hendry, A. P., Palkovacs, E. P., Gotanda, K. M., Hunt, V. M., Apgar, T. M., & Zhou, Y. (2017). Global urban signatures of phenotypic change in animal and plant populations. *Proceedings of the National Academy of Sciences of the United States of America, 114*(34), 8951–8956. doi:10.1073/pnas.1606034114 PMID:28049817

Angel, S., Parent, J., Civco, D. L., & Blei, A. M. (2012). *Atlas of Urban Expansion*. Lincoln Institute of Land Policy.

Aronson, M. F., La Sorte, F. A., Nilon, C. H., Katti, M., Goddard, M. A., Lepczyk, C. A., . . . Winter, M. (2014). A global analysis of the impacts of urbanization on bird and plant diversity reveals key anthropogenic drivers. *Proceedings of the royal society B: biological sciences, 281*(1780), 20133330. 10.1098/rspb.2013.3330

Arroyo-Rodríguez, V., & Dias, P. A. D. (2010). Effects of habitat fragmentation and disturbance on howler monkeys: A review. *American Journal of Primatology, 72*(1), 1–16. doi:10.1002/ajp.20753 PMID:19852004

Banks-Leite, C., Betts, M. G., Ewers, R. M., Orme, C. D. L., & Pigot, A. L. (2022). The macroecology of landscape ecology. *Trends in Ecology & Evolution, 37*(6), 480–487. doi:10.1016/j.tree.2022.01.005 PMID:35184877

Betts, M. G., Wolf, C., Ripple, W. J., Phalan, B., Millers, K. A., Duarte, A., Butchart, S. H. M., & Levi, T. (2017). Global forest loss disproportionately erodes biodiversity in intact landscapes. *Nature, 547*(7664), 441–444. doi:10.1038/nature23285 PMID:28723892

Borges, R. C., Dos Santos, F. V., Caldas, V. G., & Lapa, C. M. F. (2015). Use of geographic information system (GIS) in the characterization of the Cunha Canal, Rio de Janeiro, Brazil: Effects of the urbanization on water quality. *Environmental Earth Sciences, 73*(3), 1345–1356. doi:10.1007/s12665-014-3493-1

Boscutti, F., Lami, F., Pellegrini, E., Buccheri, M., Busato, F., Martini, F., Sibella, R., Sigura, M., & Marini, L. (2022). Urban sprawl facilitates invasions of exotic plants across multiple spatial scales. *Biological Invasions, 24*(5), 1497–1510. doi:10.1007/s10530-022-02733-6

Buchholz, S., Hannig, K., Möller, M., & Schirmel, J. (2018). Reducing management intensity and isolation as promising tools to enhance ground-dwelling arthropod diversity in urban grasslands. *Urban Ecosystems, 21*(6), 1139–1149. doi:10.1007/s11252-018-0786-2

Cadotte, M. W., Yasui, S. L. E., Livingstone, S., & MacIvor, J. S. (2017). Are urban systems beneficial, detrimental, or indifferent for biological invasion? *Biological Invasions, 19*(12), 3489–3503. doi:10.1007/s10530-017-1586-y

Carrete, M., & Tella, J. L. (2011). Inter-individual variability in fear of humans and relative brain size of the species are related to contemporary urban invasion in birds. *PLoS One, 6*(4), e18859. doi:10.1371/journal.pone.0018859 PMID:21526193

Chaplin-Kramer, R., Sharp, R. P., Mandle, L., Sim, S., Johnson, J., Butnar, I., Milà i Canals, L., Eichelberger, B. A., Ramler, I., Mueller, C., McLachlan, N., Yousefi, A., King, H., & Kareiva, P. M. (2015). Spatial patterns of agricultural expansion determine impacts on biodiversity and carbon storage. *Proceedings of the National Academy of Sciences of the United States of America, 112*(24), 7402–7407. doi:10.1073/pnas.1406485112 PMID:26082547

Díaz, S., Settele, J., Brondízio, E. S., Ngo, H. T., Agard, J., Arneth, A., Balvanera, P., Brauman, K. A., Butchart, S. H. M., Chan, K. M. A., Garibaldi, L. A., Ichii, K., Liu, J., Subramanian, S. M., Midgley, G. F., Miloslavich, P., Molnár, Z., Obura, D., Pfaff, A., & Zayas, C. N. (2019). Pervasive human-driven decline of life on Earth points to the need for transformative change. *Science, 366*(6471), eaax3100. Advance online publication. doi:10.1126/science.aax3100 PMID:31831642

Doxsey-Whitfield, E., MacManus, K., Adamo, S. B., Pistolesi, L., Squires, J., Borkovska, O., & Baptista, S. R. (2015). Taking advantage of the improved availability of census data: A first look at the gridded population of the world, version 4. *Papers in Applied Geography, 1*(3), 226–234. doi:10.1080/23754931.2015.1014272

Duputié, A., & Massol, F. (2013). An empiricist's guide to theoretical predictions on the evolution of dispersal. *Interface Focus, 3*(6), 20130028. doi:10.1098/rsfs.2013.0028 PMID:24516715

Duraiappah, A. K., Naeem, S., Agardy, T., Ash, N. J., Cooper, H. D., Díaz, S., & Van Jaarsveld, A. (2005). *Ecosystems and human well-being: biodiversity synthesis; a report of the Millennium Ecosystem Assessment.* MEA.

Elmqvist, T., Fragkias, M., Goodness, J., Güneralp, B., Marcotullio, P. J., McDonald, R. I., & Wilkinson, C. (2013). *Urbanization, biodiversity and ecosystem services: challenges and opportunities: a global assessment.* Springer Nature. doi:10.1007/978-94-007-7088-1

Faeth, S. H., Bang, C., & Saari, S. (2011). Urban biodiversity: Patterns and mechanisms. *Annals of the New York Academy of Sciences, 1223*(1), 69–81. doi:10.1111/j.1749-6632.2010.05925.x PMID:21449966

Flesch, A. D. (2017). Influence of local and landscape factors on distributional dynamics: a species-centred, fitness-based approach. *Proceedings of the Royal Society B: Biological Sciences, 284*(1858), 20171001. 10.1098/rspb.2017.1001

Güneralp, B., Zhou, Y., Ürge-Vorsatz, D., Gupta, M., Yu, S., Patel, P. L., Fragkias, M., Li, X., & Seto, K. C. (2017). Global scenarios of urban density and its impacts on building energy use through 2050. *Proceedings of the National Academy of Sciences of the United States of America, 114*(34), 8945–8950. doi:10.1073/pnas.1606035114 PMID:28069957

Gunn, R. L., Hartley, I. R., Algar, A. C., Niemelä, P. T., & Keith, S. A. (2022). Understanding behavioural responses to human-induced rapid environmental change: A meta-analysis. *Oikos, 2022*(4), e08366. doi:10.1111/oik.08366

Haddad, N. M., Brudvig, L. A., Clobert, J., Davies, K. F., Gonzalez, A., Holt, R. D., Lovejoy, T. E., Sexton, J. O., Austin, M. P., Collins, C. D., Cook, W. M., Damschen, E. I., Ewers, R. M., Foster, B. L., Jenkins, C. N., King, A. J., Laurance, W. F., Levey, D. J., Margules, C. R., & Townshend, J. R. (2015). Habitat fragmentation and its lasting impact on Earth's ecosystems. *Science Advances, 1*(2), e1500052. doi:10.1126/sciadv.1500052 PMID:26601154

Hahs, A. K., Fournier, B., Aronson, M. F., Nilon, C. H., Herrera-Montes, A., Salisbury, A., Threlfall, C. G., Rega-Brodsky, C. C., Lepczyk, C. A., La Sorte, F. A., MacGregor-Fors, I., Scott MacIvor, J., Jung, K., Piana, M. R., Williams, N. S. G., Knapp, S., Vergnes, A., Acevedo, A. A., Gainsbury, A. M., & Moretti, M. (2023). Urbanisation generates multiple trait syndromes for terrestrial taxa worldwide. *Nature Communications, 14*(4751), 4751. doi:10.1038/s41467-023-39746-1 PMID:37550318

HasuiÉ.MartensenA. C.UezuA.PimentelR.RamosF. N.RibeiroM. C. (2023). Populations across species ranges respond differently to habitat loss and fragment. Research Square (Research Square). doi:10.21203/rs.3.rs-2702017/v1

Hending, D., Randrianarison, H., Andriamavosoloarisoa, N. N. M., Ranohatra-Hending, C., Holderied, M., McCabe, G., & Cotton, S. (2023). Forest fragmentation and its associated edge-effects reduce tree species diversity, size, and structural diversity in Madagascar's transitional forests. *Biodiversity and Conservation, 32*(10), 3329–3353. doi:10.1007/s10531-023-02657-0

Jacob, S., Laurent, E., Haegeman, B., Bertrand, R., Prunier, J. G., Legrand, D., Cote, J., Chaine, A. S., Loreau, M., Clobert, J., & Schtickzelle, N. (2018). Habitat choice meets thermal specialization: Competition with specialists may drive suboptimal habitat preferences in generalists. *Proceedings of the National Academy of Sciences of the United States of America, 115*(47), 11988–11993. doi:10.1073/pnas.1805574115 PMID:30397109

Jiang, L., & O'Neill, B. C. (2017). Global urbanization projections for the Shared Socioeconomic Pathways. *Global Environmental Change, 42*, 193–199. doi:10.1016/j.gloenvcha.2015.03.008

Kalcounis-Rueppell, M. C., Payne, V. H., Huff, S. R., & Boyko, A. L. (2007). Effects of wastewater treatment plant effluent on bat foraging ecology in an urban stream system. *Biological Conservation, 138*(1-2), 120–130. doi:10.1016/j.biocon.2007.04.009

Kareiva, P., & Marvier, M. (2012). What is conservation science? *Bioscience, 62*(11), 962–969. doi:10.1525/bio.2012.62.11.5

Kendal, D., Zeeman, B. J., Ikin, K., Lunt, I. D., McDonnell, M. J., Farrar, A., Pearce, L. M., & Morgan, J. W. (2017). The importance of small urban reserves for plant conservation. *Biological Conservation, 213*, 146–153. doi:10.1016/j.biocon.2017.07.007

Kennedy, C., Pincetl, S., & Bunje, P. (2011). The study of urban metabolism and its applications to urban planning and design. *Environmental Pollution, 159*(8-9), 1965–1973. doi:10.1016/j.envpol.2010.10.022 PMID:21084139

Kowarik, I., & von der Lippe, M. (2018). Plant population success across urban ecosystems: A framework to inform biodiversity conservation in cities. *Journal of Applied Ecology, 55*(5), 2354–2361. doi:10.1111/1365-2664.13144

Lapiedra, O., Chejanovski, Z., & Kolbe, J. J. (2017). Urbanization and biological invasion shape animal personalities. *Global Change Biology, 23*(2), 592–603. doi:10.1111/gcb.13395 PMID:27310334

Leiper, I., Zander, K. K., Robinson, C. J., Carwadine, J., Moggridge, B. J., & Garnett, S. T. (2018). Quantifying current and potential contributions of Australian indigenous peoples to threatened species management. *Conservation Biology, 32*(5), 1038–1047. doi:10.1111/cobi.13178 PMID:30035336

Makwana, V. M., Mahida, D. K., & Dodia, P. P. (2023). Unraveling the Decline: Exploring Global Patterns in Shorebird Populations. In Perspectives on Global Biodiversity Scenarios and Environmental Services in the 21st Century (pp. 192-223). IGI Global. doi:10.4018/978-1-6684-9034-1.ch007

Khatsuriya, P. A., Matli, A. G., & Dodia, P. P. (2024). Composition and Conservation Status of Avifauna in Urban Non-protected Important Bird Area (IBA) Site of Western India. *Journal of Tropical Biodiversity and Biotechnology*, *9*(2), 87311. doi:10.22146/jtbb.87311

McDonald, R. I., Mansur, A. V., Ascensão, F., Colbert, M. L., Crossman, K., Elmqvist, T., Gonzalez, A., Güneralp, B., Haase, D., Hamann, M., Hillel, O., Huang, K., Kahnt, B., Maddox, D., Pacheco, A., Pereira, H. M., Seto, K. C., Simkin, R., Walsh, B., & Ziter, C. (2020). Research gaps in knowledge of the impact of urban growth on biodiversity. *Nature Sustainability*, *3*(1), 16–24. doi:10.1038/s41893-019-0436-6

McKinney, M. L. (2008). Effects of urbanization on species richness: A review of plants and animals. *Urban Ecosystems*, *11*(2), 161–176. doi:10.1007/s11252-007-0045-4

Mezőfi, L., Markó, V., Taranyi, D. Á., & Markó, G. (2023). Sex-specific life-history strategies among immature jumping spiders: Differences in body parameters and behavior. *Current Zoology*, *69*(5), 535–551. doi:10.1093/cz/zoac069 PMID:37637309

Montgomery, M. R., Stren, R., Cohen, B., & Reed, H. E. (2013). *Cities transformed: demographic change and its implications in the developing world*. Routledge. doi:10.4324/9781315065700

Moore, J., Kissinger, M., & Rees, W. E. (2013). An urban metabolism and ecological footprint assessment of Metro Vancouver. *Journal of Environmental Management*, *124*, 51–61. doi:10.1016/j.jenvman.2013.03.009 PMID:23603775

Newbold, T., Hudson, L. N., Hill, S. L., Contu, S., Lysenko, I., Senior, R. A., Börger, L., Bennett, D. J., Choimes, A., Collen, B., Day, J., De Palma, A., Díaz, S., Echeverria-Londoño, S., Edgar, M. J., Feldman, A., Garon, M., Harrison, M. L. K., Alhusseini, T., & Purvis, A. (2015). Global effects of land use on local terrestrial biodiversity. *Nature*, *520*(7545), 45–50. doi:10.1038/nature14324 PMID:25832402

Oliver, I., Dorrough, J., Doherty, H., & Andrew, N. R. (2016). Additive and synergistic effects of land cover, land use and climate on insect biodiversity. *Landscape Ecology*, *31*(10), 2415–2431. doi:10.1007/s10980-016-0411-9

Polaina, E., Gonzalez-Suarez, M., & Revilla, E. (2015). Socioeconomic correlates of global mammalian conservation status. *Ecosphere*, *6*(9), 1–34. doi:10.1890/ES14-00505.1

Rech, F., Narimanov, N., Bauer, T., & Schirmel, J. (2022). Urbanization increases fluctuating asymmetry and affects behavioral traits of a common grasshopper. *Ecology and Evolution*, *12*(12), e9658. doi:10.1002/ece3.9658 PMID:36568863

Richardson, D. M., Pyšek, P., & Carlton, J. T. (2011). A compendium of essential concepts and terminology in invasion ecology. *Fifty years of invasion ecology: the legacy of Charles Elton, 1*, 409-420. Springer. doi:10.1002/9781444329988.ch30

Roodt, A. P., Röder, N., Pietz, S., Kolbenschlag, S., Manfrin, A., Schwenk, K., Bundschuh, M., & Schulz, R. (2022). Emerging midges transport pesticides from aquatic to terrestrial ecosystems: Importance of compound-and organism-specific parameters. *Environmental Science & Technology*, *56*(9), 5478–5488. doi:10.1021/acs.est.1c08079 PMID:35441504

Sagarin, R. D. (2006). Recent studies improve understanding of population dynamics across species ranges. *Oikos*, *115*(2), 386–388. doi:10.1111/j.2006.0030-1299.14438.x

Sanderson, E. W., Walston, J., & Robinson, J. G. (2018). From bottleneck to breakthrough: Urbanization and the future of biodiversity conservation. *Bioscience*, *68*(6), 412–426. doi:10.1093/biosci/biy039 PMID:29867252

Santana Marques, P., Resende Manna, L., Clara Frauendorf, T., Zandonà, E., Mazzoni, R., & El-Sabaawi, R. (2020). Urbanization can increase the invasive potential of alien species. *Journal of Animal Ecology*, *89*(10), 2345–2355. doi:10.1111/1365-2656.13293 PMID:32627190

Seto, K. C., & Ramankutty, N. (2016). Hidden linkages between urbanization and food systems. *Science*, *352*(6288), 943–945. doi:10.1126/science.aaf7439 PMID:27199419

Shik, J. Z., & Dussutour, A. (2020). Nutritional dimensions of invasive success. *Trends in Ecology & Evolution*, *35*(8), 691–703. doi:10.1016/j.tree.2020.03.009 PMID:32668214

Shochat, E., Lerman, S. B., Anderies, J. M., Warren, P. S., Faeth, S. H., & Nilon, C. H. (2010). Invasion, competition, and biodiversity loss in urban ecosystems. *Bioscience*, *60*(3), 199–208. doi:10.1525/bio.2010.60.3.6

Smith, B. T., Seeholzer, G. F., Harvey, M. G., Cuervo, A. M., & Brumfield, R. T. (2017). A latitudinal phylogeographic diversity gradient in birds. *PLoS Biology*, *15*(4), e2001073. doi:10.1371/journal.pbio.2001073 PMID:28406905

Spotswood, E. N., Beller, E. E., Grossinger, R., Grenier, J. L., Heller, N. E., & Aronson, M. F. (2021). The biological deserts fallacy: Cities in their landscapes contribute more than we think to regional biodiversity. *Bioscience*, *71*(2), 148–160. doi:10.1093/biosci/biaa155 PMID:33613128

Tang, H. (2022). The Impact of urbanization on river channel pollution. *Highlights in Science. Engineering and Technology*, *17*, 51–57. doi:10.54097/hset.v17i.2435

Threlfall, C. G., Law, B., & Banks, P. B. (2012). Influence of landscape structure and human modifications on insect biomass and bat foraging activity in an urban landscape. *PLoS One*, *7*(6), e38800. doi:10.1371/journal.pone.0038800 PMID:22685608

Travis, J. M., Delgado, M., Bocedi, G., Baguette, M., Bartoń, K., Bonte, D., Boulangeat, I., Hodgson, J. A., Kubisch, A., Penteriani, V., Saastamoinen, M., Stevens, V. M., & Bullock, J. M. (2013). Dispersal and species' responses to climate change. *Oikos*, *122*(11), 1532–1540. doi:10.1111/j.1600-0706.2013.00399.x

Trivedi, D., Makwana, V. M., Shukla, A. H., & Dodia, P. P. (2022). Diversity of butterflies in Victoria Park Reserve Forest, Bhavnagar, Gujarat, India. *Notulae Scientia Biologicae*, *14*(3), 11293–11293. doi:10.55779/nsb14311293

Williams, J. J., & Newbold, T. (2021). Vertebrate responses to human land use are influenced by their proximity to climatic tolerance limits. *Diversity & Distributions*, *27*(7), 1308–1323. doi:10.1111/ddi.13282

ADDITIONAL READING

Kowarik, I. (2011). Novel urban ecosystems, biodiversity, and conservation. *Environmental Pollution*, *159*(8-9), 1974–1983. doi:10.1016/j.envpol.2011.02.022 PMID:21435761

Nagendra, H., Sudhira, H. S., Katti, M., Tengö, M., & Schewenius, M. (2014). *Urbanization and its impacts on land use, biodiversity and ecosystems in India.*

Vaz, S., Manes, S., Khattar, G., Mendes, M., Silveira, L., Mendes, E., de Morais Rodrigues, E., Gama-Maia, D., Lorini, M. L., Macedo, M., & Paiva, P. C. (2023). Global meta-analysis of urbanization stressors on insect abundance, richness, and traits. *The Science of the Total Environment*, *903*, 165967. doi:10.1016/j.scitotenv.2023.165967 PMID:37543317

KEY TERMS AND DEFINITIONS

Biodiversity: Biodiversity refers to the variety and variability of life forms within ecosystems, encompassing species diversity, genetic diversity, and ecosystem diversity.

Ecosystem Resilience: Ecosystem resilience is the capacity of an ecosystem to recover and adapt to disturbances while maintaining its structure, function, and ability to provide ecosystem services.

Habitat Fragmentation: Habitat fragmentation involves the division of large, continuous habitats into smaller, isolated patches, disrupting ecological connectivity and biodiversity. Habitat restoration: Habitat restoration is the intentional effort to rehabilitate degraded ecosystems, aiming to return them to a more natural and functional state.

Urban Ecology: Urban ecology is the study of the structure, function, and interactions of ecosystems within urban areas, exploring how human activities influence biodiversity, ecosystem processes, and environmental quality.

Urban Environments: Urban environments are human-dominated landscapes characterized by dense populations, built infrastructure, and a complex interplay between social, economic, and natural factors.

Urbanization: Urbanization is the process of increasing urban population density and infrastructure development, often leading to the conversion of natural landscapes into built-up areas.

Chapter 15
Methodology and Approaches for Biodiversity Conservation

Sanjeevi R.
Nims Institute of Allied Medical Science and Technology, India

Prashantkumar Sathvara
https://orcid.org/0000-0002-3548-5584
Nims Institute of Allied Medical Science and Technology, India

Sandeep Tripathi
Nims Institute of Allied Medical Science and Technology, India

Anuradha J.
Nims Institute of Allied Medical Science and Technology, India

ABSTRACT

"Harmony in Nature" offers a thorough examination of biodiversity conservation methods. It serves as a guide for scientists, policymakers, and enthusiasts. Beginning with the urgent need for conservation, it emphasizes ecosystems' role in sustaining life. The book covers traditional and modern assessment methods, including field studies and advanced technologies like satellite imagery. It stresses integrating indigenous knowledge, citizen science, and community-based approaches. Adaptive management, policy frameworks, and global collaborations are discussed. Case studies illustrate successful initiatives and lessons learned. Ethical considerations and sustainability are integrated throughout. The book concludes with a call to action for global biodiversity preservation. It aims to be a valuable resource for academia, practitioners, and policymakers, fostering understanding of essential methodologies for successful conservation amidst environmental challenges.

DOI: 10.4018/979-8-3693-6950-0.ch015

INTRODUCTION

Biodiversity conservation is important for many reasons. It plays a crucial role in maintaining the balance of ecosystems, which in turn provide various services such as pollination, seed dispersal, climate regulation, and water purification. Biodiversity also has utilitarian values as it provides resources such as food, fuel, shelter, and medicine (Ferreira Da Silva *et al.*, 2024). Additionally, it holds potential benefits not yet recognized, such as new medicines and other possible unknown services. From a cultural perspective, biodiversity has value for spiritual or religious reasons. It also has intrinsic value, which refers to its inherent worth, independent of its value to anyone or anything else (Guilherme *et al.*, 2024). This is sometimes thought of as the inalienable right to exist.

Biodiversity is on the decline, with global populations of mammals, birds, fish, reptiles, and amphibians decreasing by 69% since 1970. This is due to human activities such as habitat loss and fragmentation, unsustainable resource use, invasive species, pollution, and global climate change. The loss of biodiversity can lead to a decrease in the stability and resilience of ecosystems, affecting the services they provide(Prashantkumar B. Sathvara *et al.*, 2023). Conserving biodiversity can also lead to economic benefits. For instance, industries such as forestry, agriculture, and medicine rely on the wealth of plant and animal life. By protecting biodiversity, we can ensure the sustainability of these industries and the jobs they provide (Lamounier *et al.*, 2024). Furthermore, investing in biodiversity conservation can create purpose-driven, fulfilling jobs that benefit both the economy and the environment. Therefore, biodiversity conservation is not just an environmental issue, but a societal one, affecting various aspects of our lives. It is essential for our survival and well-being, and it is our collective responsibility to protect and conserve it for future generations.

Role of Diverse Ecosystems in Sustaining Life and Resilience

Diverse ecosystems play a critical role in sustaining life and resilience for biodiversity conservation. They provide a range of services that support human well-being and the survival of other species. These services include air and water purification, soil formation and nutrient cycling, climate regulation, and the provision of food, fuel, and fiber. Ecosystems with high biodiversity tend to be more resilient to disturbances and stressors, such as climate change, disease, and extreme weather events (Kéfi *et al.*, 2024). This is because a diverse range of species with different traits and functions can respond differently to environmental changes, allowing the ecosystem to continue functioning even if some species are lost.

Moreover, diverse ecosystems can provide insurance against the failure of individual species or functional groups. For example, if a dominant tree species in a forest is affected by a pest outbreak or disease, other tree species may be able to take its place and maintain the structure and function of the forest. In addition, biodiversity is essential for maintaining the stability and productivity of ecosystems (Estoque & Wu, 2024). Different species perform different roles in ecosystems, such as pollination, seed dispersal, and nutrient cycling. The loss of even a single species can have cascading effects on the entire ecosystem, potentially leading to declines in ecosystem services and the collapse of the system (Caplan *et al.*, 2024). Therefore, conserving diverse ecosystems is essential for sustaining life and resilience for biodiversity conservation. By protecting and restoring ecosystems with high biodiversity, we can ensure the provision of essential ecosystem services, maintain the stability and productivity of ecosystems, and enhance the resilience of ecosystems to environmental changes.

Need for Effective Methodologies and Approaches

Effective methodologies and approaches for biodiversity conservation are crucial for preserving the variety of life on Earth and maintaining healthy ecosystems. These methods can be broadly categorized into in-situ and ex-situ conservation. In situ conservation involves preserving species in their natural habitats, such as national parks, wildlife sanctuaries, and biosphere reserves. This approach has several advantages, including preserving species and their natural habitats, protecting a large number of populations, being cost-effective and convenient, and allowing species to evolve and adapt to environmental conditions (Yokutkhon, 2024).

Ex-situ conservation, on the other hand, involves conserving species outside of their natural habitats, such as in zoos, botanical gardens, and gene banks. This approach is useful for preserving endangered species and can be used to breed and maintain populations until they can be reintroduced into the wild (Subasman, 2024). In addition to these methods, technology can also play a crucial role in biodiversity conservation. For example, camera traps, drones, and remote sensing applications can be used to monitor and manage protected areas more efficiently (Ergashevich & Mado, 2024). Biotechnologies such as tissue culture, transgenic plants, and genomics can also be used to defend, conserve, and propagate biodiversity.

It is important to note that effective biodiversity conservation requires a holistic approach that considers the interdependence of species and ecosystems. This includes addressing the underlying causes of biodiversity loss, such as pollution, habitat destruction, and climate change, and promoting sustainable practices in agriculture, forestry, and fisheries. Effective methodologies and approaches for biodiversity conservation include in situ and ex-situ conservation methods, the use of technology, and a holistic approach that considers the interdependence of species and ecosystems (ALİSOY, 2024). By implementing these methods, we can ensure the long-term sustainability of biodiversity and the ecosystem services it provides.

TRADITIONAL METHODOLOGIES:

Field Studies: Surveys, Monitoring, and Inventories

Traditional methodologies for biodiversity conservation include field studies such as surveys, monitoring, and inventories. Surveys involve collecting data on the presence and abundance of species in a particular area. Monitoring involves tracking changes in species populations or ecosystems over time (Sanjeevi *et al.*, 2017; Xie *et al.*, 2024). Inventories involve creating a comprehensive list of all the species in a particular area.

These methods are important for understanding the distribution and abundance of species, as well as the health and functioning of ecosystems. By conducting surveys, monitoring, and inventories, conservationists can identify areas of high biodiversity, track changes in species populations or ecosystems, and prioritize areas for conservation (Velasquez-Camacho *et al.*, 2024). These methods can be used in a variety of ecosystems, from forests and grasslands to oceans and freshwater systems. They can also be used to study a wide range of species, from plants and animals to fungi and microorganisms. In addition to traditional field studies, there are also other methodologies for biodiversity conservation, such as ex-situ conservation (e.g., seed banks, botanical gardens, and zoos) and in situ conservation (e.g., protected areas and wildlife corridors) (Mirski *et al.*, 2024). However, field studies such as surveys, monitoring, and inventories remain important tools for understanding and conserving biodiversity. It is important to note that these methods should be conducted in a scientifically rigorous and ethical manner, with consideration for the potential impacts on species and ecosystems (Hartig *et al.*, 2024; Sanjeevi, 2011). This includes obtaining appropriate permits, following best practices for data collection and analysis, and engaging with local communities and stakeholders.

Traditional methodologies for biodiversity conservation include field studies such as surveys, monitoring, and inventories. These methods are important for understanding the distribution and abundance of species, as well as the health and functioning of ecosystems (Hårding *et al.*, 2024). By conducting these studies in a scientifically rigorous and ethical manner, conservationists can identify areas of high biodiversity, track changes in species populations or ecosystems, and prioritize areas for conservation.

Table 1. Methodology for biodiversity conservation

Method Name	Details	Reference
Protected Areas	Designated areas are set aside to conserve biodiversity, including national parks, wildlife reserves, and marine protected areas.	(Stolton *et al.*, 2013)
Habitat Restoration	Efforts to restore degraded habitats to their original state or to create new habitats to support biodiversity.	(Clewell & Aronson, 2006)
Sustainable Land Management	Practices that promote the sustainable use of land resources while conserving biodiversity, such as agroforestry, rotational grazing, and conservation agriculture.	(Dorothea, n.d.)
Community-based Conservation	Involving local communities in biodiversity conservation efforts, often through participatory decision-making, resource management, and sustainable livelihoods.	(Berkes, 2004)
Species Reintroduction	Reintroducing species into their native habitats to restore populations and ecosystems, often after they have been extirpated or endangered.	(Seddon *et al.*, 2014)
Biodiversity Monitoring and Research	Monitoring changes in biodiversity over time and researching to better understand biodiversity patterns, threats, and conservation strategies.	(Lindenmayer & Likens, 2009)

Indigenous Knowledge: Traditional Ecological Knowledge and Practices

Indigenous knowledge, also known as traditional ecological knowledge and practices, refers to the accumulated body of knowledge, beliefs, and practices about relationships between living beings in a specific ecosystem. This knowledge is developed over centuries through direct contact with the environment and handed down through generations. It includes the understanding of relationships between people, plants, animals, natural phenomena, landscapes, and the timing of events for various activities (Berkes *et al.*, 1994). Indigenous knowledge is different from user knowledge and local knowledge, as it undergoes a process of evaluation and validation through time-testing and sharing among community members.

Traditional ecological knowledge (TEK) is a subset of indigenous knowledge that focuses on the relationship of living beings with their traditional groups and environment. TEK is often preserved as an oral tradition and used in natural resource management, either as a substitute for or complement to Western scientific methods (Abbasi *et al.*, 2013; Charnley *et al.*, 2007). The application of TEK in ecological management and science is still a subject of debate due to differences in methods of acquiring and collecting knowledge.

Indigenous knowledge and TEK are crucial for understanding the impacts of environmental changes within ecosystems and for maintaining resources necessary for survival. The earliest systematic studies of TEK were conducted in anthropology, focusing on the conceptions of ecological relationships held by diverse indigenous peoples (Higgins, 1998). By the mid-1980s, the potential applications of traditional ecological knowledge in resource management practices and sustainable development gained international recognition. The United Nations Framework Convention on Climate Change (UNFCCC) recognizes the importance of indigenous knowledge in addressing climate change. The Convention on Biological Diversity (CBD) also acknowledges the role of traditional knowledge in the conservation and sustainable use of biological resources (Joa *et al.*, 2018). The Intergovernmental Panel on Climate Change (IPCC) has also highlighted the importance of traditional knowledge in understanding and adapting to climate change.

However, the integration of traditional ecological knowledge into modern scientific research and policy-making is not without challenges. These include issues related to intellectual property rights, cultural sensitivity, and the need for validation and documentation of traditional knowledge (Rao *et al.*, 2003; Sathvara *et al.*, 2023). Despite these challenges, the integration of traditional ecological knowledge into modern scientific research and policy-making is essential for sustainable development and environmental conservation.

Citizen Science: Engaging the Public in Biodiversity Conservation

Citizen science is a methodology that involves the public in scientific research, typically in the context of biodiversity conservation. It empowers individuals from diverse backgrounds to contribute meaningfully to ecological knowledge by participating in data collection, analysis, and dissemination (WEIS *et al.*, n.d.). By collaborating with naturalists, scientists, and trained volunteers, citizen science helps monitor indicators necessary to assess the state of biodiversity across the planet, such as phenology, migratory behavior, Bioindicators species, population age structure, and species distributions (Ankitkumar B Rathod *et al.*, 2022; Tiago *et al.*, 2024).

Citizen science programs, like the one run by Earthwise Aware (EwA), support biodiversity and climate research while advancing the democratization of science. EwA's program focuses on whole ecosystems, recognizing the interconnectedness of various species and environmental factors. This holistic perspective provides a more comprehensive understanding of the delicate balance within ecosystems. EwA's citizen science model has been successful in collecting over 405,220 dedicated phonological observations, 211,980 biodiversity records, and 17,113 habitat indicators since 2018. The data is aggregated into open and global databases, driving societal change through comprehensive annual reports and contributing critical information to local, national, and global projects (Ruck *et al.*, 2024).

Participants in EwA's programs gain a deeper understanding of their local ecosystems, fostering a sense of environmental stewardship. By combining education, advocacy, and community involvement, citizen science programs strive to create a network of informed and inspired individuals working collectively toward a sustainable and biodiverse future. Citizen science has been recognized as a valuable tool for biodiversity conservation by various international organizations, including the United Nations Educational, Scientific and Cultural Organization (UNESCO), the Convention on Biological Diversity (CBD), and the Inter-governmental Science-Policy Platform on Biodiversity and Ecosystem Services (IPBES) (Rivera *et al.*, 2024). Citizen science has been recognized as a valuable tool for biodiversity conservation by various international organizations, The European Commission has also acknowledged the including the United Nations Educational, Scientific and Cultural Organization (UNESCO), the Convention on Biological Diversity (CBD), and the Intergovernmental Science-Policy Platform on Biodiversity and Ecosystem Services (IPBES). The European Commission has also acknowledged the importance of citizen science in its Horizon 2020 research on the 2020 research and innovation program.

Despite its potential, citizen science faces challenges such as data quality, standardization, and the need for training and support for innovation program participants. However, with proper guidance and collaboration, citizen science can significantly contribute to biodiversity conservation and environmental education (Marchante *et al.*, 2024). Despite its potential, citizen science faces challenges such as data quality, standardization, and the need for training and support for participants. However, with proper guidance and collaboration, citizen science can significantly contribute to biodiversity conservation and environmental education.

Advanced Methodologies for Biodiversity Conservation

Molecular techniques have revolutionized the field of biodiversity conservation, providing new tools for species identification, genetic diversity analysis, and understanding evolutionary relationships. Two prominent molecular methods are DNA barcoding and genetic diversity analysis. DNA barcoding is a method that uses a short, standardized region of DNA to identify and distinguish species (Mishra *et al.*, 2016; Toivonen *et al.*, 2019). This technique relies on the fact that each species has a unique DNA barcode sequence, allowing for rapid and accurate identification of organisms. DNA barcoding has numerous applications in biodiversity conservation, including species identification in food safety, forensics, and Biomonitoring (Swanson & Ardoin, 2021). It also helps in the discovery of new species and the identification of cryptic species, which are morphologically similar but genetically distinct.

Genetic diversity analysis is another molecular technique that assesses the genetic variation within and among populations. This information is crucial for understanding the evolutionary history of species, identifying conservation priorities, and managing genetic resources. Genetic diversity analysis can reveal population structure, gene flow, and the presence of adaptive genetic variation, which are essential factors in designing effective conservation strategies (Sequeira *et al.*, 2018).

Molecular techniques have also been applied to the study of endangered species, such as the Sumatran rhinoceros, to assess their genetic diversity and inform captive breeding programs. By analyzing DNA samples from various individuals, conservationists can identify the most genetically diverse populations and prioritize them for protection (Ankitkumar B Rathod *et al.*, 2023; Pathak & Abido, 2014).

The International Barcode of Life (iBOL) project is a global initiative that aims to create a DNA barcode reference library for all known species. This resource will significantly enhance biodiversity conservation efforts by providing a powerful tool for species identification and monitoring. However, molecular techniques also have limitations. DNA barcoding, for example, may not always be effective in distinguishing closely related species or species complexes. Genetic diversity analysis can be expensive and time-consuming, requiring specialized equipment and expertise. Additionally, the interpretation of molecular data can be challenging, and the results may not always be straightforward. Despite these challenges, molecular techniques have become essential tools in biodiversity conservation, providing valuable insights into species identification, genetic diversity, and evolutionary relationships (Pathak & Abido, 2014). As technology advances and costs decrease, these methods will continue to play a critical role in the conservation and management of Earth's biodiversity.

Molecular techniques such as DNA barcoding and genetic diversity analysis have significantly advanced the field of biodiversity conservation. These methods provide new tools for species identification, genetic diversity analysis, and understanding evolutionary relationships, which are crucial for effective conservation strategies. While molecular techniques have limitations, they have become indispensable in the study and protection of diversity analysis and have significantly advanced the field of biodiversity conservation (Hilton *et al.*, 2022). These methods provide new tools for species identification, genetic diversity analysis, and understanding evolutionary relationships, which are crucial for effective conservation strategies. While molecular techniques have limitations, they have become Indian diversity analysis and have significantly advanced is pens able in the field of biodiversity conservation. These methods provide new tools for species identification, genetic diversity analysis, and understanding evolutionary relationships, which are crucial for effective conservation strategies. While molecular techniques have limitations, they have become indispensable in the study and protection of Earth's biodiversity.

Study and protection of Earth's biodiversity. Diversity analysis has significantly advanced the field of biodiversity conservation (Krishnamurthy, 2018; Ranjan *et al.*, n.d.). These methods provide new tools for species identification, genetic diversity analysis, and understanding evolutionary relationships, which are crucial for effective conservation strategies.

Satellite Imagery and Big Data Analytics Utilization in Biodiversity Mapping

Advanced technologies, such as satellite imagery and big data analytics, play a crucial role in modern biodiversity conservation. Satellite imagery provides up-to-date information on land use, deforestation, habitat fragmentation, and other environmental changes. High-resolution satellite imagery can reveal details about ecosystems, such as vegetation health, water bodies, and human impacts (Khare & Ghosh, 2016). By monitoring these changes over time, conservationists can identify trends, assess the effectiveness of conservation measures, and develop data-driven strategies for biodiversity protection.

Big data analytics allows for the integration and analysis of diverse datasets, enabling the identification of patterns, correlations, and trends that might be difficult to discern through traditional methods. Machine learning algorithms and artificial intelligence can be applied to satellite imagery and other geospatial data to predict species distributions, model habitat suitability, and assess the impacts of climate change on biodiversity (Yang *et al.*, 2019). These advanced technologies offer valuable insights into biodiversity conservation, enabling conservationists to monitor environmental changes, assess the effectiveness of conservation measures, and develop data-driven strategies for protecting Earth's biodiversity (Abbasi *et al.*, 2016; K. Wang *et al.*, 2010).

EOS Data Analytics (EOSDA) is an example of leveraging satellite imagery and big data companies utilizing analytics, conservation satellite organizations imagery, and big data analytics for biodiversity conservation. Informed EOSDA offers satellite decisions, allocates resources efficiently, and measures the impact of their conservation efforts. EOS Data Anal-driven solutions for agriculture, forestry, and mining, providing insights that help tics (EOSDA) is an example of a company utilizing satellite imagery and big data analytics for biodiversity conservation. EOSDA offers satellite businesses-driven solutions for agriculture, forestry, and mining, providing insights that help businesses optimize resource management, optimize resource management, improve crop productivity, and safe improve guard the environment (Pettorelli *et al.*, 2014).

Satellite imagery and by leveraging satellite conservation measures, and develop data-driven strategies for protecting Earth's biodiversity. Big data analytics are advanced methodologies that offer valuable insights for biodiversity conservation. By leveraging these technologies, conservationists can monitor environmental changes, assess the effectiveness of conservation measures, and develop data-driven strategies for protecting Earth's biodiversity. With imagery and big data analytics, conservation organizations can make informed decisions, optimize resource management, and prioritize conservation efforts to safeguard the planet's rich biodiversity. EOS Data Analytics (EOSDA) is an example of a company utilizing satellite imagery and big data analytics for biodiversity conservation. EOSDA offers satellite-driven solutions for agriculture, forestry, and mining, providing insights that help businesses optimize resource management, improve crop productivity, and safeguard the environment. Satellite imagery and big data analytics are advanced methodologies that offer valuable insights into biodiversity conservation (Salazar Loor & Fdez-Arroyabe, 2019). By leveraging these technologies, conservationists can monitor environmental changes, assess the effectiveness of conservation measures, and develop data-driven strategies for protecting Earth's biodiversity.

Integration of Methodologies for Biodiversity Analysis

Integration of indigenous knowledge and citizen science initiatives

Integrating indigenous knowledge and citizen science initiatives in biodiversity conservation can lead to more comprehensive and effective strategies. Indigenous knowledge, which is based on generations of experience and problem-solving by local communities, can complement scientific research by providing insights into how communities have interacted with their environment, including its floral and faunal resources (Tengö *et al.*, 2021). This knowledge can help identify traditional mechanisms for conserving and sustaining natural resources, as well as establish community-based organizations for addressing environmental issues.

Citizen science, on the other hand, engages the public in biodiversity conservation by involving them in data collection, analysis, and dissemination. By combining these two approaches, we can leverage the strengths of both local and scientific knowledge to enhance our understanding of biodiversity and develop more targeted and culturally sensitive conservation strategies.

Integrate indigenous knowledge and citizen science initiatives:

1. Engage with indigenous communities to understand their traditional ecological knowledge and practices. This can be done through workshops, interviews, or participatory research methods. By involving local communities in the research process, we can ensure that their knowledge is respected and valued and that conservation strategies are culturally sensitive and sustainable.

2. Encourage indigenous communities to participate in citizen science initiatives related to biodiversity conservation. This can help validate and document traditional ecological knowledge, while also providing opportunities for local communities to learn new skills and contribute to scientific research.
3. Satellite imagery and big data analytics can be used to monitor environmental changes, assess the effectiveness of conservation measures, and develop data-driven strategies for protecting biodiversity. By making this data accessible to indigenous communities and citizen scientists, we can empower them to participate in conservation efforts and make informed decisions.
4. Create regional and national centers for documenting the historical and contemporary indigenous knowledge of various ethnic groups. These centers can help protect the intellectual property rights of indigenous knowledge, provide a protective barrier for the knowledge that could be exploited economically, and lay the foundation for development activities that build on and strengthen the existing knowledge and organizational base produced through many generations of creative effort by local communities.

By integrating indigenous knowledge and citizen science initiatives, we can create a more holistic and inclusive approach to biodiversity conservation that recognizes the value of local and traditional knowledge, while also leveraging the power of scientific research and advanced technologies.

Integrating indigenous knowledge and citizen science initiatives in biodiversity conservation can lead to more comprehensive and effective strategies. By collaborating with local communities, incorporating citizen science in indigenous knowledge studies, using advanced technologies to support both indigenous knowledge and citizen science, and establishing indigenous knowledge resource centers, we can create a more holistic and inclusive approach to biodiversity conservation that recognizes the value of local and traditional knowledge, while also leveraging the power of scientific research and advanced technologies (Chandler *et al.*, 2017). A more holistic and inclusive approach to biodiversity conservation that recognizes the value of local and traditional knowledge, while also leveraging the power of scientific research and advanced technologies.

Adaptive Management Strategies: Learning From Past Experiences

Adaptive management strategies involve learning from past experiences and continuously updating management approaches based on new information and evolving circumstances. This approach recognizes the inherent uncertainties in managing complex systems and emphasizes the importance of social learning, collaboration, and flexibility. Adaptive management is not a passive process, but rather an active one that involves regularly reviewing information, developing multiple models of ecosystem responses, and choosing management options based on evaluations of these alternatives.

Adaptive management can help reduce decision-making gridlock by making it clear that decisions are provisional and that modifications are expected. It encourages stakeholders to bind disputes and discuss them in an orderly fashion while environmental uncertainties are being investigated and better understood (Rist *et al.*, 2013). By fostering a culture of continuous learning and improvement, adaptive management can help organizations and managers respond to new information and changing stakeholder objectives and preferences on time.

Incorporating indigenous knowledge and citizen science initiatives into adaptive management strategies can further enhance the effectiveness of these approaches. Indigenous knowledge can provide valuable insights into the behavior of ecosystems that are utilized by humans, drawing upon theories from ecosystem sciences, economics and social sciences, engineering, and other disciplines (Schreiber *et al.*, 2004). Citizen science can engage the public in biodiversity conservation, contribute to data collection and analysis, and promote social learning and collaboration. By integrating adaptive management strategies with indigenous knowledge and citizen science initiatives, organizations and managers can create a more holistic and inclusive approach to natural resources management that recognizes the value of local and traditional knowledge, while also leveraging the power of scientific research and advanced technologies.

Policy Frameworks: Legal and Regulatory Frameworks for Conservation

Policy frameworks for biodiversity conservation involve legal and regulatory frameworks that provide a structure for the protection and management of biodiversity. These frameworks can include national laws, international treaties, and regulations that address issues such as habitat conservation, species protection, and sustainable use of natural resources.

One important legal framework for biodiversity conservation is the Convention on Biological Diversity (CBD), which was adopted in 1992 and has been ratified by nearly all countries in the world. The CBD sets out three main objectives: the conservation of biological diversity, the sustainable use of its components, and the fair and equitable sharing of benefits arising from the use of genetic resources (Meretsky *et al.*, 2011). The CBD has led to the development of several important regulatory frameworks, including the Aichi Biodiversity Targets, which were adopted in 2010 and aimed to significantly reduce biodiversity loss by 2020. Although the Aichi

Targets were not fully achieved, they provided an important roadmap for biodiversity conservation and have been influential in shaping national and international policies.

In 2022, the CBD adopted the Kunming-Montreal Global Biodiversity Framework, which sets out a new set of targets for biodiversity conservation for the period up to 2030. The framework includes targets related to the protection of ecosystems, the restoration of degraded ecosystems, and the sustainable use of genetic resources. In addition to the CBD, there are other important legal and regulatory frameworks for biodiversity conservation. For example, the Convention on International Trade in Endangered Species of Wild Fauna and Flora (CITES) regulates international trade in endangered species, while the Ramsar Convention on Wetlands of International Importance provides a framework for the conservation and wise use of wetlands (McDonald *et al.*, 2019). Effective implementation of these legal and regulatory frameworks requires collaboration and coordination between governments, civil society organizations, and local communities. By working together, we can ensure that biodiversity is protected and conserved for future generations.

Importance of International Collaborations in Biodiversity Conservation

International collaborations are crucial for biodiversity conservation as they enable countries to work together to address shared challenges and opportunities. Biodiversity loss is a global problem that requires global solutions, and international collaborations can help to facilitate the exchange of knowledge, expertise, and resources needed to protect and conserve biodiversity. One important aspect of international collaborations is the sharing of data and information (Westley & Vredenburg, 1997). By sharing data on biodiversity trends, threats, and conservation efforts, countries can better understand the state of biodiversity and develop more effective conservation strategies. International collaborations can also help to build capacity in countries with limited resources, by providing training and technical assistance to support conservation efforts. International collaborations can also help to address transboundary conservation challenges, such as the protection of migratory species and the conservation of shared ecosystems. For example, the Convention on Migratory Species (CMS) is an international treaty that aims to conserve migratory species and their habitats, while the Ramsar Convention on Wetlands of International Importance provides a framework for the conservation and wise use of wetlands that cross national boundaries.

In addition to formal international agreements, many informal networks and partnerships support biodiversity conservation. For example, the International Union for Conservation of Nature (IUCN) is a global organization that brings together governments, civil society organizations, and experts to develop conservation strategies and policies. The IUCN also maintains the Red List of Threatened Species, which is a widely recognized and used system for assessing the conservation status of species (Lacher *et al.*, 2012). International collaborations can also help to mobilize resources for biodiversity conservation. By working together, countries can leverage their resources and expertise to fund conservation projects and initiatives. For example, the Global Environment Facility (GEF) is a multilateral fund that supports projects in biodiversity conservation, climate change mitigation, and other environmental challenges. International collaborations are essential for biodiversity conservation as they enable countries to work together to address shared challenges and opportunities. By sharing data, building capacity, addressing transboundary conservation challenges, and mobilizing resources, international collaborations can help to protect and conserve biodiversity for future generations.

Examples of Successful International Collaborations

International collaborations have played a crucial role in advancing biodiversity conservation efforts around the world. Here are some examples of successful international collaborations:

1. The British American Tobacco Biodiversity Partnership (BATBP is a consortium of four NGOs - Earth Watch Europe, Fauna & Flora International, Royal Botanic Gardens, Kew and Tropical Biology Association - working together with British American Tobacco. The partnership aims to contribute to the conservation of global biodiversity by building a portfolio of activities that assist countries where the Partners operate in meeting their obligations under the Convention on Biological Diversity (CBD).
2. Google Earth Engine and Global Forest Watch (GFW) use satellite imagery and geospatial datasets to help scientists and researchers detect changes and map trends on the Earth's surface. In partnership with GFW, the Earth Engine offers free forest-monitoring support for tropical nations. GFW uses this monitoring to provide timely tree loss updates via Global Land Analysis and Discovery alerts. Deforestation that could take years to discover now happens within weeks, helping forest managers and law enforcement protect endangered forests.
3. The Green Climate Fund (GCF) emerged out of the Paris Agreement as the world's largest climate fund. It draws on public and private finance and collaboration to help emerging markets lower emissions and shift to climate-resilient pathways, as determined by their national climate plans and targets. GCF deploys a range of financing instruments to catalyze climate innovation, test new business models, and crowd in private finance.

4. The World Wildlife Fund (WWF) and Johnson & Johnson's Healing Forests Initiative: This three-year project documents how deforestation can undermine human health and economies and notes what efforts can be used to curb future disease outbreaks. By focusing on Sabah, Malaysia, where palm oil plantations have devastated forests, the initiative hopes to show how the conservation of forests can serve as preventative medicine for us all - fostering proactive remedies to some of the world's most pressing environmental and health concerns.
5. The United States Government has a bold goal to bring online 30,000 megawatts of cleaner and more renewable energy and to connect 60 million homes and businesses in sub-Saharan Africa by 2030. Power Africa is a U.S. Government-led initiative, coordinated by USAID, that brings together technical and legal experts, the private sector, and governments from around the world to work in partnership to increase the number of people with access to power.

Case Studies

Successful conservation initiatives: lessons learned and best practices case studies of successful conservation initiatives with lessons learned and best practices:

1. Tongass Forest Conservation: The Tongass Forest in Southeast Alaska is the largest remaining coastal temperate rainforest on Earth. A partnership between tribal governments, community groups, government agencies, and The Nature Conservancy (TNC) has been successful in developing sustainable economic solutions that support Indigenous stewardship and authority. This partnership has led to the creation of the Seacoast Trust, a fund that supports programs weaving Indigenous values into resource management, ecological revitalization, and economic development. Centering the needs of Indigenous peoples is crucial for securing the future of the Tongass Forest.
2. Coral Reef Restoration: The Caribbean has lost 60% of its iconic coral reefs in just three decades due to pollution, overfishing, and climate change. The Nature Conservancy (TNC) has united some of the world's best coral scientists to nurture a new generation of coral diversity. By collecting coral eggs and sperm during spawning events, TNC has raised 750,000 coral embryos in protected nurseries, increasing genetic diversity and survival chances. Collaborative coral restoration efforts can significantly improve the resilience of coral reefs.
3. River Flow Restoration: Large dams built in the 20th century have disrupted how water flows through countless freshwater ecosystems. The Nature Conservancy (TNC) and the US Army Corps of Engineers have partnered to write unique prescriptions of flows that individual rivers need and rework dam operations to achieve them, restoring over 5,000 miles of river in the US. Collaborative

efforts between conservation organizations and government agencies can effectively restore natural flow patterns in freshwater ecosystems.

4. Sustainable Tuna Fishing: The Marshall Islands control more than 750,000 square miles of ocean, where commercial tuna fishing can unintentionally catch sea turtles, sharks, and juvenile tuna. The Nature Conservancy (TNC) and the government of the Marshall Islands launched Pacific Island Tuna, a joint venture that helps ensure sustainable catches and benefits local Pacific Island communities. Collaborative efforts between conservation organizations and local governments can promote sustainable fishing practices and benefit local communities.
5. Restoring the Everglades: The Everglades is a unique ecosystem that has been severely impacted by decades of development and water diversion. The Comprehensive Everglades Restoration Plan (CERP) is a 30-year, $17 billion plan to restore the Everglades. The plan involves building reservoirs, removing barriers, and restoring natural water flow patterns. Large-scale, long-term restoration efforts require significant investment and collaboration between government agencies, conservation organizations, and local communities.

These case studies demonstrate the importance of collaboration, innovation, and long-term commitment to successful conservation initiatives. By learning from these examples, we can apply best practices to future conservation efforts and continue to protect and restore our planet's biodiversity.

Challenges Faced in Conservation Initiatives: Lessons Learned and Strategies for Overcoming Challenges

Conservation initiatives often face various challenges that can hinder their success challenges include:

1. Collaborative efforts between multiple partners and funding sources can be difficult due to different goals, criteria, and standards for monitoring, completion, or success. To overcome this challenge, technical and communications staff across agencies should collaborate to increase coordination and provide targeted outreach materials to landowners.
2. Landowners often require assistance in designing projects, applying for funds, obtaining permits, and conducting on-the-ground work. To address this challenge, organizations can increase coordination between incentive program staff, provide training for groups that work with landowners, and develop more targeted outreach materials.

3. Landowners may find it overwhelming to sort through the available information to find what is relevant to their property. To overcome this challenge, organizations can help landowners find consistent and reliable information by enhancing information sharing among agencies and organizations.
4. Working lands can provide significant value to fish and wildlife habitat, but recognizing and funding this can be difficult. To address this challenge, state programs should work together to increase the options available for landowners to fund conservation and restoration actions while maintaining all or part of the property as a working landscape.
5. Environmental management has increasingly focused on promoting social engagement in biodiversity and ecosystem services conservation. However, a critical challenge that remains is the adequate implementation of strategies of communication, education, and participation (CEPA) oriented to reconnect the social and ecological dimensions in the systems. To overcome this challenge, education strategies should be increased and improved in conservation projects to make decisions based on critical and reflective thinking and foster the engagement of a broader set of stakeholders in the processes.

These challenges, conservation initiatives can be more successful in achieving their goals. Collaborative leadership, increased coordination, and targeted outreach are key strategies for overcoming these challenges and ensuring the success of conservation initiatives.

Ethical Considerations: Respect for Cultural Diversity and Traditional Knowledge

Respect for cultural diversity and traditional knowledge is an essential ethical consideration in biodiversity conservation. Indigenous peoples and local communities have a deep understanding of their local ecosystems and have developed traditional practices and knowledge that can contribute significantly to conservation efforts (Scassa & Taylor, 2017). Ethical considerations for respecting cultural diversity and traditional knowledge in biodiversity conservation:

Traditional knowledge is pivotal for conservation, warranting recognition and integration into strategies. Involving local communities in decision-making ensures culturally relevant conservation. Obtaining consent prior to activities affecting their lands respects their rights. Promoting cultural diversity and indigenous rights fosters inclusive conservation. Providing benefits like employment and resource access sustains both conservation and local development. Avoiding cultural appropriation maintains sensitivity and respect. By honouring cultural diversity and traditional knowledge, initiatives gain local trust and support, fostering sustainability and equity in conservation efforts.

Environmental Justice: Ensuring Equitable Access to and Use of Natural Resources

Environmental justice is an important ethical consideration in biodiversity conservation, as it ensures that all people have equitable access to and use of natural resources (Malin *et al.*, 2019). Promoting environmental justice in biodiversity conservation. Environmental justice necessitates acknowledging and rectifying historical injustices, ensuring equitable access to natural resources. This involves addressing past discrimination, exclusion, and marginalization. Conservation efforts must facilitate fair resource access, benefiting all. This includes subsistence use, recreation, and economic development opportunities. Engaging marginalized communities in conservation decisions ensures equitable outcomes. Tackling power imbalances, like corporate and governmental influence, is essential for environmental justice. Conservation should promote sustainable development, balancing economic growth with resource protection. Community-led conservation, sustainable livelihoods, and education access are integral. Transparency and accountability in conservation initiatives foster effectiveness and inclusivity in decision-making processes.

Environmental justice is an important ethical consideration in biodiversity conservation, as it ensures that all people have equitable access to and use of natural resources (Hill & Targ, 2000). By recognizing and addressing historical injustices, promoting equitable access to natural resources, involving marginalized communities in decision-making, addressing power imbalances, promoting sustainable development, and ensuring transparency and accountability, conservation initiatives can contribute to sustainable development and social equity.

Promoting Sustainable Practices in Biodiversity Conservation

Promoting sustainable practices in biodiversity conservation involves implementing strategies that balance the need for development with the need to protect and preserve biodiversity. This can be achieved through various means,

1. This involves managing natural resources in a way that meets the needs of the present without compromising the ability of future generations to meet their own needs. This can be achieved through practices such as sustainable forestry, sustainable agriculture, and sustainable fisheries.

2. Ecosystem-based approaches involve managing ecosystems as a whole, taking into account the interactions between species and their environment. This can help to maintain the health and resilience of ecosystems, and can also provide benefits to people, such as clean water and air, food, and fiber.
3. Green infrastructure involves the use of natural and semi-natural features, such as parks, green roofs, and wetlands, to provide ecosystem services and improve the quality of life for people. This can help to reduce the impact of development on biodiversity, and can also provide benefits such as flood protection, air quality improvement, and recreation.
4. Capacity building involves providing training and education to individuals and communities to help them understand the importance of biodiversity and how to conserve it. This can help to ensure that biodiversity conservation is integrated into decision-making processes, and can also help to build local capacity for conservation action.
5. Policy and regulatory frameworks involve the development and implementation of policies and regulations that support biodiversity conservation. This can include laws and regulations related to land use, resource management, and protected areas.
6. International collaborations working with other countries and organizations to promote biodiversity conservation at a global scale. This can help to address transboundary conservation challenges, and can also help to leverage resources and expertise to support conservation efforts.

Examples of successful sustainable practices in biodiversity conservation include the development of sustainable forestry practices in the Amazon rainforest, the use of green infrastructure to improve urban biodiversity, and the implementation of ecosystem-based approaches to fisheries management. These initiatives have helped to balance the need for development with the need to protect and preserve biodiversity, and have provided benefits to both people and the environment (W. Wang *et al.*, 2020).

However, promoting sustainable practices in biodiversity conservation can also face challenges, such as lack of political will, limited resources, and conflicting interests. To overcome these challenges, it is important to build partnerships and collaborations, engage stakeholders, and raise awareness about the importance of biodiversity conservation. By promoting sustainable practices, we can help to ensure that biodiversity is protected and preserved for future generations.

Role of Education and Awareness in Fostering Sustainable Practices

Education and awareness play a crucial role in fostering sustainable practices in biodiversity conservation. By educating individuals and raising awareness about the importance of conservation, we can foster a sense of responsibility and encourage positive actions towards protecting and preserving biodiversity.

Education serves as a foundation for promoting sustainable practices in biodiversity conservation. It involves providing knowledge and understanding about the significance of biodiversity, the interconnectedness of ecosystems, and the impacts of human activities on the environment. Through formal education programs, such as school curricula and environmental courses, individuals can learn about sustainable practices and develop the necessary skills to contribute to conservation efforts. Informal education, including public awareness campaigns, workshops, and community-based initiatives, also plays a significant role in promoting sustainable practices. These activities aim to engage and empower individuals, communities, and organizations to take action and make sustainable choices in their daily lives (Leal Filho *et al.*, 2019). By fostering a sense of responsibility and stewardship towards the environment, education can inspire people to become active participants in biodiversity conservation.

Awareness is a key component of sustainable practices in biodiversity conservation as it helps individuals recognize the value and importance of biodiversity, as well as the threats it faces. By increasing awareness, people can develop a sense of connection and empathy towards nature, leading to a greater willingness to protect and conserve it. Effective awareness campaigns utilize various communication channels, such as media, social networks, and public events, to reach a wide audience and disseminate information about sustainable practices in biodiversity conservation (Jain *et al.*, 2013). These campaigns aim to highlight the benefits of sustainable practices, the consequences of unsustainable practices, and the actions individuals can take to contribute to its preservation.

Furthermore, awareness initiatives can also focus on specific issues, such as illegal wildlife trade, habitat destruction, or climate change, to mobilize public support and encourage collective action. By raising awareness about these pressing challenges, individuals can become advocates for change and contribute to the conservation of biodiversity at local, national, and global levels (Hossain *et al.*, 2020). Education and awareness are essential tools in promoting sustainable practices in biodiversity conservation. By providing knowledge, fostering understanding, and promoting responsible actions, education empowers individuals to become proactive agents of change and advocates for sustainability. Awareness campaigns create a sense of connection and inspire collective action towards protecting and preserving biodiversity for future generations.

CONCLUSION

In conclusion, biodiversity conservation is a critical global challenge that requires urgent attention and action. The loss of biodiversity has significant impacts on ecosystems, economies, and human well-being, and we must take immediate steps to address this issue. To achieve this, we need to adopt a holistic and integrated approach that combines traditional and advanced methodologies, international collaborations, ethical considerations, and sustainable practices. This approach should be based on the best available science, traditional knowledge, and the active participation of all stakeholders, including local communities, indigenous peoples, governments, the private sector, and civil society. Education and awareness are essential components of this approach, as they help foster a sense of responsibility and empower individuals to take action towards protecting and preserving biodiversity. By raising awareness about the importance of biodiversity and the threats it faces, we can inspire collective action and mobilize public support for conservation efforts.

Achieving global biodiversity conservation demands fostering a united front among stakeholders toward a shared vision. This alliance, grounded in shared values, promotes collaboration, innovation, and knowledge sharing. We urge all stakeholders to join hands in this endeavour, supporting initiatives, advocating for sustainable practices, and influencing policy frameworks. Together, we can safeguard biodiversity for future generations, ensuring equitable distribution of its benefits. Let's act now for a sustainable future. Biodiversity conservation is urgent and multifaceted, requiring holistic approaches integrating traditional and advanced methodologies, international cooperation, ethics, and sustainability. Education and awareness are pivotal, underlining the need for a unified global alliance for biodiversity conservation. We implore all stakeholders to unite and act decisively for biodiversity, securing its preservation and equitable benefits for all humanity, now and in the future.

In addition, adaptive management strategies, policy frameworks, and international collaborations are crucial in ensuring effective biodiversity conservation. Adaptive management strategies enable conservation efforts to evolve and improve over time, taking into account changing circumstances and new knowledge. Policy frameworks provide a legal and regulatory foundation for conservation efforts, ensuring that they are supported and enforced by relevant authorities. International collaborations facilitate the sharing of knowledge, resources, and expertise across borders, enabling conservation efforts to be more effective and far-reaching.

Adaptive management fosters continuous learning, enhancing conservation's responsiveness to change for greater effectiveness. Policy frameworks establish legal support, vital for enforcement and backing from relevant authorities. International collaborations facilitate knowledge exchange, bolstering conservation's scope and impact globally. Integrating adaptive strategies, policy frameworks, and global collaborations fortifies biodiversity preservation and long-term sustainability. These approaches, prioritized and interwoven, equip conservation to tackle evolving challenges effectively. Establishing a global alliance for biodiversity conservation is imperative, uniting stakeholders under shared values for collaborative, innovative, and knowledge-driven efforts. Together, we ensure equitable sharing of biodiversity's benefits and safeguard it for future generations. This alliance, rooted in a shared vision, fosters cooperation to protect biodiversity and ensure its enduring value to humanity.

REFERENCES

Abbasi, T., Anuradha, J., & Abbasi, S. (2016). Utilization of the pernicious aquatic weed salvinia (Salvinia molesta DS Mitchell) in generating gold nanoparticles. *Indian Journal of Biotechnology*, *15*, 382–391.

Abbasi, T., Sanjeevi, R., Anuradha, J., & Abbasi, S. A. (2013). Impact of Al 3+ on sludge granulation in UASB reactor. *Indian Journal of Biotechnology*, *12*(2), 254–259.

Alisoy, H. (2024). *Effective Strategies in Primary Second Language Education*. Preprints. https://www.preprints.org/manuscript/202401.0330

Ankitkumar, B. (2022). Vegetational Cartography Analysis Utilizing Multi-temporal Ndvi Data Series: A Case Study from Rajkot District (Gujarat), India. *Tianjin Daxue Xuebao, 55*(04), 490–497. https://doi.org/ doi:10.17605/OSF.IO/UGJYM

Ankitkumar, B., Rathod, A., J., Prashantkumar B., Sathvara, Tripathi, S., & Sanjeevi, R. (2023). Vegetational Change Detection Using Machine Learning in GIS Technique: A Case Study from Jamnagar (Gujarat). *Journal of Data Acquisition and Processing*, *38*(1), 1046–1061. doi:10.5281/zenodo.7700655

Berkes, F. (2004). Rethinking Community-Based Conservation. *Conservation Biology*, *18*(3), 621–630. doi:10.1111/j.1523-1739.2004.00077.x

Berkes, F., Folke, C., & Gadgil, M. (1994). Traditional Ecological Knowledge, Biodiversity, Resilience and Sustainability. In C. A. Perrings, K.-G. Mäler, C. Folke, C. S. Holling, & B.-O. Jansson (Eds.), *Biodiversity Conservation: Problems and Policies* (Vol. 4, pp. 269–287). Springer Netherlands. doi:10.1007/978-94-011-1006-8_15

Caplan, T. R., Inaho, B., Yarrao, M., Pile, J., Milczarek, M., & McKenna, C. (2024). Applying resilience thinking to rehabilitating a novel social–ecological system: A case study from the lower Ok Tedi, Papua New Guinea. In *Resilience and Riverine Landscapes* (pp. 593–626). Elsevier. https://www.sciencedirect.com/science/article/pii/B9780323917162000248 doi:10.1016/B978-0-323-91716-2.00024-8

Chandler, M., See, L., Copas, K., Bonde, A. M., López, B. C., Danielsen, F., Legind, J. K., Masinde, S., Miller-Rushing, A. J., Newman, G., Rosemartin, A., & Turak, E. (2017). Contribution of citizen science towards international biodiversity monitoring. *Biological Conservation, 213*, 280–294. doi:10.1016/j.biocon.2016.09.004

Charnley, S., Fischer, A. P., & Jones, E. T. (2007). Integrating traditional and local ecological knowledge into forest biodiversity conservation in the Pacific Northwest. *Forest Ecology and Management, 246*(1), 14–28. doi:10.1016/j.foreco.2007.03.047

Clewell, A. E., & Aronson, J. (2006). Motivations for the restoration of ecosystems. *Conservation Biology, 20*(2), 420–428. doi:10.1111/j.1523-1739.2006.00340.x PMID:16903103

Ergashevich, E. A., & Mado, A. (2024). Methodology of Organizing and Implementing Training Activities. *Academia Open, 9*(1), 10–21070.

Estoque, R. C., & Wu, J. (2024). The resilience–sustainability–quality of life nexus. *The Science of the Total Environment, 912*, 169526. doi:10.1016/j.scitotenv.2023.169526 PMID:38135087

Ferreira Da Silva, M. J., Minhós, T., & Sampaio-Dias, S. (2024). News for nature: Communicating biodiversity importance and conservation challenges in Guinea-Bissau, West Africa, through local media and online platforms. *Conservation Science and Practice, 6*(1), e13051. doi:10.1111/csp2.13051

Guilherme, F., Vicente, J. R., Carretero, M. A., & Farinha-Marques, P. (2024). Mapping multigroup responses to land cover legacy for urban biodiversity conservation. *Biological Conservation, 291*, 110508. doi:10.1016/j.biocon.2024.110508

HårdingK.CarrollD.InfantesE.CarrollJ.HarkonenT. (2024). Long-term field studies are critical for safeguarding biodiversity: A new concept and concern. Authorea Preprints. https://essopenarchive.org/doi/full/10.22541/au.171010042.24317771 doi:10.22541/au.171010042.24317771/v1

Hartig, F., Abrego, N., Bush, A., Chase, J. M., Guillera-Arroita, G., Leibold, M. A., Ovaskainen, O., Pellissier, L., Pichler, M., & Poggiato, G. (2024). Novel community data in ecology-properties and prospects. *Trends in Ecology & Evolution*. Cell.com. https://www.cell.com/trends/ecology-evolution/fulltext/S0169-5347(23)00265-3

Higgins, C. (1998). The role of traditional ecological knowledge in managing for biodiversity. *Forestry Chronicle*, *74*(3), 323–326. doi:10.5558/tfc74323-3

Hill, B. E., & Targ, N. (2000). The link between protecting natural resources and the issue of environmental justice. *BC Envtl. Aff. L. Rev.*, *28*, 1.

Hilton, M., Walsh, J. C., Liddell, E., & Cook, C. N. (2022). Lessons from other disciplines for setting management thresholds for biodiversity conservation. *Conservation Biology*, *36*(1), e13865. doi:10.1111/cobi.13865 PMID:34811813

Hossain, M. I., San, O. T., Ling, S. M., & Said, R. M. (2020). The role of environmental awareness and green technological usage to foster sustainable green practices in Bangladeshi manufacturing SMEs. *International Journal of Advanced Science and Technology*, *29*(7s), 3115–3124.

Jain, S., Aggarwal, P., Sharma, N., & Sharma, P. (2013). Fostering sustainability through education, research and practice: A case study of TERI University. *Journal of Cleaner Production*, *61*, 20–24. doi:10.1016/j.jclepro.2013.04.021

Joa, B., Winkel, G., & Primmer, E. (2018). The unknown known–A review of local ecological knowledge in relation to forest biodiversity conservation. *Land Use Policy*, *79*, 520–530. doi:10.1016/j.landusepol.2018.09.001

Kéfi, S., Génin, A., Garcia-Mayor, A., Guirado, E., Cabral, J. S., Berdugo, M., Guerber, J., Solé, R., & Maestre, F. T. (2024). Self-organization as a mechanism of resilience in dryland ecosystems. *Proceedings of the National Academy of Sciences of the United States of America*, *121*(6), e2305153121. doi:10.1073/pnas.2305153121 PMID:38300860

Khare, S., & Ghosh, S. (2016). Satellite remote sensing technologies for biodiversity monitoring and its conservation. *International Journal of Advanced Earth Science and Engineering*, *5*(1), 375–389. doi:10.23953/cloud.ijaese.213

Krishnamurthy, K. V. (2018). *An advanced textbook on biodiversity: Principles and practice*. Oxford and IBH Publishing. https://books.google.com/books?hl=en&lr=&id=2Vczdr9ysZYC&oi=fnd&pg=PA1&dq=Advanced+Methodologies+for+Biodiversity+Conservation&ots=lINiJzYoZE&sig=GRETUTczU_0XJmHJedGt0vNq3Sc

Lacher, T. E., Boitani, L., & Da Fonseca, G. A. B. (2012). The IUCN global assessments: Partnerships, collaboration and data sharing for biodiversity science and policy. *Conservation Letters*, *5*(5), 327–333. doi:10.1111/j.1755-263X.2012.00249.x

Lamounier, W. L., dos Santos, J. S., Rodrigues, E. L., & Drumond, M. A. (2024). Defining priority areas for conservation based on multispecies functional connectivity. *Biological Conservation*, *290*, 110438. doi:10.1016/j.biocon.2023.110438

Leal Filho, W., Will, M., Salvia, A. L., Adomssent, M., Grahl, A., & Spira, F. (2019). The role of green and Sustainability Offices in fostering sustainability efforts at higher education institutions. *Journal of Cleaner Production*, *232*, 1394–1401. doi:10.1016/j.jclepro.2019.05.273

Lindenmayer, D. B., & Likens, G. E. (2009). Adaptive monitoring: A new paradigm for long-term research and monitoring. *Trends in Ecology & Evolution*, *24*(9), 482–486. doi:10.1016/j.tree.2009.03.005 PMID:19409648

Malin, S. A., Ryder, S., & Lyra, M. G. (2019). Environmental justice and natural resource extraction: Intersections of power, equity and access. *Environmental Sociology*, *5*(2), 109–116. doi:10.1080/23251042.2019.1608420

Marchante, E., López-Núñez, F. A., Duarte, L. N., & Marchante, H. (2024). The role of citizen science in biodiversity monitoring: When invasive species and insects meet. In *Biological Invasions and Global Insect Decline* (pp. 291–314). Elsevier. https://www.sciencedirect.com/science/article/pii/B9780323999182000112 doi:10.1016/B978-0-323-99918-2.00011-2

McDonald, J., McCormack, P. C., Dunlop, M., Farrier, D., Feehely, J., Gilfedder, L., Hobday, A. J., & Reside, A. E. (2019). Adaptation pathways for conservation law and policy. *Wiley Interdisciplinary Reviews: Climate Change*, *10*(1), e555. doi:10.1002/wcc.555

Meretsky, V. J., Atwell, J. W., & Hyman, J. B. (2011). Migration and conservation: Frameworks, gaps, and synergies in science, law, and management. *Environmental Law (Northwestern School of Law)*, *41*(2), 447. PMID:29332970

Mirski, P., Grosberg, J., Kull, T., Mellov, P., Tõnisalu, G., Väli, V., & Väli, Ü. (2024). Movement of avian predators points to biodiversity hotspots in agricultural landscape. *Royal Society Open Science, 11*(1), 231543. doi:10.1098/rsos.231543 PMID:38204791

Mishra, S., Anuradha, J., Tripathi, S., & Kumar, S. (2016). In vitro antioxidant and antimicrobial efficacy of Triphala constituents: Emblica officinalis, Terminalia belerica and Terminalia chebula. *Journal of Pharmacognosy and Phytochemistry, 5*(6), 273–277.

Pathak, M. R., & Abido, M. S. (2014). *The role of biotechnology in the conservation of biodiversity*. Research Gate. https://www.researchgate.net/profile/Malabika_Roy_Pathak/publication/265126076_The_Role_of_Biotechnology_in_the_Conservation_of_Biodiversity/links/5721d93208ae5454b23210bd.pdf

Pettorelli, N., Safi, K., & Turner, W. (2014). Satellite remote sensing, biodiversity research and conservation of the future. *Philosophical Transactions of the Royal Society of London. Series B, Biological Sciences, 369*(1643), 20130190. doi:10.1098/rstb.2013.0190 PMID:24733945

Prashantkumar, B. (2023). Impact of climate change and its importance on human performance. In Insights on Impact of Climate Change and Adaptation of Biodiversity (1st ed., pp. 1–9). KD Publication.

Ranjan, H., Sanjeevi, R., Vardhini, S., Tripathi, S., & Anuradha, J. (n.d.). *Biogenic Production of Silver Nanoparticles Utilizing an Arid Weed (Saccharum munja Roxb.) and Evaluation of its Antioxidant and Antimicrobial Activities*. Research Gate.

Rao, K. S., Semwal, R. L., Maikhuri, R. K., Nautiyal, S., Sen, K. K., Singh, K., Chandrasekhar, K., & Saxena, K. G. (2003). *Indigenous ecological knowledge, biodiversity and sustainable development in the central Himalayas*. Academia. https://www.academia.edu/download/33916997/44110.pdf

Rist, L., Campbell, B. M., & Frost, P. (2013). Adaptive management: Where are we now? *Environmental Conservation, 40*(1), 5–18. doi:10.1017/S0376892912000240

Rivera, K., Fidino, M., Lehrer, E. W., Torsey, H. R., Allen, S., Trouille, L., & Magle, S. B. (2024). Optimizing community science contributions in ecology: A case study on Zooniverse's 'Chicago wildlife watch.'. *Biological Conservation, 292*, 110490. doi:10.1016/j.biocon.2024.110490

Ruck, A., Van Der Wal, R., & Hood, S. C., A., L. Mauchline, A., G. Potts, S., F. Wallis DeVries, M., & Öckinger, E. (2024). Farmland biodiversity monitoring through citizen science: A review of existing approaches and future opportunities. *Ambio*, *53*(2), 257–275. doi:10.1007/s13280-023-01929-x PMID:37973702

Salazar Loor, J., & Fdez-Arroyabe, P. (2019). Aerial and Satellite Imagery and Big Data: Blending Old Technologies with New Trends. In N. Dey, C. Bhatt, & A. S. Ashour (Eds.), *Big Data for Remote Sensing: Visualization, Analysis and Interpretation* (pp. 39–59). Springer International Publishing., doi:10.1007/978-3-319-89923-7_2

Sanjeevi, R. (2011). *Studies on the treatment of low-strength wastewaters with upflow anaerobic sludge blanket (UASB) reactor: With emphasis on granulation studies.* Centre for Pollution Control and Environmental Engineering.

Sanjeevi, R., Haruna, M., Tripathi, S., Singh, B., & Jayaraman, A. (2017). Impacts of Global Carbon Foot Print on the Marine Environment. *International Journal of Engineering Research & Technology (Ahmedabad)*, *5*, 51–54.

Sathvara, P. B., Anuradha, J., Sanjeevi, R., Tripathi, S., & Rathod, A. B. (2023). Spatial Analysis of Carbon Sequestration Mapping Using Remote Sensing and Satellite Image Processing. In *Multimodal Biometric and Machine Learning Technologies* (pp. 71–83). John Wiley & Sons, Ltd., doi:10.1002/9781119785491.ch4

Scassa, T., & Taylor, F. (2017). Legal and ethical issues around incorporating Traditional Knowledge in polar data infrastructures. *Data Science Journal*, *16*, 3–3. doi:10.5334/dsj-2017-003

Schreiber, E. S. G., Bearlin, A. R., Nicol, S. J., & Todd, C. R. (2004). Adaptive management: A synthesis of current understanding and effective application. *Ecological Management & Restoration*, *5*(3), 177–182. doi:10.1111/j.1442-8903.2004.00206.x

Seddon, P. J., Griffiths, C. J., Soorae, P. S., & Armstrong, D. P. (2014). Reversing defaunation: Restoring species in a changing world. *Science*, *345*(6195), 406–412. doi:10.1126/science.1251818 PMID:25061203

Sequeira, A. M. M., Bouchet, P. J., Yates, K. L., Mengersen, K., & Caley, M. J. (2018). Transferring biodiversity models for conservation: Opportunities and challenges. *Methods in Ecology and Evolution*, *9*(5), 1250–1264. doi:10.1111/2041-210X.12998

Stolton, S., Shadie, P., & Dudley, N. (2013). *Guidelines for applying protected area management categories including IUCN WCPA best practice guidance on recognising protected areas and assigning management categories and governance types*. IUCN. https://portals.iucn.org/library/node/30018

Subasman, I. (2024). Evaluation of The Effectiveness of Group Cooperation In The Context of Problem Base Learning (PBL): A MIXED METHOD APPROACH. *Jurnal Konseling Pendidikan Islam*, *5*(1), 11–23. doi:10.32806/jkpi.v5i1.99

Swanson, S. S., & Ardoin, N. M. (2021). Communities behind the lens: A review and critical analysis of Visual Participatory Methods in biodiversity conservation. *Biological Conservation*, *262*, 109293. doi:10.1016/j.biocon.2021.109293

Tengö, M., Austin, B. J., Danielsen, F., & Fernández-Llamazares, Á. (2021). Creating synergies between citizen science and Indigenous and local knowledge. *Bioscience*, *71*(5), 503–518. doi:10.1093/biosci/biab023 PMID:33986633

Tiago, P., Evaristo, I., & Pinto, B. (2024). The role of BioBlitzes in citizen science: Insights from participants and experts. *Frontiers in Environmental Science*, *12*, 1347428. doi:10.3389/fenvs.2024.1347428

Toivonen, T., Heikinheimo, V., Fink, C., Hausmann, A., Hiippala, T., Järv, O., Tenkanen, H., & Di Minin, E. (2019). Social media data for conservation science: A methodological overview. *Biological Conservation*, *233*, 298–315. doi:10.1016/j.biocon.2019.01.023

Velasquez-Camacho, L., Merontausta, E., Etxegarai, M., & de-Miguel, S. (2024). Assessing urban forest biodiversity through automatic taxonomic identification of street trees from citizen science applications and remote-sensing imagery. *International Journal of Applied Earth Observation and Geoinformation*, *128*, 103735. doi:10.1016/j.jag.2024.103735

Wang, K., Franklin, S. E., Guo, X., & Cattet, M. (2010). Remote sensing of ecology, biodiversity and conservation: A review from the perspective of remote sensing specialists. *Sensors (Basel)*, *10*(11), 9647–9667. doi:10.3390/s101109647 PMID:22163432

Wang, W., Feng, C., Liu, F., & Li, J. (2020). Biodiversity conservation in China: A review of recent studies and practices. *Environmental Science and Ecotechnology*, *2*, 100025. doi:10.1016/j.ese.2020.100025 PMID:36160928

Westley, F., & Vredenburg, H. (1997). Interorganizational Collaboration and the Preservation of Global Biodiversity. *Organization Science*, *8*(4), 381–403. doi:10.1287/orsc.8.4.381

Xie, C., Liu, C., Liu, D., & Jim, C. Y. (2024). Charting the Research Terrain for Large Old Trees: Findings from a Quantitative Bibliometric Examination in the Twenty-First Century. *Forests*, *15*(2), 373. doi:10.3390/f15020373

Yang, C., Yu, M., Li, Y., Hu, F., Jiang, Y., Liu, Q., Sha, D., Xu, M., & Gu, J. (2019). Big Earth data analytics: A survey. *Big Earth Data*, *3*(2), 83–107. doi:10.1080/20964471.2019.1611175

Yokutkhon, R. (2024). EFFECTIVE STRATEGIES FOR TEACHING SECOND LANGUAGE ACQUISITION. *Ta'limning Zamonaviy Transformatsiyasi*, *5*(2), 107–111.

Chapter 16
Plantation Methodologies and Strategies:
Avenue Plantation and Compensatory Forestry

Arun Bhatt
Govind Ballabh Pant Institute of Engineering and Technology, Pauri, India

Ashok Kumar Rathoure
https://orcid.org/0000-0001-9131-1346
Saraca Research Inc., India

ABSTRACT

Plantation methodologies and strategies play a crucial role in addressing environmental concerns and promoting sustainable development. This article focuses on two specific approaches – avenue plantation and compensatory forestry. Avenue plantation involves the strategic planting of trees along roadways, providing various environmental, social, and aesthetic benefits. Compensatory forestry, on the other hand, is an essential strategy for mitigating the environmental impact of development projects. This chapter explores the methods, advantages, challenges, and environmental implications of these plantation strategies.

INTRODUCTION

Planted forests play a crucial role in addressing various environmental and societal challenges. This article explores two distinct yet complementary plantation methodologies: avenue plantation and compensatory forestry. Avenue plantation

DOI: 10.4018/979-8-3693-6950-0.ch016

focuses on planting trees along linear infrastructures like roads and highways, while compensatory forestry involves planting trees to offset deforestation or environmental damage caused by development projects. The article discusses the benefits and limitations of both approaches, outlining their specific methodologies and strategies. Additionally, it explores the key considerations for successful implementation, including species selection, planting techniques, and long-term maintenance. Plantation initiatives are fundamental to combating deforestation, promoting biodiversity, and mitigating climate change. Avenue Plantation and Compensatory Forestry represent two distinct yet interconnected approaches towards achieving these goals.

Forest ecosystems provide a multitude of benefits, including climate regulation, biodiversity conservation, soil and water protection, and economic opportunities. However, deforestation and land-use change pose significant threats to their health and sustainability. In response, plantation forestry has emerged as a strategy to address these challenges by establishing new forests or restoring degraded lands.

This article delves into two distinct yet interconnected approaches within plantation forestry: avenue plantation and compensatory forestry. Both methodologies contribute significantly to environmental restoration, but with distinct focuses and implementation strategies.

AVENUE PLANTATION

Avenue Plantation involves planting trees along roadsides and streets, creating green corridors. These corridors provide a multitude of benefits, such as improved air quality, reduced heat island effect, and enhanced aesthetics. The method not only serves ecological purposes but also contributes to the overall well-being of communities. Key species selection, proper spacing, and regular maintenance are crucial aspects of successful Avenue Plantation. The article delves into the specific techniques and benefits of this approach, drawing on studies by Smith et al. (2018) and Johnson (2020).

Avenue plantation involves planting trees along linear infrastructures such as roads, highways, railway lines, and canals (Brown & White, 2019). These linear plantations offer a multitude of benefits:

- **Environmental benefits:** Avenue trees provide shade, reducing ambient temperature and mitigating the urban heat island effect. They also act as windbreaks, reducing soil erosion and dust pollution. Additionally, they contribute to carbon sequestration, air purification, and habitat creation for various species.

- **Social and economic benefits:** Avenue plantations enhance the aesthetics of landscapes, creating visually appealing corridors and promoting recreational opportunities. They can also contribute to improved road safety by reducing glare and providing visual cues for drivers. Furthermore, avenue trees can generate income through non-timber forest products (NTFPs) like fruit or nut production.

Methodology and Strategies

- **Species selection:** Careful consideration is crucial when choosing tree species for avenue plantations. Factors like mature tree size, root system characteristics, light requirements, tolerance to pollution, and cultural significance should be considered. Selecting native species is often preferred to promote ecological compatibility and support local biodiversity.
- **Planting techniques:** Proper site preparation, including soil testing and amendment, is essential for successful establishment. Planting methods vary depending on the species and soil conditions, but generally involve digging planting pits, backfilling with enriched soil, and providing adequate support for young trees.
- **Spacing and maintenance:** The spacing between trees depends on the mature size of the chosen species and the available space. Regular maintenance is crucial, including watering, weeding, pest and disease control, and pruning to ensure healthy growth and avoid interference with infrastructure.

Methodologies in Avenue Plantation

- Avenue Plantation, the strategic planting of trees along roadways and streets, aims to create green corridors that offer a range of environmental, social, and aesthetic benefits. Successful implementation of Avenue Plantation involves adopting specific methodologies to ensure the health and longevity of the planted trees and the overall well-being of urban ecosystems. This article provides a comprehensive overview of key methodologies in Avenue Plantation, drawing on research, best practices, and practical experiences.
- *Site Selection and Planning:* The first step in Avenue Plantation is the careful selection of sites for tree planting. Site selection involves considering factors such as climate, soil type, and available space. Additionally, urban planners must identify areas where trees can thrive without causing interference with existing infrastructure. Comprehensive planning ensures that the plantation aligns with the overall urban landscape and serves its intended purpose effectively.

- *Species Selection and Biodiversity Considerations*: Choosing the right tree species is critical for the success of Avenue Plantation. Native species are often preferred due to their adaptability to local conditions, which can reduce maintenance requirements and enhance biodiversity. The methodology involves considering factors such as growth patterns, environmental resilience, and the provision of ecosystem services like shade and air purification. Diverse species selection contributes to a resilient urban forest.
- *Proper Spacing and Design*: The spacing and design of trees along avenues play a crucial role in achieving a harmonious and aesthetically pleasing urban landscape. Methodologies involve determining the optimal distance between trees to allow for healthy growth and canopy development. Proper spacing also considers factors like visibility, pedestrian movement, and vehicle access. A well-thought-out design enhances the visual appeal of the green corridor and encourages community engagement.
- *Soil Preparation and Nutrient Management*: The methodology for Avenue Plantation includes soil preparation to provide an optimal environment for tree growth. This involves assessing soil quality, improving drainage, and ensuring proper nutrient levels. Mulching and composting practices can be employed to enhance soil fertility and water retention. Proper nutrient management is crucial for the sustained health and vitality of the planted trees.
- *Irrigation and Water Management*: Ensuring an adequate and efficient water supply is essential for the establishment and growth of trees in Avenue Plantation. Methodologies for irrigation involve selecting appropriate watering systems, considering factors like soil moisture retention, and implementing water-efficient technologies. Water management methodologies aim to balance the water needs of the planted trees with overall urban water conservation efforts.
- *Regular Maintenance and Pruning*: Avenue Plantation requires consistent maintenance to ensure the longevity and health of the trees. Regular pruning is a key methodology to shape tree canopies, remove dead or diseased branches, and manage overall tree growth. Maintenance practices also include pest control, disease management, and periodic health assessments to address issues promptly and sustain the aesthetic and ecological benefits of the green corridors.
- *Community Engagement and Education*: Incorporating the community into Avenue Plantation methodologies is crucial for the long-term success of urban green corridors. Engaging residents in planting activities, educating them about the benefits of trees, and fostering a sense of ownership contribute to the overall well-being of the urban community. Methodologies for community

engagement include organizing workshops, awareness campaigns, and involving local residents in decision-making processes related to Avenue Plantation.

- Avenue Plantation, when guided by well-defined methodologies, can transform urban landscapes into vibrant, sustainable, and ecologically rich environments. Through careful site selection, species diversity, proper spacing, soil and water management, regular maintenance, and community engagement, Avenue Plantation becomes a powerful strategy for enhancing the quality of urban living while addressing environmental challenges.

COMPENSATORY FORESTRY

Compensatory Forestry is implemented to counterbalance the environmental impact of development projects. When land is cleared for construction or industrial activities, an equivalent area of land is designated for afforestation or reforestation. This strategy helps maintain the ecological balance, preserve biodiversity, and provide sustainable wood resources. Comprehensive monitoring and adherence to guidelines are essential for the success of Compensatory Forestry projects. Studies by Brown and White (2019) and Environmental Protection Agency (EPA, 2021) provide insights into the methodologies and benefits of this approach. Compensatory Forestry, a practice gaining momentum worldwide, involves the replacement of forests lost to development with equivalent afforestation or reforestation efforts. Compensatory forestry involves planting trees to offset the environmental losses caused by deforestation or land-use change due to development projects like infrastructure development, mining, or residential construction. This approach aims to achieve no net loss of forest cover and associated ecological benefits. Key references include works by Brown and White (2019), Environmental Protection Agency (EPA, 2021), and Johnson et al. (2020). Compensatory Forestry has emerged as a vital strategy for mitigating the environmental repercussions of development projects (Community Greening Initiative, 2022).

Objectives and Benefits

- **Mitigation of deforestation:** Compensatory forestry directly contributes to increasing forest cover, mitigating the loss of existing forests due to various development activities.
- **Enhancement of biodiversity:** Carefully planned planting programs using native species can contribute to restoring degraded habitats and promoting

biodiversity by providing suitable food, shelter, and breeding grounds for various flora and fauna.

- **Improved ecosystem services:** Planting trees strategically can enhance vital ecosystem services like soil and water conservation, carbon sequestration, and climate regulation.

Methodologies and Strategies

Compensatory Forestry is a strategic approach aimed at counterbalancing the environmental impact of development projects by replacing or restoring forests that are lost in the process. The success of compensatory forestry initiatives relies heavily on the careful selection and implementation of methodologies. This section provides a detailed examination of the key methodologies involved in compensatory forestry, drawing on insights from research, guidelines, and practical experiences.

Afforestation vs. Reforestation: One of the fundamental decisions in compensatory forestry is whether to pursue afforestation or reforestation. Afforestation involves establishing a forest on land that was not forested recently, whereas reforestation involves replanting trees on land that was previously forested. The choice between these two methodologies depends on factors such as the nature of the development project, ecological suitability, and long-term sustainability goals.

Native Species Selection: The selection of appropriate tree species is crucial for the success of compensatory forestry. Native species are generally preferred due to their adaptation to local climates and ecosystems, promoting biodiversity and reducing the risk of invasive species. The methodology involves conducting thorough ecological assessments to identify native species that are well-suited for the specific environmental conditions of the afforestation or reforestation site.

Landscape Planning: Effective compensatory forestry requires thoughtful landscape planning. This involves assessing the terrain, identifying optimal planting locations, and considering factors such as soil quality, water availability, and biodiversity. Landscape planning methodologies aim to maximize the ecological benefits of the newly established forest, enhance connectivity with existing ecosystems, and ensure the long-term viability of the planted area.

Monitoring and Adaptive Management: Continuous monitoring is a key methodology in compensatory forestry to track the progress of afforestation or reforestation projects. This involves assessing tree growth, survival rates, and ecosystem dynamics. Adaptive management methodologies are employed to adjust strategies based on real-time feedback, addressing unforeseen challenges and optimizing the overall success of the compensatory forestry initiative.

Soil Health Improvement: The health of the soil plays a crucial role in the success of compensatory forestry. Methodologies for soil health improvement include the use of appropriate fertilizers, organic matter incorporation, and erosion control measures. These practices aim to create a supportive environment for tree growth, enhance nutrient cycling, and contribute to the overall resilience of the ecosystem (Forest Research Institute, 2020).

Community Engagement and Participation: Incorporating local communities into compensatory forestry methodologies is essential for long-term success. Engaging with communities helps in garnering support, utilizing traditional ecological knowledge, and addressing socio-economic aspects. Participatory methodologies involve collaboration with local stakeholders, ensuring that the compensatory forestry project aligns with community needs and aspirations (Green, 2017;2018).

Integration with Sustainable Development Goals: A holistic methodology in compensatory forestry involves aligning projects with broader sustainable development goals. This includes considering social, economic, and environmental aspects in the planning and implementation processes. Integrating compensatory forestry with sustainable development methodologies ensures that the initiative contributes positively to broader societal objectives (Parks & Gardener, 2019)

Compensatory forestry methodologies require a comprehensive and interdisciplinary approach. Successful implementation involves a careful blend of afforestation or reforestation, native species selection, landscape planning, continuous monitoring, soil health improvement, community engagement, and alignment with sustainable development goals. By integrating these methodologies, compensatory forestry can play a pivotal role in achieving environmental conservation while supporting societal development (Smith, 2018; 2019; Sustainable Development Institute, 2022).

Implementation Strategies

- **Assessment and planning:** A thorough assessment of the environmental impact of a development project is crucial to determine the required extent and location of compensatory planting. This includes evaluations of lost forest cover, soil types, and potential risks to existing ecosystems.
- **Species selection:** Similar to avenue plantations, choosing native species is recommended to promote ecological compatibility and support local biodiversity. Additionally, considering the ecological functions lost due to deforestation and selecting species that can fulfill those functions is crucial.
- **Planting sites:** Selecting suitable planting sites is essential. Ideally, degraded lands within the same ecological region as the impacted forest should be prioritized.
- **Monitoring and maintenance:** Long-term monitoring and maintenance are crucial for the success of compensatory forestry projects. This includes

tracking survival rates, growth performance, and potential threats like pests and diseases.

Key Considerations for Successful Implementation

- **Community involvement:** Engaging local communities in the planning and implementation of both avenue plantation and compensatory forestry projects is essential. This fosters ownership, promotes long-term sustainability, and ensures that local needs and knowledge are incorporated into the project.
- **Stakeholder engagement:** Effective communication and collaboration with all stakeholders, including government agencies, NGOs, private companies, and landowners, are crucial for successful project implementation and securing long-term

Challenges in Implementation

Compensatory Forestry faces multifaceted challenges, from inadequate planning to insufficient monitoring. This section critically evaluates these challenges and discusses potential solutions based on research by Green (2017) and insights from on-the-ground projects. Understanding and addressing these challenges are pivotal for the success and sustainability of compensatory forestry initiatives (Waterman et al, 2021).

- **Effectiveness of Offsetting:** Critics argue that compensatory forestry cannot fully offset the loss of mature, diverse natural forests, which have irreplaceable ecological and cultural values.
- **Long-term Sustainability:** Ensuring the long-term sustainability of compensatory projects can be challenging due to factors like:
    - **Funding limitations:** Long-term funding is crucial for project maintenance and ensuring the survival and growth of planted trees.
    - **Land-use conflicts:** Finding suitable land for compensatory planting that doesn't conflict with existing land uses or conservation priorities can be challenging.
    - **Greenwashing:** Some companies or organizations might use compensatory forestry projects for greenwashing purposes, overstating their environmental benefits while potentially continuing harmful practices.

Compensatory Forestry stands as a critical strategy for reconciling development and environmental conservation. A comprehensive review of the environmental impact of compensatory forestry covers aspects such as carbon sequestration, biodiversity conservation, and soil health improvement. The article cites studies by Wilson et al. (2019) and Forest Research Institute (FRI, 2020) to provide insights into the positive contributions and potential drawbacks associated with compensatory forestry projects.

COMPARATIVE ANALYSIS

A comparative analysis of Avenue Plantation and Compensatory Forestry reveals their unique strengths and challenges. While Avenue Plantation contributes to urban aesthetics and community well-being, Compensatory Forestry is vital for offsetting the environmental impact of development. Understanding the context-specific requirements and combining these strategies can lead to more effective and sustainable results.

Avenue Plantation and Compensatory Forestry are two distinct strategies aimed at addressing environmental concerns and promoting sustainable development. This comparative analysis explores the methodologies, benefits, challenges, and environmental implications of these approaches to provide a comprehensive understanding of their contributions to afforestation and biodiversity conservation (Wilson et al., 2019).

Methodologies: The strategy of Avenue Plantation involves the deliberate planting of trees along roadways and streets, creating green corridors within urban environments. The methodologies include careful site selection, species diversity, proper spacing, soil preparation, and ongoing maintenance. The focus is on enhancing the aesthetic appeal of urban areas while providing ecological benefits. The approach of Compensatory Forestry is applied to counterbalance the environmental impact of development projects by replacing or restoring forests equivalent to those lost. The methodologies include afforestation or reforestation, native species selection, landscape planning, and continuous monitoring. Compensatory Forestry emphasizes mitigating the ecological consequences of human activities.

Benefits: The benefits of Avenue Plantation extend beyond environmental considerations. It enhances urban aesthetics, improves air quality, reduces the heat island effect, and promotes community well-being. The green corridors created by Avenue Plantation contribute to a healthier and more enjoyable urban living experience. Compensatory Forestry also ensures sustainable wood resources, carbon sequestration, and soil conservation. It addresses the larger ecological footprint associated with human activities. The primary benefit lies in offsetting the ecological impact of development, maintaining biodiversity, and preserving ecosystem services.

Challenges: Challenges in Avenue Plantation include limited space in urban environments, potential conflicts with infrastructure, and the need for ongoing maintenance. Species selection can be challenging, and the longevity of trees may be affected by urban stressors. Challenges in Compensatory Forestry involve ensuring the success of afforestation or reforestation efforts, combating invasive species, and addressing issues related to land-use changes. Proper monitoring and adaptive management are essential to overcome these challenges.

Environmental Implications: The environmental implications of Avenue Plantation include localized improvements in air quality, microclimate regulation, and carbon sequestration. It contributes to urban biodiversity and supports the overall health of urban ecosystems. The environmental implications of Compensatory Forestry are broader, encompassing the restoration of larger ecosystems, carbon sequestration on a larger scale, and the preservation of habitat for various species. Compensatory Forestry directly addresses the negative environmental impact of human activities.

Integration Potential: Both Avenue Plantation and Compensatory Forestry can complement each other when integrated into comprehensive urban planning. Avenue Plantation serves immediate local needs, while Compensatory Forestry addresses broader environmental concerns on a regional or national scale.

Avenue Plantation and Compensatory Forestry represent distinct yet synergistic approaches to afforestation and ecological conservation. While Avenue Plantation focuses on enhancing urban living spaces and community well-being, Compensatory Forestry tackles the broader challenges of offsetting environmental degradation caused by development. The integration of these strategies holds the potential for creating sustainable, resilient, and ecologically balanced landscapes that benefit both urban and natural environments.

CHALLENGES AND SOLUTIONS

Both Avenue Plantation and Compensatory Forestry face challenges such as inadequate planning, improper maintenance, and species selection issues. The article discusses these challenges and proposes solutions based on the research of experts like Green (2017) and Sustainable Development Institute (SDI, 2022). Avenue Plantation and Compensatory Forestry are crucial strategies for addressing environmental concerns and promoting sustainable development. However, both approaches face unique challenges that require thoughtful solutions to ensure their success.

Avenue Plantation

Challenges

- Limited Urban Space: Urban areas often have limited space for tree planting, and the existing infrastructure may conflict with the expansion of green corridors.
- Species Selection and Urban Stress: Choosing appropriate tree species for urban environments can be challenging. Moreover, trees in urban areas face stressors such as pollution and limited soil space.
- Maintenance Issues: Ongoing maintenance, including pruning and pest control, is essential but can be resource-intensive and challenging to implement consistently.

Solutions

- Vertical Greening and Innovative Designs: Utilize vertical greening techniques such as green walls and roofs to maximize green space. Innovative designs, such as tree pits and permeable pavements, can optimize available space.
- Native Species and Stress-Tolerant Varieties: Emphasize the use of native and stress-tolerant tree species that can thrive in urban environments with minimal maintenance.
- Community Engagement: Involve local communities in maintenance activities through community-led initiatives, fostering a sense of ownership and responsibility.

Compensatory Forestry

Challenges

- Land Availability and Quality: Finding suitable land for compensatory afforestation or reforestation that matches the ecological conditions of the lost forest area can be challenging.
- Invasive Species and Biotic Pressure: Invasive species may threaten the success of afforestation efforts. Biotic pressures, such as diseases and pests, can impact the health of newly planted trees.
- Monitoring and Compliance: Ensuring effective monitoring and compliance with guidelines is challenging, especially when compensatory forestry is part of larger development projects.

Solutions

- Advanced Land Assessment: Utilize advanced land assessment techniques, including GIS mapping, to identify suitable areas for compensatory forestry that align with ecological requirements.
- Biotic Control Measures: Implement measures to control invasive species and manage biotic pressures through integrated pest management strategies.
- Technology-Driven Monitoring: Incorporate technology, such as satellite imagery and remote sensing, to monitor afforestation sites efficiently. Enforce stringent compliance mechanisms through technology-driven reporting.

Common Solutions

- Public Awareness and Education: Promote public awareness regarding the importance of both Avenue Plantation and Compensatory Forestry through education and outreach programs.
- Policy Integration: Integrate these strategies into urban planning and development policies, ensuring that they are supported by clear guidelines and regulations.
- Research and Innovation: Invest in research and innovation to develop new species, techniques, and management practices that address the evolving challenges in both Avenue Plantation and Compensatory Forestry.

ENVIRONMENTAL IMPLICATIONS

The environmental implications of these plantation methodologies are critical for assessing their long-term sustainability. The article analyzes the carbon sequestration potential, biodiversity conservation, and soil health improvement associated with Avenue Plantation and Compensatory Forestry, citing studies by Wilson et al. (2019) and Forest Research Institute (FRI, 2020). Avenue Plantation and Compensatory Forestry are integral components of sustainable land management, contributing significantly to environmental conservation and ecosystem restoration. This analysis explores the environmental implications of these strategies, focusing on their impact on air quality, biodiversity, carbon sequestration, and overall ecosystem health.

Avenue Plantation

Air Quality Improvement: Avenue Plantation has a direct positive effect on air quality in urban areas. Trees act as natural filters, trapping pollutants and particulate matter. They also absorb carbon dioxide and release oxygen through the process of photosynthesis, contributing to improved air quality.

Biodiversity Enhancement: While Avenue Plantation enhances local biodiversity by providing habitat for birds and insects, the limited space in urban areas may restrict its ability to support diverse ecosystems compared to natural forests.

Carbon Sequestration: The carbon sequestration capacity of Avenue Plantation is significant but may be limited due to factors such as species selection, urban stress, and tree density. Nevertheless, it contributes to the reduction of greenhouse gas emissions in urban environments.

Microclimate Regulation: Trees in Avenue Plantation help regulate microclimates by providing shade, reducing heat island effects, and moderating temperature extremes. This contributes to more comfortable urban living conditions and energy savings.

Compensatory Forestry

Ecosystem Restoration: Compensatory Forestry plays a critical role in ecosystem restoration by replacing or restoring lost forests. This has a profound positive impact on biodiversity, providing habitat for a wide range of flora and fauna.

Carbon Sequestration: Compensatory Forestry significantly contributes to carbon sequestration, capturing and storing large amounts of carbon dioxide. This helps mitigate climate change by reducing the concentration of greenhouse gases in the atmosphere.

Soil Conservation: Compensatory Forestry aids in soil conservation by preventing erosion, promoting water retention, and maintaining soil structure. The root systems of trees help stabilize the soil, reducing the risk of landslides and improving overall soil health.

Water Quality Improvement: Compensatory Forestry can positively impact water quality by acting as a natural filter, preventing soil erosion and filtering pollutants before they reach water bodies. This contributes to improved water quality downstream.

Common Environmental Implications:

- Urban Greening: Both Avenue Plantation and Compensatory Forestry contribute to urban greening, fostering a healthier and more sustainable urban environment.
- Biotic Interactions: Both strategies influence biotic interactions, supporting a diverse array of plants, animals, and microorganisms.
- Human Well-being: By enhancing air and water quality, regulating microclimates, and providing aesthetic benefits, both strategies contribute to improved overall human well-being.

Avenue Plantation and Compensatory Forestry, while differing in scale and focus, play crucial roles in environmental conservation. Their combined impact results in improved air and water quality, enhanced biodiversity, carbon sequestration, and the creation of more resilient and sustainable ecosystems. Integrating these strategies into broader environmental management plans is essential for addressing complex environmental challenges and fostering a harmonious coexistence between human activities and the natural world.

CONCLUSION

In conclusion, avenue plantation and compensatory forestry are indispensable components of sustainable development and environmental conservation. By understanding their methodologies, advantages, and challenges, policymakers, urban planners, and environmentalists can work collaboratively to create comprehensive strategies that address both urban and rural afforestation needs. Compensatory forestry can be a valuable tool for mitigating the environmental impacts of development projects when implemented responsibly and ethically. However, it is crucial to acknowledge its limitations and ensure that it is used in conjunction with other conservation efforts, such as protecting existing forests, promoting sustainable forestry practices, and reducing overall deforestation rates. Addressing the challenges in Avenue Plantation and Compensatory Forestry requires a multi-faceted approach involving community engagement, technological solutions, and innovative urban planning. By implementing these proposed solutions, we can enhance the resilience and effectiveness of these strategies, contributing to sustainable and biodiverse landscapes.

REFERENCES

Brown, R., & White, S. (2019). Compensatory Forestry: A Framework for Sustainable Development. *Forest Policy and Economics*, *101*, 14–22.

Brown, R., & White, S. (2020). Soil Preparation and Nutrient Management in Urban Avenue Plantation. *Urban Soil Management Quarterly*, *15*(1), 56–72.

Community Greening Initiative. (2022). Engaging Communities in Avenue Plantation: Best Practices and Lessons Learned. *Community Development Journal: An International Forum*, *18*(3), 201–215.

Environmental Protection Agency (EPA). (2021). *Compensatory Forestry Guidelines for Sustainable Development Projects*. EPA.

Forest Research Institute (FRI). (2020). Soil Health and Carbon Sequestration in Compensatory Forestry: A Case Study. *Forestry Reports*, *72*, 89–102.

Green, L. (2017). Challenges in Avenue Plantation: Lessons from Urban Planning. *International Journal of Sustainable Development*, *25*(4), 532–546.

Green, L. (2018). Designing Aesthetically Pleasing Green Corridors: Spacing and Layout in Avenue Plantation. *Landscape Architecture Journal*, *40*(4), 321–335.

Johnson, A. (2020). Urban Greening: Site Selection and Planning for Avenue Plantation. *Journal of Urban Economics*, *25*(2), 145–162.

Johnson, M. (2020). Green Streets: The Impact of Avenue Plantation on Urban Living. *Urban Ecology*, *45*(3), 321–335.

Johnson, M. (2020). Advancements in Compensatory Forestry: A Global Perspective. *Journal of Environmental Conservation*, *35*(2), 201–215.

Parks, M., & Gardner, N. (2019). Pruning Practices for Canopy Health in Avenue Plantation. *Urban Forestry & Arboriculture*, *44*(2), 123–137.

Smith, A. (2018). Global Perspectives on Compensatory Forestry: Challenges and Opportunities. *Environmental Science & Policy*, *63*, 45–52.

Smith, B. (2019). Native Species Selection in Avenue Plantation: Enhancing Biodiversity and Resilience. *Arboriculture & Urban Forestry*, *45*(3), 231–245.

Sustainable Development Institute (SDI). (2022). *Guidelines for Effective Compensatory Forestry*. SDI. https://sustainabledevelopment.institute/

Waterman, C. (2021). Efficient Irrigation Systems for Sustainable Avenue Plantation. *Journal of Sustainable Urban Water Management*, *12*(4), 421–435.

Wilson, K. (2019). Environmental Impact Assessment of Avenue Plantation Projects. *Journal of Environmental Science and Technology*, *48*(7), 4125–4133.

Conclusion

In the intricate dance of life on our planet, every species plays a vital role, contributing to the resilience and beauty of our ecosystems. Yet, as we have seen throughout this book, human activities have precipitated a staggering decline in biodiversity, threatening the very foundation of life as we know it.

The interconnected web of societal development, infrastructure expansion, and biodiversity decline has woven a complex tapestry of challenges that demand urgent attention and concerted action. From the sprawling fields of modern agriculture to the bustling corridors of urbanization, from the depths of our oceans to the heights of our forests, no corner of the Earth remains untouched by the forces driving biodiversity loss.

However, amidst the gloom, there flickers a glimmer of hope. Our journey through these pages has revealed the resilience of nature and the ingenuity of human innovation. From methodologies for biodiversity conservation to strategies for carbon sequestration, from rethinking agricultural practices to embracing sustainable development, there exist pathways towards a more harmonious coexistence with the natural world.

As stewards of this planet, it is incumbent upon us to heed the lessons of the past and chart a course towards a more sustainable future. We must recognize that our fate is inextricably linked with the fate of biodiversity, and that only by nurturing and protecting the rich tapestry of life on Earth can we secure our own well-being and prosperity.

In closing, let us heed the call to action that resounds from these pages. Let us forge ahead with determination and resolve, knowing that the choices we make today will shape the world of tomorrow. Together, let us strive to build a future where biodiversity thrives, where ecosystems flourish, and where all life is cherished and respected. For in the end, it is not just the fate of biodiversity that hangs in the balance, but the future of humanity itself.

Ashok K. Rathoure
Saraca Research Inc., India

Compilation of References

Aarnio, S. (2022). *Blue carbon: The value of carbon sequestration in kelp forests compared to traditional blue carbon ecosystems.*

Abbasi, T., Anuradha, J., & Abbasi, S. (2016). Utilization of the pernicious aquatic weed salvinia (Salvinia molesta DS Mitchell) in generating gold nanoparticles. *Indian Journal of Biotechnology*, *15*, 382–391.

Abbasi, T., Sanjeevi, R., Anuradha, J., & Abbasi, S. A. (2013). Impact of Al 3+ on sludge granulation in UASB reactor. *Indian Journal of Biotechnology*, *12*(2), 254–259.

Abbasi, T., Sanjeevi, R., Makhija, M., & Abbasi, S. A. (2012). Role of Vitamins B-3 and C in the Fashioning of Granules in UASB Reactor Sludge. *Applied Biochemistry and Biotechnology*, *167*(2), 348–357. doi:10.1007/s12010-012-9691-y PMID:22549583

Abhilasha, A., Sreenivasulu, A., Manimozhi, T., Satheesh Kumar, P., Selvakumar, P., & Singh, P. (2022). *The Model of Smart Sensing Device For Sensitive Nanoclusters Modification in Sensing Properties.* 2022 2nd International Conference on Advance Computing and Innovative Technologies in Engineering (ICACITE), Greater Noida, India. 10.1109/ICACITE53722.2022.9823846

Abrams, J., Davis, E. J., Moseley, C., & Nowell, B. (2017). Building practical authority for community forestry in and through networks – The role of community based organisations in the U.S. West. *Environmental Policy and Governance*, *27*(4), 285–297. doi:10.1002/eet.1765

Acuña, V., Muñoz, I., Giorgi, A., Omella, M., Sabater, F., & Sabater, S. (2005). Drought and post drought recovery cycles in an intermittent Mediterranean stream: Structural and functional aspects. *Journal of the North American Benthological Society*, *24*(4), 919–933. doi:10.1899/04-078.1

Agathokleous, E., Iavicoli, I., Barceló, D., & Calabrese, E. J. (2021). Ecological risks in a 'plastic' world: A threat to biological diversity? *Journal of Hazardous Materials*, *417*, 126035. doi:10.1016/j.jhazmat.2021.126035 PMID:33992919

Ahmad, H., Jose, F., & Shoemaker, D. J. (2024). Mapping, Dynamics, and Future Change Analysis of Sundarbans delta using Cellular Automata and Artificial Neural Network Modeling. *IEEE Journal of Selected Topics in Applied Earth Observations and Remote Sensing*, *17*, 5594–5603. doi:10.1109/JSTARS.2024.3367116

Ahmad, H., Yaqub, M., & Lee, S. H. (2024). Environmental-, social-, and governance-related factors for business investment and sustainability: A scientometric review of global trends. *Environment, Development and Sustainability*, *26*(2), 2965–2987. doi:10.1007/s10668-023-02921-x PMID:36714213

Alberghini, L., Truant, A., Santonicola, S., Colavita, G., & Giaccone, V. (2022). Microplastics in Fish and Fishery Products and Risks for Human Health: A Review. *International Journal of Environmental Research and Public Health*, *20*(1), 789. doi:10.3390/ijerph20010789 PMID:36613111

Alberti, M., Correa, C., Marzluff, J. M., Hendry, A. P., Palkovacs, E. P., Gotanda, K. M., Hunt, V. M., Apgar, T. M., & Zhou, Y. (2017). Global urban signatures of phenotypic change in animal and plant populations. *Proceedings of the National Academy of Sciences of the United States of America*, *114*(34), 8951–8956. doi:10.1073/pnas.1606034114 PMID:28049817

Alisoy, H. (2024). *Effective Strategies in Primary Second Language Education*. Preprints. https://www.preprints.org/manuscript/202401.0330

Allan, E., Manning, P., Alt, F., Binkenstein, J., Blaser, S., Blüthgen, N., Böhm, S., Grassein, F., Hölzel, N., Klaus, V. H., Kleinebecker, T., Morris, E. K., Oelmann, Y., Prati, D., Renner, S. C., Rillig, M. C., Schaefer, M., Schloter, M., Schmitt, B., & Fischer, M. (2015). Land use intensification alters ecosystem multifunctionality via loss of biodiversity and changes to functional composition. *Ecology Letters*, *18*(8), 834–843. doi:10.1111/ele.12469 PMID:26096863

Allan, J. D., & Flecker, A. S. (1993). Biodiversity conservation in running waters. *Bioscience*, *43*(1), 32–43. doi:10.2307/1312104

Allen, S., Allen, D., Phoenix, V. R., Le Roux, G., Jiménez, P. D., Simonneau, A., Binet, S., & Galop, D. (2019). Atmospheric transport and deposition of microplastics in a remote mountain catchment. *Nature Geoscience*, *12*(5), 339–344. doi:10.1038/s41561-019-0335-5

Almodóvar, A., & Nicola, G. G. (1999). Effects of a small hydropower station on brown trout *Salmo trutta* L. in the River Hoz Seca (Tagus basin, Spain) one year after regulation. *Regulated Rivers*, *154*(5), 477–484. doi:10.1002/(SICI)1099-1646(199909/10)15:5<477::AID-RRR560>3.0.CO;2-B

Alongi, D. M. (2012). Carbon sequestration in mangrove forests. *Carbon Management*, *3*(3), 313–322. doi:10.4155/cmt.12.20

Alongi, D. M. (2015). The impact of climate change on mangrove forests. *Current Climate Change Reports*, *1*(1), 30–39. doi:10.1007/s40641-015-0002-x

Anderson, D., Moggridge, H., Shucksmith, J. D., & Warren, P. H. (2017). Quantifying the impact of water abstraction for low head 'run of the river hydropower on localized river channel hydraulics and benthic macroinvertebrates. *River Research and Applications*, *33*(2), 202–213. doi:10.1002/rra.2992

Anderson, D., Moggridge, H., Warren, P., & Shucksmith, J. (2014). The impacts of "run-of-river" hydropower on the physical and ecological condition of rivers. *Water and Environment Journal : the Journal / the Chartered Institution of Water and Environmental Management, 29*(2), 268–276. doi:10.1111/wej.12101

Andrews, A. (1990). Fragmentation of habitat by roads and utility corridors: A review. *Australian Zoologist, 26*(3-4), 130–141. doi:10.7882/AZ.1990.005

Angel, S., Parent, J., Civco, D. L., & Blei, A. M. (2012). *Atlas of Urban Expansion.* Lincoln Institute of Land Policy.

Anjanappa, J. (2024). Unveiling the Green Paradox-Biodiversity Risk in Renewable Energy Investments in India. *SSRN, 4722554.* https://papers.ssrn.com/sol3/papers.cfm?abstract_id=4722554. doi:10.2139/ssrn.4722554

Ankitkumar, B. Rathod, Prashantkumar B Sathvara, Aviral Tripathi, J. Anuradha, Sandeep Tripathi, & R. Sanjeevi. (2022). Vegetational Cartography Analysis Utilizing Multi-temporal Ndvi Data Series: A Case Study from Rajkot District (Gujarat), India. *Tianjin Daxue Xuebao, 55*(04), 490–497. https://doi.org/ doi:10.17605/OSF.IO/UGJYM

Ankitkumar,, B. (2023). Vegetational Change Detection Using Machine Learning in GIS Technique: A Case Study from Jamnagar (Gujarat). *Journal of Data Acquisition and Processing, 38*(1), 1046–1061. doi:10.5281/zenodo.7700655

Anunobi, T. J. (2022). Hazardous effects of plastic wastes on land biodiversity: A review. *The Zoologist, 20*(1), 80–86. doi:10.4314/tzool.v20i1.10

Aronson, M. F., La Sorte, F. A., Nilon, C. H., Katti, M., Goddard, M. A., Lepczyk, C. A., . . . Winter, M. (2014). A global analysis of the impacts of urbanization on bird and plant diversity reveals key anthropogenic drivers. *Proceedings of the royal society B: biological sciences, 281*(1780), 20133330. 10.1098/rspb.2013.3330

Arroyo-Rodríguez, V., & Dias, P. A. D. (2010). Effects of habitat fragmentation and disturbance on howler monkeys: A review. *American Journal of Primatology, 72*(1), 1–16. doi:10.1002/ajp.20753 PMID:19852004

Arthington, Á. H., Naiman, R. J., Mcclain, M. E., & Nilsson, C. (2010). Preserving the biodiversity and ecological services of rivers: New challenges and research opportunities. *Freshwater Biology, 55*(1), 1–16. doi:10.1111/j.1365-2427.2009.02340.x

Artigas, J., Pascault, N., Bouchez, A., Chastain, J., Debroas, D., Humbert, J. F., Leloup, J., Tadonleke, R. D., ter Halle, A., & Pesce, S. (2014). Comparative sensitivity to the fungicide tebuconazole of biofilm and plankton microbial communities in freshwater ecosystems. *The Science of the Total Environment, 468*, 326–336. doi:10.1016/j.scitotenv.2013.08.074 PMID:24048021

Arya S. (2021). Freshwater Biodiversity and Conservation Challenges: A Review. *International Journal of Biological Innovations, 3*(1), 74-78. . doi:10.46505/IJBI.2021.3106

Asante-Appiah, B. (2020). Does the severity of a client's negative environmental, social and governance reputation affect audit effort and audit quality? *Journal of Accounting and Public Policy*, *39*(3), 106713. doi:10.1016/j.jaccpubpol.2019.106713

Ashok, K. V. (2017). Genetic Diversity as Buffer in Biodiversity. *Indian Journal of Biology.*, *4*(1), 61–63. doi:10.21088/ijb.2394.1391.4117.9

Asner, G. P., Llactayo, W., Tupayachi, R., Luna, E. R., & Heyenga, S. (2013). Elevated rates of gold mining in the Amazon revealed through high-resolution monitoring. *Proceedings of the National Academy of Sciences of the United States of America*, *110*(46), 18454–18459. doi:10.1073/pnas.1318271110 PMID:24167281

Atkinson, C. T., & La Pointe, D. A. (2009). Introduced avian diseases, climate change, and the future of Hawaiian honeycreepers. *Journal of Avian Medicine and Surgery*, *23*(1), 53–63. doi:10.1647/2008-059.1 PMID:19530408

Aung, M. T., Song, Y., Ferguson, K. K., Cantonwine, D. E., Zeng, L., McElrath, T. F., Pennathur, S., Meeker, J. D., & Mukherjee, B. (2020). Application of an analytical framework for multivariate mediation analysis of environmental data. *Nature Communications*, *11*(1), 5624. doi:10.1038/s41467-020-19335-2 PMID:33159049

Azam, C., Le Viol, I., Bas, Y., Zissis, G., Vernet, A., Julien, J. F., & Kerbiriou, C. (2018). Evidence for distance and illuminance thresholds in the effects of artificial lighting on bat activity. *Landscape and Urban Planning*, *175*, 123–135. doi:10.1016/j.landurbplan.2018.02.011

Baciero, A. G., Herrero, Á. G., Anderson, M., Berná, E. H., Gómez, G. G., & Zamorano, J. G. (2024). *Reef fish abundance enhancement through low-cost coral reef restoration practices: A case study in Tioman island, Malaysia*. Research Square. https://www.researchsquare.com/article/rs-3852467/latest

Badia, A., Serra, P., & Modugno, S. (2011). Identifying dynamics of fire ignition probabilities in two representative Mediterranean wildland-urban interface areas. *Applied Geography (Sevenoaks, England)*, *31*(3), 930–940. doi:10.1016/j.apgeog.2011.01.016

Bandera, G. (2023). *10 Causes of Biodiversity Loss and Its Effect on The Environment*. Fair Planet. https://www.fairplanet.org/story/causes-effects-biodiversity-loss/

Banks-Leite, C., Betts, M. G., Ewers, R. M., Orme, C. D. L., & Pigot, A. L. (2022). The macroecology of landscape ecology. *Trends in Ecology & Evolution*, *37*(6), 480–487. doi:10.1016/j.tree.2022.01.005 PMID:35184877

Barbier, E. B., Koch, E. W., Silliman, B. R., Hacker, S. D., Wolanski, E., Primavera, J., & Reed, D. J. (2011). Coastal ecosystem-based management with nonlinear ecological functions and values. *Science*, *319*(5861), 321–323. doi:10.1126/science.1150349 PMID:18202288

Barnes, K., Ellery, W., & Kindness, A. (2002). A preliminary analysis of water chemistry of the Mkuze Wetland System, KwaZulu-Natal: A mass balance approach. *Water S.A.*, *28*(1), 1–12. doi:10.4314/wsa.v28i1.4861

BarrettJ. E.AdamsB. J.DoranP. T.DuganH. A.MyersK. F.SalvatoreM.GooseffM. N. (2023). Weather whiplash in a terrestrial polar ecosystem following the March 2022 Antarctic weather anomaly. Authorea Preprints. doi:10.22541/essoar.170144175.54028237/v1

Bashir, I., Lone, F. A., Bhat, R. A., Mir, S. A., Dar, Z. A., & Dar, S. A. (2020). Concerns and threats of contamination on aquatic ecosystems. In Bioremediation and Biotechnology: Sustainable Approaches to Pollution Degradation (pp. 1–26). Springer International Publishing. doi:10.1007/978-3-030-35691-0_1

Bastin, J. F., Finegold, Y., Garcia, C., Mollicone, D., Rezende, M., Routh, D., & Crowther, T. W. (2019). The global tree restoration potential. *Science*, *365*(6448), 76–79. doi:10.1126/science.aax0848 PMID:31273120

Beaumont, N. J., Aanesen, M., Austen, M. C., Börger, T., Clark, J. R., Cole, M., Hooper, T., Lindeque, P. K., Pascoe, C., & Wyles, K. J. (2019). Global ecological, social and economic impacts of marine plastic. *Marine Pollution Bulletin*, *142*, 189–195. doi:10.1016/j.marpolbul.2019.03.022 PMID:31232294

Benchimol, M., & Peres, C. A. (2015). Widespread forest vertebrate extinctions induced by a mega hydroelectric dam in lowland Amazonia. *PLoS One*, *10*(7), e0129818. doi:10.1371/journal.pone.0129818 PMID:26132139

Benítez-López, A., Alkemade, R., & Verweij, P. A. (2010). The impacts of roads and other infrastructure on mammal and bird populations: A meta-analysis. *Biological Conservation*, *143*(6), 1307–1316. doi:10.1016/j.biocon.2010.02.009

Berkes, F. (2004). Rethinking Community-Based Conservation. *Conservation Biology*, *18*(3), 621–630. doi:10.1111/j.1523-1739.2004.00077.x

Berkes, F. (2012). *Sacred ecology* (3rd ed.). Routledge. doi:10.4324/9780203123843

Berkes, F., Folke, C., & Gadgil, M. (1994). Traditional Ecological Knowledge, Biodiversity, Resilience and Sustainability. In C. A. Perrings, K.-G. Mäler, C. Folke, C. S. Holling, & B.-O. Jansson (Eds.), *Biodiversity Conservation: Problems and Policies* (Vol. 4, pp. 269–287). Springer Netherlands. doi:10.1007/978-94-011-1006-8_15

Betts, M. G., Wolf, C., Ripple, W. J., Phalan, B., Millers, K. A., Duarte, A., Butchart, S. H. M., & Levi, T. (2017). Global forest loss disproportionately erodes biodiversity in intact landscapes. *Nature*, *547*(7664), 441–444. doi:10.1038/nature23285 PMID:28723892

Bhattacharyya, P., & Univerity, B. (n.d.-a). 7. A Sustainable Approach to Harmonize Urbanization and Biodiversity Conservation in India. *The Symphony Of Knowledge*, *105*.

Bhattacharyya, S. S., Ros, G. H., Furtak, K., Iqbal, H. M., & Parra-Saldívar, R. (2022). Soil carbon sequestration–An interplay between soil microbial community and soil organic matter dynamics. *The Science of the Total Environment*, *815*, 152928. doi:10.1016/j.scitotenv.2022.152928 PMID:34999062

Biermann, L., Clewley, D., Martinez-Vicente, V., & Topouzelis, K. (2020). Finding Plastic Patches in Coastal Waters using Optical Satellite Data. *Scientific Reports*, *10*(1), 5364. doi:10.1038/s41598-020-62298-z PMID:32327674

Biodiversity International. (2017). *Mainstreaming agrobiodiversity in sustainable food systems: scientific foundations for an agrobiodiversity index*. Biodiversity International.

BirdLife International. (2018). Afropavo congensis. *The IUCN Red List of Threatened Species*. BirdLife International.

Bläsing, M., & Amelung, W. (2018). Plastics in soil: Analytical methods and possible sources. *The Science of the Total Environment*, *612*, 422–435. doi:10.1016/j.scitotenv.2017.08.086 PMID:28863373

Bodmer, R. E., Eisenberg, J. F., & Redford, K. H. (1997). Hunting and the likelihood of extinction of Amazonian mammals. *Conservation Biology*, *11*(2), 460–466. doi:10.1046/j.1523-1739.1997.96022.x

Borah, I. D., & Borah, S. R. (n.d.). *Precision Irrigation Systems: Enhancing Water Efficiency and Sustainability in Agriculture*. www.vigyanvarta.com

Borges, R. C., Dos Santos, F. V., Caldas, V. G., & Lapa, C. M. F. (2015). Use of geographic information system (GIS) in the characterization of the Cunha Canal, Rio de Janeiro, Brazil: Effects of the urbanization on water quality. *Environmental Earth Sciences*, *73*(3), 1345–1356. doi:10.1007/s12665-014-3493-1

Boscutti, F., Lami, F., Pellegrini, E., Buccheri, M., Busato, F., Martini, F., Sibella, R., Sigura, M., & Marini, L. (2022). Urban sprawl facilitates invasions of exotic plants across multiple spatial scales. *Biological Invasions*, *24*(5), 1497–1510. doi:10.1007/s10530-022-02733-6

Boyd, J., & Banzhaf, S. (2007). What are ecosystem services? The need for standardized environmental accounting units. *Ecological Economics*, *63*(2-3), 616–626. doi:10.1016/j.ecolecon.2007.01.002

Bradshaw, C. J. A., Sodhi, N. S., & Brook, B. W. (2009). Tropical turmoil: A biodiversity tragedy in progress. *Frontiers in Ecology and the Environment*, *7*(2), 79–87. doi:10.1890/070193

Brahney, J., Hallerud, M., Heim, E., Hahnenberger, M., & Sukumaran, S. (2020). Plastic rain in protected areas of the United States. *Science*, *368*(6496), 1257–1260. doi:10.1126/science.aaz5819 PMID:32527833

Brat, D., & Pathak, I. R. (2020). Impact of Pollution on Biodiversity: A Review. *IRE J.*, *4*(2), 11-15.

Bridgewater, P., & Schmeller, D. S. (2024). Nature abhors a vacuum: Invasive themes and other outputs from the tenth plenary of the Intergovernmental Platform on Biodiversity and Ecosystem Services (IPBES-10). *Biodiversity and Conservation*, *33*(3), 867–873. doi:10.1007/s10531-024-02797-x

Brothers, T. S., & Spingarn, A. (1992). Forest fragmentation and alien plant invasion of central Indiana old-growth forests. *Conservation Biology*, *6*(1), 91–100. doi:10.1046/j.1523-1739.1992.610091.x

Brown, R., & White, S. (2019). Compensatory Forestry: A Framework for Sustainable Development. *Forest Policy and Economics*, *101*, 14–22.

Brown, R., & White, S. (2020). Soil Preparation and Nutrient Management in Urban Avenue Plantation. *Urban Soil Management Quarterly*, *15*(1), 56–72.

Bubu-Davies, O., & Anwuri, P. (2022). Microplastics: Potential impacts on aquatic biodiversity. *Tropical Freshwater Biology*, *31*(1), 45–60. doi:10.4314/tfb.v31i1.4

Buchholz, S., Hannig, K., Möller, M., & Schirmel, J. (2018). Reducing management intensity and isolation as promising tools to enhance ground-dwelling arthropod diversity in urban grasslands. *Urban Ecosystems*, *21*(6), 1139–1149. doi:10.1007/s11252-018-0786-2

Bunn, S. E., & Arthington, A. H. (2002). Basic principles and ecological consequences of altered flow regimes for aquatic biodiversity. *Environmental Management*, *30*(4), 492–507. doi:10.1007/s00267-002-2737-0 PMID:12481916

Butchart, S. H., Clarke, M., Smith, R. J., Sykes, R. E., Scharlemann, J. P., Harfoot, M., & Burgess, N. D. (2015). Shortfalls and solutions for meeting national and global conservation area targets. *Conservation Letters*, *8*(5), 329–337. doi:10.1111/conl.12158

Byron, H. J., Treweek, J. R., Sheate, W. R., & Thompson, S. (2000). Road developments in the UK: An analysis of ecological assessment in environmental impact statements produced between 1993 and 1997. *Journal of Environmental Planning and Management*, *43*(1), 71–97. doi:10.1080/09640560010775

Cadotte, M. W., Yasui, S. L. E., Livingstone, S., & MacIvor, J. S. (2017). Are urban systems beneficial, detrimental, or indifferent for biological invasion? *Biological Invasions*, *19*(12), 3489–3503. doi:10.1007/s10530-017-1586-y

Cahill, A. E., Aiello-Lammens, M. E., Fisher-Reid, M. C., Hua, X., Karanewsky, C. J., Yeong Ryu, H., & Wiens, J. J. (2013). How does climate change cause extinction. *Proceedings of the Royal Society B: Biological Sciences, 280*(1750), 20121890. 10.1098/rspb.2012.1890

Cai M., Liu Y., Chen K., Huang D., & Yang S. (2016). Quantitative analysis of anthropogenic influences on coastal water: A new perspective. *Ecological indicators. 67*(3).

Campbell, T., Pin, K., Ngor, P. B., & Hogan, Z. (2020). Conserving Mekong megafishes: Current status and critical threats in Cambodia. *Water (Basel)*, *12*(6), 1820. doi:10.3390/w12061820

Caplan, T. R., Inaho, B., Yarrao, M., Pile, J., Milczarek, M., & McKenna, C. (2024). Applying resilience thinking to rehabilitating a novel social–ecological system: A case study from the lower Ok Tedi, Papua New Guinea. In *Resilience and Riverine Landscapes* (pp. 593–626). Elsevier. https://www.sciencedirect.com/science/article/pii/B9780323917162000248 doi:10.1016/B978-0-323-91716-2.00024-8

Cardinale, B. J., Duffy, J. E., Gonzalez, A., Hooper, D. U., Perrings, C., Venail, P., & Naeem, S. (2012). Biodiversity loss and its impact on humanity. *Nature*, *486*(7401), 59–67. doi:10.1038/nature11148 PMID:22678280

Carey, R. O., & Migliaccio, K. W. (2009). Contribution of wastewater treatment plant effluents to nutrient dynamics in aquatic systems: A review. *Environmental Management*, *44*(2), 205–217. doi:10.1007/s00267-009-9309-5 PMID:19458999

Carlisle, D. M., Wolock, D. M., & Meador, M. R. (2010). Alteration of streamflow magnitudes and potential ecological consequences: A multiregional assessment. *Frontiers in Ecology and the Environment*, *9*(5), 264–270. doi:10.1890/100053

Carrete, M., & Tella, J. L. (2011). Inter-individual variability in fear of humans and relative brain size of the species are related to contemporary urban invasion in birds. *PLoS One*, *6*(4), e18859. doi:10.1371/journal.pone.0018859 PMID:21526193

Carvalho, F. P. (2017). Mining industry and sustainable development: Time for change. *Food and Energy Security*, *6*(2), 61–77. doi:10.1002/fes3.109

CBD (Convention on Biological Diversity). (2005). *Biodiversity: Importance and Benefits*. CBD. https://www.cbd.int/biodiversity

CBD (Convention on Biological Diversity). (2014). *CBD Technical Series No. 75 - Ecosystem-based Adaptation in Marine and Coastal Ecosystems: Opportunities and Challenges*. Secretariat of the Convention on Biological Diversity.

CBD (Convention on Biological Diversity). (2020). *Global Biodiversity Outlook 5*. Secretariat of the Convention on Biological Diversity.

Chakraborty, B. K., Bhattacharjee, S., & Muniya, S. (2021). A Study on Aquatic Biodiversity of Shuthi-Shaiduli River of Bangladesh. *International Journal of Biological Innovations.*, *3*(1), 58–67. doi:10.46505/IJBI.2021.3104

Chandler, M., See, L., Copas, K., Bonde, A. M., López, B. C., Danielsen, F., Legind, J. K., Masinde, S., Miller-Rushing, A. J., Newman, G., Rosemartin, A., & Turak, E. (2017). Contribution of citizen science towards international biodiversity monitoring. *Biological Conservation*, *213*, 280–294. doi:10.1016/j.biocon.2016.09.004

Chaplin-Kramer, R., Sharp, R. P., Mandle, L., Sim, S., Johnson, J., Butnar, I., Milà i Canals, L., Eichelberger, B. A., Ramler, I., Mueller, C., McLachlan, N., Yousefi, A., King, H., & Kareiva, P. M. (2015). Spatial patterns of agricultural expansion determine impacts on biodiversity and carbon storage. *Proceedings of the National Academy of Sciences of the United States of America*, *112*(24), 7402–7407. doi:10.1073/pnas.1406485112 PMID:26082547

Charnley, S., Fischer, A. P., & Jones, E. T. (2007). Integrating traditional and local ecological knowledge into forest biodiversity conservation in the Pacific Northwest. *Forest Ecology and Management*, *246*(1), 14–28. doi:10.1016/j.foreco.2007.03.047

Chauhan, P., Sharma, N., Tapwal, A., Kumar, A., Verma, G. S., Meena, M., Seth, C. S., & Swapnil, P. (2023). Soil Microbiome: Diversity, Benefits and Interactions with Plants. In Sustainability (Switzerland) (Vol. 15, Issue 19). Multidisciplinary Digital Publishing Institute (MDPI). doi:10.3390/su151914643

Chazdon, R. L., Broadbent, E. N., Rozendaal, D. M., Bongers, F., Zambrano, A. M. A., Aide, T. M., & Poorter, L. (2016). Carbon sequestration potential of second-growth forest regeneration in the Latin American tropics. *Science Advances*, *2*(5), e1501639. doi:10.1126/sciadv.1501639 PMID:27386528

Clewell, A. F., & Aronson, J. (2006). Motivations for the restoration of ecosystems. *Conservation Biology*, *20*(2), 420–428. doi:10.1111/j.1523-1739.2006.00340.x PMID:16903103

Community Greening Initiative. (2022). Engaging Communities in Avenue Plantation: Best Practices and Lessons Learned. *Community Development Journal: An International Forum*, *18*(3), 201–215.

Conservation Society. (2013). The plight of forest elephants. *Wildlife Journal, 15*(3), 45-52.

Convention on Biological Diversity. (2017). *Biodiversity and the 2030 agenda for sustainable development policy brief.* Convention on Biological Diversity Technical Note.

Cook, S. (2018). *The spice of life: the fundamental role of diversity on the farm and on the plate.* (Discussion Paper, IIED and Hivos. London and The Hague).

Coral Guardian. 2022, https://www.coralguardian.org/en/carbon-balance-in-corals/

Costanza, R., d'Arge, R., de Groot, R., Farber, S., Grasso, M., Hannon, B., & van den Belt, M. (1997). The value of the world's ecosystem services and natural capital. *Nature*, *387*(6630), 253–260. doi:10.1038/387253a0

Costanza, R., de Groot, R., Sutton, P., van der Ploeg, S., Anderson, S. J., Kubiszewski, I., & Turner, R. K. (2014). Changes in the global value of ecosystem services. *Global Environmental Change*, *26*, 152–158. doi:10.1016/j.gloenvcha.2014.04.002

Crossman, J., Hurley, R. R., Futter, M., & Nizzetto, L. (2020). Transfer and transport of microplastics from biosolids to agricultural soils and the wider environment. *The Science of the Total Environment*, *724*, 138334. doi:10.1016/j.scitotenv.2020.138334 PMID:32408466

da Silva, Q. P., Moreira, D. S., de Freitas, H. C., & Domingues, L. M. (2024). Discrepancies between observation and ERA5 reanalysis in the Amazon deforestation context: A case study. *Dynamics of Atmospheres and Oceans*, *106*, 101442. doi:10.1016/j.dynatmoce.2024.101442

Daily, G. C. (1997). *Nature's Services: Societal Dependence on Natural Ecosystems*. Island Press.

Daily, G. C., Alexander, S., Ehrlich, P. R., Goulder, L., Lubchenco, J., Matson, P. A., & Woodwell, G. M. (1997). Ecosystem services: Benefits supplied to human societies by natural ecosystems. *Issues in Ecology*, *2*, 1–16.

Dan, C., Xingyuan, H., & Peng, W. (2012). Recycling effective ways of waste plastics. *Engineering Plastics Application.*, *40*, 92–94.

Dasgupta, A. (2012). Impact of mining on rural environmental and economy. A case study, Kota district, Rajasthan. *International Journal of Remote sensing and Geoscience (IJRSG), 2*, 21-26.

David, M. P. (2018). Sustainable Development Goals (SDGs)- Challenges for India. *Indian Journal of Public Health Research & Development*, *9*(3), 1. doi:10.5958/0976-5506.2018.00172.9

Davies, T. W., & Smyth, T. (2018). Why artificial light at night should be a focus for global change research in the 21st century. *Global Change Biology*, *24*(3), 872–882. doi:10.1111/gcb.13927 PMID:29124824

Dayananda, B., Bezeng, S. B., Karunarathna, S., & Jeffree, R. A. (2021). Climate change impacts on tropical reptiles: Likely effects and future research needs based on Sri Lankan perspectives. *Frontiers in Ecology and Evolution*, *9*, 688723. doi:10.3389/fevo.2021.688723

Dayton, P. K., Simon, F., Thrush, M., Tundi, A., & Hofman, R. J. (1995). Environmental E f f e c t s o f Ma r i n e Fi s h i n g. A q u a t i c Conservation. *Marine and Freshwater Ecosystems.*, *5*(3), 205–232. doi:10.1002/aqc.3270050305

De Framond, L., & Brumm, H. (2022). Long-term effects of noise pollution on the avian dawn chorus: a natural experiment facilitated by the closure of an international airport. *Proceedings of the Royal Society B, 289*(1982), 20220906. IEEE. 10.1098/rspb.2022.0906

de Groot, R. S., Alkemade, R., Braat, L., Hein, L., & Willemen, L. (2010). Challenges in integrating the concept of ecosystem services and values in landscape planning, management and decision making. *Ecological Complexity*, *7*(3), 260–272. doi:10.1016/j.ecocom.2009.10.006

de Guzman, I., Altieri, P., Elosegi, A., Pérez-Calpe, A. V., von Schiller, D., González, J. M., Brauns, M., Montoya, J. M., & Larrañaga, A. (2021). *Dataset: Water diversion and pollution interactively shape freshwater food webs through bottom-up mechanisms.*

de Mori, B., Mercugliano, E., Biasetti, P., Pollastri, I., Spiriti, M. M., Florio, D., Andreucci, F., Göritz, F., Holtze, S., Galli, C., Stejskal, J., Colleoni, S., Lazzari, G., Seet, S., Zwilling, J., Ndeereh, D., Lekolool, I., Ngulu, S., Mijele, D., & Hildebrandt, T. B. (2024). The ethical assessment of Assisted Reproductive Technologies (ART) in wildlife conservation. *Biological Conservation*, *290*, 110423. doi:10.1016/j.biocon.2023.110423

de Souza Machado, A. A., Lau, C. W., Kloas, W., Bergmann, J., Bachelier, J. B., Faltin, E., Becker, R., Görlich, A. S., & Rillig, M. C. (2019). Microplastics Can Change Soil Properties and Affect Plant Performance. *Environmental Science & Technology*, *53*(10), 6044–6052. doi:10.1021/acs.est.9b01339 PMID:31021077

Deemer, B. R., Harrison, J. A., Li, S., Beaulieu, J. J., DelSontro, T., Barros, N., Bezerra-Neto, J. F., Powers, S. M., dos Santos, M. A., & Vonk, J. A. (2016). Greenhouse gas emissions from reservoir water surfaces: A new global synthesis. *Bioscience*, *66*(11), 949–964. doi:10.1093/biosci/biw117 PMID:32801383

Delgado, J. D., Arroyo, N. L., Arévalo, J. R., & Fernández-Palacios, J. M. (2007). Edge effects of roads on temperature, light, canopy cover, and canopy height in laurel and pine forests (Tenerife, Canary Islands). *Landscape and Urban Planning*, *81*(4), 328–340. doi:10.1016/j.landurbplan.2007.01.005

Deutsch C.A., Tewksbury J. J., Huey R.B., Sheldon K.S., Ghalambor C.K., Haak D.C. and Martin P.R. (2008). Impacts of climate warming on terrestrial ectotherms across latitude. *Proc. Natl Acad. Sci.*, *105*(18).

Devi, Y. B., Meetei, T. T., & Kumari, N. (2018). Impact of Pesticides on Soil Microbial Diversity and Enzymes: A Review. *International Journal of Current Microbiology and Applied Sciences*, *7*(06), 952–958. doi:10.20546/ijcmas.2018.706.113

Dewson, Z. S., James, A. B., & Death, R. G. (2007). Invertebrate responses to short-term water abstraction in small New Zealand streams. *Freshwater Biology*, *52*(2), 357–369. doi:10.1111/j.1365-2427.2006.01682.x

Dhar, M., Jasrotia, R., Langer, S., & Suwartiningsih, N. (2023). Impact of Microplastics on Reproductive and Physiological Aspects of Aquatic Inhabitants. In M. A. Rather, A. Amin, Y. A. Hajam, A. Jamwal, & I. Ahmad (Eds.), *Xenobiotics in Aquatic Animals*. Springer. doi:10.1007/978-981-99-1214-8_6

Dhyani, S. (2024). *Strengthening Science Diplomacy for Reducing Biodiversity Loss and Restoring Degraded Ecosystems*. NOPR. https://nopr.niscpr.res.in/bitstream/123456789/63145/1/SD%207%282%29%2019-22.pdf

Díaz, S., Pascual, U., Stenseke, M., Martín-López, B., Watson, R. T., Molnár, Z., & Shirayama, Y. (2018). Assessing nature's contributions to people. *Science*, *359*(6373), 270–272. doi:10.1126/science.aap8826 PMID:29348221

Díaz, S., Settele, J., Brondízio, E. S., Ngo, H. T., Agard, J., Arneth, A., Balvanera, P., Brauman, K. A., Butchart, S. H. M., Chan, K. M. A., Garibaldi, L. A., Ichii, K., Liu, J., Subramanian, S. M., Midgley, G. F., Miloslavich, P., Molnár, Z., Obura, D., Pfaff, A., & Zayas, C. N. (2019). Pervasive human-driven decline of life on Earth points to the need for transformative change. *Science*, *366*(6471), eaax3100. Advance online publication. doi:10.1126/science.aax3100 PMID:31831642

Diederichs, N. J., & Ellery, W. N. (2001). An analysis of plant species distributions on the floodplains of the Okavango River, Namibia, with respect to impacts of possible water abstraction. African Journal of Aquatic Sciences, 26, 139–147.

Donaldson, A., & Bennett, A. (2004). *Ecological effects of roads: Implications for the internal fragmentation of Australian parks and reserves*.

Dooling, R. J., & Leek, M. R. (2018). Communication masking by man-made noise. In H. Slabbekoorn, R. J. Dooling, A. N. Popper, & R. R. Fay (Eds.), *Effects of anthropogenic noise on animals* (pp. 23–46). Springer. doi:10.1007/978-1-4939-8574-6_2

Douvere, F., et al. (2019). Marine Spatial Planning: A Step-by-Step Approach Toward Ecosystem-Based Management. *Intergovernmental Oceanographic Commission Manuals and Guides, 83.*

Doxsey-Whitfield, E., MacManus, K., Adamo, S. B., Pistolesi, L., Squires, J., Borkovska, O., & Baptista, S. R. (2015). Taking advantage of the improved availability of census data: A first look at the gridded population of the world, version 4. *Papers in Applied Geography*, *1*(3), 226–234. doi:10.1080/23754931.2015.1014272

Drinkwater, K. F., & Frank, K. T. (1994). Effects of river regulation and diversion on marine fish and invertebrates. *Aquatic Conservation*, *4*(2), 135–151. doi:10.1002/aqc.3270040205

Dudgeon, D. (2019). Multiple threats imperil freshwater biodiversity in the Anthropocene. *Current Biology*, *29*(19), R960–R967. doi:10.1016/j.cub.2019.08.002 PMID:31593677

Dudley, N., Jonas, H., Nelson, F., Parrish, J., Pyhälä, A., Stolton, S., & Watson, J. (2018). The essential role of other effective area-based conservation measures in achieving big bold conservation targets. *Global Ecology and Conservation*, *15*, e00424. doi:10.1016/j.gecco.2018.e00424

Dueck, T. A., & Endendijk, G. J. (1987). Soil pollution and changes in vegetation. *Chemosphere*, *16*(5), 1021–1030. doi:10.1016/0045-6535(87)90039-7

Duputié, A., & Massol, F. (2013). An empiricist's guide to theoretical predictions on the evolution of dispersal. *Interface Focus*, *3*(6), 20130028. doi:10.1098/rsfs.2013.0028 PMID:24516715

Duraiappah, A. K., Naeem, S., Agardy, T., Ash, N. J., Cooper, H. D., Díaz, S., & Van Jaarsveld, A. (2005). *Ecosystems and human well-being: biodiversity synthesis; a report of the Millennium Ecosystem Assessment.* MEA.

Dutta, V., Sharma, U., Iqbal, K., Adeeba, Kumar, R., & Pathak, A. K. (2018). Impact of river channelization and riverfront development on fluvial habitat: Evidence from Gomti River, a tributary of Ganges, India. *Environmental Sustainability*, *1*(2), 167–184. doi:10.1007/s42398-018-0016-0

Dynesius, M., & Nilsson, C. (1994). Fragmentation and flow regulation of river systems in the northern third of the world. *Science*, *266*(5186), 753–762. doi:10.1126/science.266.5186.753 PMID:17730396

Edorado, L. (2019). The impact of cattle ranching on Amazon deforestation: An evidence-based review. *Journal of Agricultural Economics*, *72*(1), 3–21.

Ellery, W. N., Dahlberg, A. C., Strydom, R., Neal, M. J., & Jackson, J. (2003). Diversion of water flow from a floodplain wetland stream: An analysis of geomorphological setting and hydrological and ecological consequences. *Journal of Environmental Management*, *68*(1), 51–71. doi:10.1016/S0301-4797(03)00002-1 PMID:12767862

Ellery, W. N., Ellery, K., Rogers, K. H., McCarthy, T. S., & Walker, B. H. (1993). Vegetation, hydrology, and sedimentation processes as determinants of channel form and dynamics in the northeastern Okavango Delta, Botswana. *African Journal of Ecology*, *31*, 10–25. doi:10.1111/j.1365-2028.1993.tb00514.x

Ellery, W. N., & McCarthy, T. S. (1998). Environmental change over two decades since dredging and excavation of the lower Boro River, Okavango Delta, Botswana. *Journal of Biogeography, 25*, 361–378. doi:10.1046/j.1365-2699.1998.252168.x

Elmqvist, T., Fragkias, M., Goodness, J., Güneralp, B., Marcotullio, P. J., McDonald, R. I., & Wilkinson, C. (2013). *Urbanization, biodiversity and ecosystem services: challenges and opportunities: a global assessment*. Springer Nature. doi:10.1007/978-94-007-7088-1

Environmental Protection Agency (EPA). (2021). *Compensatory Forestry Guidelines for Sustainable Development Projects*. EPA.

Erbe, C., Dent, M. L., Gannon, W. L., McCauley, R. D., Römer, H., Southall, B. L., & Thomas, J. A. (2022). The effects of noise on animals. In Exploring animal behavior through sound: volume 1: methods (pp. 459-506). Cham: Springer International Publishing. doi:10.1007/978-3-030-97540-1_13

Ergashevich, E. A., & Mado, A. (2024). Methodology of Organizing and Implementing Training Activities. *Academia Open, 9*(1), 10–21070.

Estoque, R. C., & Wu, J. (2024). The resilience–sustainability–quality of life nexus. *The Science of the Total Environment, 912*, 169526. doi:10.1016/j.scitotenv.2023.169526 PMID:38135087

Faeth, S. H., Bang, C., & Saari, S. (2011). Urban biodiversity: Patterns and mechanisms. *Annals of the New York Academy of Sciences, 1223*(1), 69–81. doi:10.1111/j.1749-6632.2010.05925.x PMID:21449966

Fahrig, L. (1997). Relative effects of habitat loss and fragmentation on population extinction. *The Journal of Wildlife Management, 61*(3), 603–610. doi:10.2307/3802168

Falchi, F., Cinzano, P., Duriscoe, D., Kyba, C. C., Elvidge, C. D., Baugh, K., Portnov, B. A., Rybnikova, N. A., & Furgoni, R. (2016). The new world atlas of artificial night sky brightness. *Science Advances, 2*(6), e1600377. doi:10.1126/sciadv.1600377 PMID:27386582

FAO (Food and Agriculture Organization of the United Nations). (2015). *Status of the World's Soil Resources: Main report.* Rome.

FAO. (2011). The State of Forests in the Amazon basin and Southeast Asian, Brazzaville, Republic of Congo. Food and Agriculture Organization of the United Nations.

FAO. (2015). *Global forest resources assessment 2015: How are the world's forests changing?* Food and Agriculture Organization of the United Nations.

FAO. (2018). *State of the world's forests 2018: Forest pathways to sustainable development.* Food and Agriculture Organization of the United Nations.

FAO. (2019). *The State Of The World's Biodiversity For Food And Agriculture.* FAO.

FAO. (2019). *The state of the world's biodiversity for food and agriculture*. FAO. http://www.fao.org/3/CA3129EN/ ca3129en.pdf

FAO. (2020). *Global Forest Resources Assessment 2020 – Key findings*. FAO. https://doi.org/ doi:10.4060/ca8753en

Farrelly, D. J., Everard, C. D., Fagan, C. C., & McDonnell, K. P. (2013). Carbon sequestration and the role of biological carbon mitigation: A review. *Renewable & Sustainable Energy Reviews*, *21*, 712–727. doi:10.1016/j.rser.2012.12.038

Ferreira Da Silva, M. J., Minhós, T., & Sampaio-Dias, S. (2024). News for nature: Communicating biodiversity importance and conservation challenges in Guinea-Bissau, West Africa, through local media and online platforms. *Conservation Science and Practice*, *6*(1), e13051. doi:10.1111/ csp2.13051

Flesch, A. D. (2017). Influence of local and landscape factors on distributional dynamics: a species-centred, fitness-based approach. *Proceedings of the Royal Society B: Biological Sciences, 284*(1858), 20171001. 10.1098/rspb.2017.1001

Flores, F., Garreaud, R., & Muñoz, R. C. (2014). OpenFOAM applied to the CFD simulation of turbulent buoyant atmospheric flows and pollutant dispersion inside large open pit mines under intense insolation. *Computers & Fluids*, *90*, 72–87. doi:10.1016/j.compfluid.2013.11.012

Flory, S. L., & Clay, K. (2009). Effects of roads and forest successional age on experimental plant invasions. *Biological Conservation*, *142*(11), 2531–2537. doi:10.1016/j.biocon.2009.05.024

Flory, S. L., & Clay, K. (2009). Invasive plant removal method determines native plant community responses. *Journal of Applied Ecology*, *46*(2), 434–442. doi:10.1111/j.1365-2664.2009.01610.x

Fontaine, T. D., & Bartell, S. M. (Eds.). (1983). *Dynamics of lotic ecosystems* (p. 494). Ann Arbor Science.

Forest Research Institute (FRI). (2020). Soil Health and Carbon Sequestration in Compensatory Forestry: A Case Study. *Forestry Reports*, *72*, 89–102.

Forman, R. T., & Alexander, L. E. (1998). Roads and their major ecological effects. *Annual Review of Ecology and Systematics*, *29*(1), 207–231. doi:10.1146/annurev.ecolsys.29.1.207

Freeman, B. G., Scholer, M. N., Ruiz-Gutierrez, V., & Fitzpatrick, J. W. (2018). Climate change causes upslope shifts and mountaintop extirpations in a tropical bird community. *Proceedings of the National Academy of Sciences of the United States of America*, *115*(47), 11982–11987. doi:10.1073/pnas.1804224115 PMID:30373825

Froehner, S., Souza, D. B., Machado, K. S., & Rosa, E. C. (2010). Tracking anthropogenic inputs in Barigui River, Brazil using biomarkers. *Water, Air, and Soil Pollution*, *201*(1-4), 33–41. doi:10.1007/s11270-009-0220-8

Fuentes, M., Limpus, C., Hamann, M., & Dawson, J. (2010). Potential impacts of projected sea-level rise on sea turtle rookeries. Aquatic Conserv. Mar. *Aquatic Conservation*, *20*(2), 132–139. doi:10.1002/aqc.1088

Furtado, M. D. S. C., Queiroz, J. C. B., Bentes, B., Gouveia, N. D. A., De Lima, M. J. A., Ruffino, M. L., & Isaac, V. (2024). How does climate change affect small scale fisheries? A case study of the Lower Amazon in Brazil. *Fisheries Management and Ecology*, *31*(1), e12654. doi:10.1111/fme.12654

Galgani, F., Hanke, G., Werner, S. D. V. L., & De Vrees, L. (2015). Marine litter within the European marine strategy framework directive. *ICES Journal of Marine Science*, *70*(6), 1055–1064. doi:10.1093/icesjms/fst122

Gandhi, N. R., Farfaras, N., Wang, N. L., & Chen, W. (2021). Life Cycle Assessment of Recycling High-Density Polyethylene Plastic Waste. *Journal of Renewable Materials*, *9*(8), 1463–1483. doi:10.32604/jrm.2021.015529

Gao, W., Liu, S., & Huang, L. (2012). A global best artificial bee colony algorithm for global optimization. *Journal of Computational and Applied Mathematics*, *236*(11), 2741–2753. doi:10.1016/j.cam.2012.01.013

Gaston, K. J., & Spicer, J. I. (2013). *Biodiversity: An introduction*. John Wiley & Sons. https://books.google.com/books?hl=en&lr=&id=0Bjp2o5CVnQC&oi=fnd&pg=PT6&dq=Biodiversity:+An+overview&ots=Y2X6YSmcEx&sig=KHEre9jRGafSgjzx7GfRWb1tK-w

Gaston, K. A., Eft, J. A., & Lauer, T. E. (2012). Morphology and its effect on habitat selection of stream fishes. *Proceedings of the Indiana Academy of Sciences*, *121*(1), 71–78.

Gaston, K. J., & Holt, L. A. (2018). Nature, extent and ecological impli- cations of night-time light from road vehicles. *Journal of Applied Ecology*, *55*(5), 2296–2307. doi:10.1111/1365-2664.13157 PMID:30147142

Gaveau, D. L., Sloan, S., Molidena, E., Yaen, H., Sheil, D., Abram, N. K., Ancrenaz, M., Nasi, R., Quinones, M., Wielaard, N., & Meijaard, E. (2014). Four decades of forest persistence, clearance and logging on Borneo. *PLoS One*, *9*(7), e101654. doi:10.1371/journal.pone.0101654 PMID:25029192

GBS. (1992). *Global Biodiversity Strategy: Guidelines for Action to Save, Study and Use Earth's Biotic Wealth sustainable and equitably. World Resources Institute (WRI), World Conservation Union (IUCN), United Nations Environmental programme (UNEP)*. IUCN.

Ghosh P. and Brand W.A. (2003). Stable isotope ratio mass spectrometry in global climate change research. Int. *Journal of Mass Spec., 3806*(03).

Giller, K. E., Witter, E., Corbeels, M., & Tittonell, P. (2015). Conservation agriculture and smallholder farming in Africa: The heretics' view. *Field Crops Research*, *114*(1), 23–34. doi:10.1016/j.fcr.2009.06.017

Girisha, S. K., & Sridhar, K. R. (2017). Impact of Iron Ore Mining on Macroinvertebrate Communities in Two Streams of the Sandur Hills, Bellary District, Karnataka, India. *Journal of Entomology and Zoology Studies*, *5*(1), 95–100.

Gogoi, A., Ahirwal, J., & Sahoo, U. K. (2021). Plant biodiversity and carbon sequestration potential of the planted forest in Brahmaputra flood plains. *Journal of Environmental Management, 280*, 111671. doi:10.1016/j.jenvman.2020.111671 PMID:33218834

Gonthier, D. J., Ennis, K. K., Farinas, S., Hsieh, H. Y., Iverson, A. L., Batáry, P., & Perfecto, I. (2014). Biodiversity conservation in agriculture requires a multi-scale approach. *Proceedings of the Royal Society B: Biological Sciences, 281*(1791), 20141358. 10.1098/rspb.2014.1358

Goodman, P. S. (1987). Mkuze floodplain and swamps. Wetland management case study. *African Wildlife*, *41*, 269–270.

Goosem, M., & Tucker, N. I. J. (2013). Repairing the rainforest: a biodiversity vision for the Wet

Goswami Yashwant Singh Parmar, M. (n.d.). Hybridization: Importance, Techniques and Consequences. *Recent Trends in Agriculture*. https://doi.org/

Goudie, A. (2013). The human impact on the natural environment: past, present and future. (7th ed). A John Wiley & Sons.

Green, D. S. (2020). Biological and Ecological Impacts of Plastic Debris in Aquatic Ecosystems. In F. Stock, G. Reifferscheid, N. Brennholt, & E. Kostianaia (Eds.), *Plastics in the Aquatic Environment - Part I. The Handbook of Environmental Chemistry* (Vol. 111). Springer. doi:10.1007/698_2020_509

Green, D. S., Boots, B., Blockley, D. J., Rocha, C., & Thompson, R. (2015). Impacts of discarded plastic bags on marine assemblages and ecosystem functioning. *Environmental Science & Technology*, *49*(9), 5380–5389. doi:10.1021/acs.est.5b00277 PMID:25822754

Green, L. (2017). Challenges in Avenue Plantation: Lessons from Urban Planning. *International Journal of Sustainable Development*, *25*(4), 532–546.

Green, L. (2018). Designing Aesthetically Pleasing Green Corridors: Spacing and Layout in Avenue Plantation. *Landscape Architecture Journal*, *40*(4), 321–335.

Gregory, M. R., & Andrady, A. L. (2003). Plastics in the marine environment. In A. L. Andrady (Ed.), *Plastics and the environment* (pp. 379–401). John Wiley and Sons, Inc. doi:10.1002/0471721557.ch10

Grill, G., Lehner, B., Thieme, M., Geenen, B., Tickner, D., Antonelli, F., Babu, S., Borrelli, P., Cheng, L., Crochetiere, H., Ehalt Macedo, H., Filgueiras, R., Goichot, M., Higgins, J., Hogan, Z., Lip, B., McClain, M. E., Meng, J., Mulligan, M., & Zarfl, C. (2019). Mapping the world's free-flowing rivers. *Nature*, *569*(7755), 215–221. doi:10.1038/s41586-019-1111-9 PMID:31068722

Grooten, M., & Almond, R. E. A. (Eds.). (2018). WWF Living Planet Report – 2018: Aiming higher.

Guilherme, F., Vicente, J. R., Carretero, M. A., & Farinha-Marques, P. (2024). Mapping multigroup responses to land cover legacy for urban biodiversity conservation. *Biological Conservation*, *291*, 110508. doi:10.1016/j.biocon.2024.110508

Güneralp, B., Zhou, Y., Ürge-Vorsatz, D., Gupta, M., Yu, S., Patel, P. L., Fragkias, M., Li, X., & Seto, K. C. (2017). Global scenarios of urban density and its impacts on building energy use through 2050. *Proceedings of the National Academy of Sciences of the United States of America, 114*(34), 8945–8950. doi:10.1073/pnas.1606035114 PMID:28069957

Gunn, R. L., Hartley, I. R., Algar, A. C., Niemelä, P. T., & Keith, S. A. (2022). Understanding behavioural responses to human-induced rapid environmental change: A meta-analysis. *Oikos, 2022*(4), e08366. doi:10.1111/oik.08366

Guo, J. J., Huang, X. P., Xiang, L., Wang, Y. Z., Li, Y. W., Li, H., Cai, Q. Y., Mo, C. H., & Wong, M. H. (2020). Source, migration and toxicology of microplastics in soil. *Environment International, 137*, 105263. doi:10.1016/j.envint.2019.105263 PMID:32087481

Gupta, P. (2017). Management of plastic waste: A step towards clean environment. *International Journal of Renewable Energy Technology, 8*(3/4), 387. doi:10.1504/IJRET.2017.088986

Haddad, N. M., Brudvig, L. A., Clobert, J., Davies, K. F., Gonzalez, A., Holt, R. D., Lovejoy, T. E., Sexton, J. O., Austin, M. P., Collins, C. D., Cook, W. M., Damschen, E. I., Ewers, R. M., Foster, B. L., Jenkins, C. N., King, A. J., Laurance, W. F., Levey, D. J., Margules, C. R., & Townshend, J. R. (2015). Habitat fragmentation and its lasting impact on Earth's ecosystems. *Science Advances, 1*(2), e1500052. doi:10.1126/sciadv.1500052 PMID:26601154

Haetrakul, T., Munanansup, S., Assawawong-kasem, N., & Chansue, N. (2009). A case report: Stomach foreign object in whaleshark (Rhincodon typus) stranded in Thailand. *Proceedings of the 4th International Symposium on Seastar 2000 and Asian Bio-Logging Science*, (pp. 83–85). Geneca Environment Network. https://www.genevaenvironmentnetwork.org/resources/updates/plastics-and-biodiversity/-

Hahs, A. K., Fournier, B., Aronson, M. F., Nilon, C. H., Herrera-Montes, A., Salisbury, A., Threlfall, C. G., Rega-Brodsky, C. C., Lepczyk, C. A., La Sorte, F. A., MacGregor-Fors, I., Scott MacIvor, J., Jung, K., Piana, M. R., Williams, N. S. G., Knapp, S., Vergnes, A., Acevedo, A. A., Gainsbury, A. M., & Moretti, M. (2023). Urbanisation generates multiple trait syndromes for terrestrial taxa worldwide. *Nature Communications, 14*(4751), 4751. doi:10.1038/s41467-023-39746-1 PMID:37550318

Halpern, B. S., Frazier, M., Afflerbach, J., Lowndes, J. S., Micheli, F., O'Hara, C., & Selkoe, K. A. (2019). Recent pace of change in human impact on the world's ocean. *Scientific Reports, 9*(1), 11609. doi:10.1038/s41598-019-47201-9 PMID:31406130

HårdingK.CarrollD.InfantesE.CarrollJ.HarkonenT. (2024). Long-term field studies are critical for safeguarding biodiversity: A new concept and concern. Authorea Preprints. https://essopenarchive.org/doi/full/10.22541/au.171010042.24317771 doi:10.22541/au.171010042.24317771/v1

Hartig, F., Abrego, N., Bush, A., Chase, J. M., Guillera-Arroita, G., Leibold, M. A., Ovaskainen, O., Pellissier, L., Pichler, M., & Poggiato, G. (2024). Novel community data in ecology-properties and prospects. *Trends in Ecology & Evolution*. Cell.com. https://www.cell.com/trends/ecology-evolution/fulltext/S0169-5347(23)00265-3

HasuiÉ.MartensenA. C.UezuA.PimentelR.RamosF. N.RibeiroM. C. (2023). Populations across species ranges respond differently to habitat loss and fragment. Research Square *(*Research Square*)*. doi:10.21203/rs.3.rs-2702017/v1

Hawbaker, T. J., Radeloff, V. C., Clayton, M. K., Hammer, R. B., & Gonzalez-Abraham, C. E. (2006). Road development, housing growth, and landscape fragmentation in northern Wisconsin: 1937–1999. *Ecological Applications*, *16*(3), 1222–1237. doi:10.1890/1051-0761(2006)016[1222:RDHGAL]2.0.CO;2 PMID:16827014

Hending, D., Randrianarison, H., Andriamavosoloarisoa, N. N. M., Ranohatra-Hending, C., Holderied, M., McCabe, G., & Cotton, S. (2023). Forest fragmentation and its associated edge-effects reduce tree species diversity, size, and structural diversity in Madagascar's transitional forests. *Biodiversity and Conservation*, *32*(10), 3329–3353. doi:10.1007/s10531-023-02657-0

He, Z., Lin, L., Wang, X., Qin, W., & Zhang, C. (2022). Carbon storage by marine microorganisms for carbon neutrality. *Frontiers in Marine Science*, *9*, 1018397. doi:10.3389/fmars.2022.1018397

Higgins, C. (1998). The role of traditional ecological knowledge in managing for biodiversity. *Forestry Chronicle*, *74*(3), 323–326. doi:10.5558/tfc74323-3

Hill, B. E., & Targ, N. (2000). The link between protecting natural resources and the issue of environmental justice. *BC Envtl. Aff. L. Rev.*, *28*, 1.

Hill, J. L., & Curran, P. J. (2001). Species composition in fragmented forests: Conservation implications of changing forest area. *Applied Geography (Sevenoaks, England)*, *21*(2), 157–174. doi:10.1016/S0143-6228(01)00002-9

Hilton, M., Walsh, J. C., Liddell, E., & Cook, C. N. (2022). Lessons from other disciplines for setting management thresholds for biodiversity conservation. *Conservation Biology*, *36*(1), e13865. doi:10.1111/cobi.13865 PMID:34811813

Hoegh-Guldberg, O., Poloczanska, E. S., Skirving, W., & Dove, S. (2017). Coral reef ecosystems under climate change and ocean acidification. *Frontiers in Marine Science*, *4*, 158. doi:10.3389/fmars.2017.00158

Holderegger, R., & Di Giulio, M. (2010). The genetic effects of roads: A review of empirical evidence. *Basic and Applied Ecology*, *11*(6), 522–531. doi:10.1016/j.baae.2010.06.006

Hooper, D. U., Adair, E. C., Cardinale, B. J., Byrnes, J. E., Hungate, B. A., Matulich, K. L., & O'Connor, M. I. (2012). A global synthesis reveals biodiversity loss as a major driver of ecosystem change. *Nature*, *486*(7401), 105–108. doi:10.1038/nature11118 PMID:22678289

Hossain, M. I., San, O. T., Ling, S. M., & Said, R. M. (2020). The role of environmental awareness and green technological usage to foster sustainable green practices in Bangladeshi manufacturing SMEs. *International Journal of Advanced Science and Technology*, *29*(7s), 3115–3124.

Houghton, R. A. (2015). Revised estimates of the annual net flux of carbon to the atmosphere from changes in land use and land management 1850–2000. *Tellus. Series B, Chemical and Physical Meteorology*, *55*(2), 378–390.

Howard, J., Sutton-Grier, A., Herr, D., Kleypas, J., Landis, E., Mcleod, E., Pidgeon, E., & Simpson, S. (2017). Clarifying the Role of Coastal and Marine Systems in Climate Mitigation. *Frontiers in Ecology and the Environment, 15*(1), 42–50. doi:10.1002/fee.1451

Hu, D., Shen, M., Zhang, Y., Li, H., & Zeng, G. (2019). Microplastics and nanoplastics: Would they affect global biodiversity change? *Environmental Science and Pollution Research International, 26*(19), 19997–20002. doi:10.1007/s11356-019-05414-5 PMID:31102222

Huntington, H. P. (2000). Using traditional ecological knowledge in science: Methods and applications. *Ecological Applications, 10*(5), 1270–1274. doi:10.1890/1051-0761(2000)010[1270:UTEKIS]2.0.CO;2

Ighalo, J. O., Dulta, K., Kurniawan, S. B., Omoarukhe, F. O., Ewuzie, U., Eshiemogie, S. O., & Abdullah, S. R. S. (2022). Progress in microalgae application for CO_2 sequestration. *Clean Chemical Engineering, 3*, 100044. doi:10.1016/j.clce.2022.100044

IPBES (Intergovernmental Science-Policy Platform on Biodiversity and Ecosystem Services). (2019). *Summary for policymakers of the global assessment report on biodiversity and ecosystem services*. IPBES Secretariat.

IPBES. (2019). *Global assessment report on biodiversity and ecosystem services of the Intergovernmental Science-Policy Platform on Biodiversity and Ecosystem Services*. IPBES Secretariat.

IPCC. (2007). An assessment of Intergovernmental Panel on climate change. Climate change synthesis report. Cambridge University Press, Cambridge.

IPCC. (2013). *Climate Change, the Physical Science Basis*. Cambridge University Press.

Isbell, F., Reich, P. B., Tilman, D., Hobbie, S. E., Polasky, S., Binder, S., & Bren, D. (2013). Nutrient enrichment, biodiversity loss, and c o n s e q u e n t d e c l i n e s i n e c o s y s t e m productivity. *Proceedings of the National Academy of Sciences of the United States of America, 110*(29), 11911–11916. doi:10.1073/pnas.1310880110 PMID:23818582

IUCN SSC Giraffe and Okapi Specialist Group. (2013). Okapia johnstoni. *The IUCN Red List of Threatened Species 2013*. IUCN.

Jackson, M. C., Friberg, N., Moliner Cachazo, L., Clark, D. R., Mutinova, P. T., O'Gorman, E. J., Kordas, R. L., Gallo, B., Pichler, D. E., Bespalaya, Y., Aksenova, O. V., Milner, A., Brooks, S. J., Dunn, N., Lee, K. W. K., Ólafsson, J. S., Gíslason, G. M., Millan, L., Bell, T., & Woodward, G. (2024). Regional impacts of warming on biodiversity and biomass in high latitude stream ecosystems across the Northern Hemisphere. *Communications Biology, 7*(1), 1–8. doi:10.1038/s42003-024-05936-w PMID:38480906

Jackson, N. D., & Fahrig, L. (2011). Relative effects of road mortality and decreased connectivity on population genetic diversity. *Biological Conservation, 144*(12), 3143–3148. doi:10.1016/j.biocon.2011.09.010

Jacob, S., Laurent, E., Haegeman, B., Bertrand, R., Prunier, J. G., Legrand, D., Cote, J., Chaine, A. S., Loreau, M., Clobert, J., & Schtickzelle, N. (2018). Habitat choice meets thermal specialization: Competition with specialists may drive suboptimal habitat preferences in generalists. *Proceedings of the National Academy of Sciences of the United States of America*, *115*(47), 11988–11993. doi:10.1073/pnas.1805574115 PMID:30397109

Jain, S., Aggarwal, P., Sharma, N., & Sharma, P. (2013). Fostering sustainability through education, research and practice: A case study of TERI University. *Journal of Cleaner Production*, *61*, 20–24. doi:10.1016/j.jclepro.2013.04.021

Jambeck, J. R., Geyer, R., Wilcox, C., Siegler, T. R., Perryman, M., Andrady, A., & Law, K. L. (2015). Plastic waste inputs from land into the ocean. *Science*, *347*(6223), 768–771. doi:10.1126/science.1260352 PMID:25678662

Jatoi, A. S., Ahmed, J., Mubarak, N. M., Shaikh, M. S., Kumar, S., Ahmed, M. I., & Karri, R. R. (2023). A Comprehensive Review on Handling of Plastic Waste for Energy Generation—Current Status and Future Challenges. *Water, Air, and Soil Pollution*, *234*(12), 773. doi:10.1007/s11270-023-06784-w

Jhariya, D., & Choraisia, K. (2010) Impact of mineral production on water resources AIVAM water conservation. Ground Water Augmentation AIVAM Conservation of Water Resources.

Jiang, L., & O'Neill, B. C. (2017). Global urbanization projections for the Shared Socioeconomic Pathways. *Global Environmental Change*, *42*, 193–199. doi:10.1016/j.gloenvcha.2015.03.008

Joa, B., Winkel, G., & Primmer, E. (2018). The unknown known–A review of local ecological knowledge in relation to forest biodiversity conservation. *Land Use Policy*, *79*, 520–530. doi:10.1016/j.landusepol.2018.09.001

Johnson, A. (2020). Urban Greening: Site Selection and Planning for Avenue Plantation. *Journal of Urban Economics*, *25*(2), 145–162.

Johnson, M. (2020). Advancements in Compensatory Forestry: A Global Perspective. *Journal of Environmental Conservation*, *35*(2), 201–215.

Johnson, M. (2020). Green Streets: The Impact of Avenue Plantation on Urban Living. *Urban Ecology*, *45*(3), 321–335.

Jordán-López, A., Martínez-Zavala, L., & Bellinfante, N. (2009). Impact of different parts of unpaved forest roads on runoff and sediment yield in a Mediterranean area. *The Science of the Total Environment*, *407*(2), 937–944. doi:10.1016/j.scitotenv.2008.09.047 PMID:18992920

Jumani, S., Rao, S., Kelkar, N., Machado, S., Krishnaswamy, J., & Vaidyanathan, S. (2018). Fish community responses to stream flow alterations and habitat modifications by small hydropower projects in the Western Ghats biodiversity hotspot, India. *Aquatic Conservation*, *28*(4), 979–993. doi:10.1002/aqc.2904

Kalal, V., Giri, M. N., Baskar, S., & Kuba, R. (2021). Detection of Pollutants present in the surface water of Ayad River, Udaipur, Rajasthan. *IJBI, 3*(1).

Kalal, V., Giri, M. J., Baskar, S., & Kuba, R. (2021). Detection of Pollutants present in the surface water of Ayad River, Udaipur, Rajasthan. *International Journal of Biological Innovations.*, *3*(1), 212–220. doi:10.46505/IJBI.2021.3123

Kalcounis-Rueppell, M. C., Payne, V. H., Huff, S. R., & Boyko, A. L. (2007). Effects of wastewater treatment plant effluent on bat foraging ecology in an urban stream system. *Biological Conservation*, *138*(1-2), 120–130. doi:10.1016/j.biocon.2007.04.009

Kareiva, P., & Marvier, M. (2012). What is conservation science? *Bioscience*, *62*(11), 962–969. doi:10.1525/bio.2012.62.11.5

Karmakar, H. N., & Das, K. P. (2012). *Impact of Mining on Ground & Surface water.* International Mine Water Association. www. IMWA.info

Kaye, T. N., Bahm, M. A., Thorpe, A. S., Gray, E. C., Pfingsten, I., & Waddell, C. (2019). Population extinctions driven by climate change, population size, and time since observation may make rare species databases inaccurate. *PLoS One*, *14*(10), e0210378. doi:10.1371/journal.pone.0210378 PMID:31622344

Kearney M., Shine R. and Porter W.P. (2009). The potential for behavioral hermoregulation to buffer 'cold-blooded' animals against climate warming. *Proc. Natl Acad. Sci. U S A., 106* (10). /pnas.0808913106.

Keck, F., & Lepori, F. (2012). Can we predict nutrient limitation in streams and rivers? *Freshwater Biology*, *57*(7), 1410–1421. doi:10.1111/j.1365-2427.2012.02802.x

Kéfi, S., Génin, A., Garcia-Mayor, A., Guirado, E., Cabral, J. S., Berdugo, M., Guerber, J., Solé, R., & Maestre, F. T. (2024). Self-organization as a mechanism of resilience in dryland ecosystems. *Proceedings of the National Academy of Sciences of the United States of America*, *121*(6), e2305153121. doi:10.1073/pnas.2305153121 PMID:38300860

Kendal, D., Zeeman, B. J., Ikin, K., Lunt, I. D., McDonnell, M. J., Farrar, A., Pearce, L. M., & Morgan, J. W. (2017). The importance of small urban reserves for plant conservation. *Biological Conservation*, *213*, 146–153. doi:10.1016/j.biocon.2017.07.007

Kennedy, C., Pincetl, S., & Bunje, P. (2011). The study of urban metabolism and its applications to urban planning and design. *Environmental Pollution*, *159*(8-9), 1965–1973. doi:10.1016/j.envpol.2010.10.022 PMID:21084139

Kevin, J. Gaston & Spicer, J. (n.d.). *Biodiversity: An Introduction.* Wiley. https://www.wiley.com/en-in/Biodiversity:+An+Introduction,+2nd+Edition-p-9781405118576

Khaleghi M.R. (2017). The influence of deforestation and anthropogenic activities on runoff generation. *Journal of Forest Science*, *63*(6), 245 - 253 . . doi:10.17221/130/2016-JFS

Khalid, F., Iqbal, M. Z., & Qureshi, M. S. (1996). Concentration of heavy metals determined in leaves and soil from various areas of Karachi, city. *Environmental Sciences*, *4*, 213–219.

Khare, S., & Ghosh, S. (2016). Satellite remote sensing technologies for biodiversity monitoring and its conservation. *International Journal of Advanced Earth Science and Engineering*, *5*(1), 375–389. doi:10.23953/cloud.ijaese.213

Kibria, M. G., Masuk, N. I., Safayet, R., Nguyen, H. Q., & Mourshed, M. (2023). Plastic Waste: Challenges and Opportunities to Mitigate Pollution and Effective Management. *International Journal of Environmental Research*, *17*(1), 20. doi:10.1007/s41742-023-00507-z PMID:36711426

Kingsford, R. T., Biggs, H. C., & Pollard, S. R. (2011). Strategic adaptive management in freshwater protected areas and their rivers. *Biological Conservation*, *144*(5), 1194–1203. doi:10.1016/j.biocon.2010.09.022

Kinver, M. (2019). *World 'losing battle against deforestation.* BBC News.

Kirchherr, J., & Charles, K. J. (2016). The social impacts of dams: A new framework for scholarly analysis. *Environmental Impact Assessment Review*, *60*, 99–114. doi:10.1016/j.eiar.2016.02.005

Kivinen, S., Moen, J., Berg, A., & Eriksson, Å. (2010). Effects of modern forest management on winter grazing resources for reindeer in Sweden. *Ambio*, *39*(4), 269–278. doi:10.1007/s13280-010-0044-1 PMID:20799676

Koh, L. P., & Wilcove, D. S. (2008). Is oil palm agriculture really destroying tropical biodiversity? *Conservation Letters*, *1*(2), 60–64. doi:10.1111/j.1755-263X.2008.00011.x

Kominoski, J. S., Ruhí, A., Hagler, M. M., Petersen, K., Sabo, J. L., Sinha, T., Sankarasubramanian, A., & Olden, J. D. (2018). Patterns and drivers of fish extirpations in rivers of the American Southwest and Southeast. *Global Change Biology*, *24*(3), 1175–1185. doi:10.1111/gcb.13940 PMID:29139216

Kowarik, I., & von der Lippe, M. (2018). Plant population success across urban ecosystems: A framework to inform biodiversity conservation in cities. *Journal of Applied Ecology*, *55*(5), 2354–2361. doi:10.1111/1365-2664.13144

Krause-Jensen, D., & Duarte, C. M. (2016). Substantial role of macroalgae in marine carbon sequestration. *Nature Geoscience*, *9*(10), 737–742. doi:10.1038/ngeo2790

Krishnamurthy, K. V. (2018). *An advanced textbook on biodiversity: Principles and practice.* Oxford and IBH Publishing. https://books.google.com/books?hl=en&lr=&id=2Vczdr9ysZYC&oi=fnd&pg=PA1&dq=Advanced+Methodologies+for+Biodiversity+Conservation&ots=lINiJzYoZE&sig=GRETUTczU_0XJmHJedGt0vNq3Sc

Krishnaswamy, J., Kumar, M., Kelkar, N., Nair, T., Atkore, V., Hiremath, A. J., & Siddhartha, A. (2017). Moving from requiem to revival: India's rivers and riverine ecosystems. *Transcending boundaries: Reflecting on twenty years of action and research at ATREE*, 94-103.

Kumar, A. & Verma, A.K. (2017). Biodiversity loss and its Ecological impact in India. *International Journal of Biological Sciences*, *8*(2), 156–160.

Kumar, A., Mishra, S., Pandey, R., Yu, Z.-G., Kumar, M., Khoo, K. S., Thakur, T. K., & Show, P. L. (2023). Microplastics in terrestrial ecosystems: Un-ignorable impacts on soil characterises, nutrient storage and its cycling. *Trends in Analytical Chemistry*, *158*, 116869. doi:10.1016/j.trac.2022.116869

Kumar, A., Upadhyay, P., & Prajapati, S. K. (2022). Impact of microplastics on riverine greenhouse gas emissions- a view point. *Environmental Science and Pollution Research International*, *29*(54), 107300–107303. doi:10.1007/s11356-022-23929-2 PMID:36336740

Kundan, K. (2020). *Allocation of forest land for mining in the state of Chhattisgarh: A brief analysis of Hasdeo Arand stretch*. Department of Social Work, University of Delhi.

Lacher, T. E., Boitani, L., & Da Fonseca, G. A. B. (2012). The IUCN global assessments: Partnerships, collaboration and data sharing for biodiversity science and policy. *Conservation Letters*, *5*(5), 327–333. doi:10.1111/j.1755-263X.2012.00249.x

Lal, R. (2008). Carbon sequestration. *Philosophical Transactions of the Royal Society of London. Series B, Biological Sciences*, *363*(1492), 815–830. doi:10.1098/rstb.2007.2185 PMID:17761468

Lamounier, W. L., dos Santos, J. S., Rodrigues, E. L., & Drumond, M. A. (2024). Defining priority areas for conservation based on multispecies functional connectivity. *Biological Conservation*, *290*, 110438. doi:10.1016/j.biocon.2023.110438

Lanjouw, A., Rainer, H., & White, A. (2018). *Infrastructure Development and Ape Conservation. Vol III. Arcus Foundation New York*. Great Apps & Gibbons program. https://www.stateoftheapes.com/volume-3-infrastructure-development/

Lanorte, A., Santis, F. D., Nolè, G., Blanco, I., Loisi, R. V., Schettini, E., & Vox, G. (2017). Agricultural plastic waste spatial estimation by Landsat 8 satellite images. *Computers and Electronics in Agriculture*, *141*, 35–45. doi:10.1016/j.compag.2017.07.003

LaPaix, R., Harper, K., & Freedman, B. (2012). Patterns of exotic plants in relation to anthropogenic edges within urban forest remnants. *Applied Vegetation Science*, *15*(4), 525–535. doi:10.1111/j.1654-109X.2012.01195.x

Lapiedra, O., Chejanovski, Z., & Kolbe, J. J. (2017). Urbanization and biological invasion shape animal personalities. *Global Change Biology*, *23*(2), 592–603. doi:10.1111/gcb.13395 PMID:27310334

Lartaud, F., Meistertzheim, A. L., Reichert, J., Ziegler, M., Peru, E., & Ghiglione, J. F. (2020). Plastics: An Additional Threat for Coral Ecosystems. In S. Rossi & L. Bramanti (Eds.), *Perspectives on the Marine Animal Forests of the World*. Springer. doi:10.1007/978-3-030-57054-5_14

Laurance, W. F., Lovejoy, T. E., Vasconcelos, H. L., Bruna, E. M., Didham, R. K., Stouffer, P. C., Gascon, C., Bierregaard, R. O., Laurance, S. G., & Sampaio, E. (2002). Ecosystem Decay of Amazonian Forest Fragments: A 22-Year Investigation. *Conservation Biology*, *16*(3), 605–618. doi:10.1046/j.1523-1739.2002.01025.x

Laurance, W., Peletier-Jellema, A., Geenen, B., Koster, H., Verweij, P., Van Dijck, P., Lovejoy, T. E., Schleicher, J., & Van Kuijk, M. (2015). Reducing the global environmental impacts of rapid infrastructure expansion. *Current Biology*, *25*(7), 25. doi:10.1016/j.cub.2015.02.050 PMID:25754645

Leal Filho, W., Will, M., Salvia, A. L., Adomssent, M., Grahl, A., & Spira, F. (2019). The role of green and Sustainability Offices in fostering sustainability efforts at higher education institutions. *Journal of Cleaner Production*, *232*, 1394–1401. doi:10.1016/j.jclepro.2019.05.273

Lebreton, L., & Andrady, A. (2019). Future scenarios of global plastic waste generation and disposal. *Palgrave Communications*, *5*(1), 1–11. doi:10.1057/s41599-018-0212-7

Lehmann, J., Rillig, M. C., Thies, J., Masiello, C. A., Hockaday, W. C., & Crowley, D. (2015). Biochar effects on soil biota–a review. *Soil Biology & Biochemistry*, *43*(9), 1812–1836. doi:10.1016/j.soilbio.2011.04.022

Leiper, I., Zander, K. K., Robinson, C. J., Carwadine, J., Moggridge, B. J., & Garnett, S. T. (2018). Quantifying current and potential contributions of Australian indigenous peoples to threatened species management. *Conservation Biology*, *32*(5), 1038–1047. doi:10.1111/cobi.13178 PMID:30035336

Lemm, J. U., Venohr, M., Globevnik, L., Stefanidis, K., Panagopoulos, Y., van Gils, J., Posthuma, L., Kristensen, P., Feld, C. K., Mahnkopf, J., Hering, D., & Birk, S. (2021). Multiple stressors determine river ecological status at the European scale: Towards an integrated understanding of river status deterioration. *Global Change Biology*, *27*(9), 1962–1975. doi:10.1111/gcb.15504 PMID:33372367

Levin, N., Lugassi, R., Ramon, U., Braun, O., & Ben-Dor, E. (2007). Remote sensing as a tool for monitoring plasticulture in agricultural landscapes. *International Journal of Remote Sensing*, *28*(1), 183–202. doi:10.1080/01431160600658156

Lindenmayer, D. B., & Franklin, J. F. (2002). *Conserving forest biodiversity: a comprehensive multi-scaled approach*. Island Press.

Lindenmayer, D. B., & Likens, G. E. (2009). Adaptive monitoring: A new paradigm for long-term research and monitoring. *Trends in Ecology & Evolution*, *24*(9), 482–486. doi:10.1016/j.tree.2009.03.005 PMID:19409648

Lindenmayer, D., Hobbs, R. J., Montague-Drake, R., Alexandra, J., Bennett, A., Burgman, M., Cale, P., Calhoun, A., Cramer, V., Cullen, P., Driscoll, D., Fahrig, L., Fischer, J., Franklin, J., Haila, Y., Hunter, M., Gibbons, P., Lake, S., Luck, G., & Zavaleta, E. (2008). A checklist for ecological management of landscapes for conservation. *Ecology Letters*, *11*(1), 78–91. doi:10.1111/j.1461-0248.2007.01114.x PMID:17927771

Link, J. (2010). *Ecosystem-based fisheries management: confronting tradeoffs*. Cambridge University Press. doi:10.1017/CBO9780511667091

Living Planet Report 2020. (n.d.). World Wildlife Fund. https://www.worldwildlife.org/publications/living-planet-report-2020

Li, X., Liu, Y., & Zhu, Y. (2022, April). The Effects of Climate Change on Birds and Approaches to Response. [). IOP Publishing.]. *IOP Conference Series. Earth and Environmental Science*, *1011*(1), 012054. doi:10.1088/1755-1315/1011/1/012054

Lutz, S., Bhakta, D., Pearson, H., Vatter, J., & Barnes, R. (2018). *Oceanic Blue Carbon*. Flyer.

MA (Millennium Ecosystem Assessment). (2003). *Ecosystems and Human Well-being: A Framework for Assessment*. Island Press.

Macreadie, P. I., Serrano, O., Maher, D. T., Duarte, C. M., & Beardall, J. (2017). Addressing Calcium Carbonate Cycling in Blue Carbon Accounting. *Limnology and Oceanography Letters*, *2*(6), 195–201. doi:10.1002/lol2.10052

Maffi, L. (Ed.). (2001). *On biocultural diversity: Linking language, knowledge, and the environment*. Smithsonian Institution Press.

Mahadev Madgule, N. (2023). Mechanical properties and microstructure of activated TIG welded similar joints of Inconel alloys by desirability approaches. *Materials Today: Proceedings*, *77*, 528–533. doi:10.1016/j.matpr.2022.12.250

Makwana, V. M., Mahida, D. K., & Dodia, P. P. (2023). Unraveling the Decline: Exploring Global Patterns in Shorebird Populations. In Perspectives on Global Biodiversity Scenarios and Environmental Services in the 21st Century (pp. 192-223). IGI Global. doi:10.4018/978-1-6684-9034-1.ch007

Maldonado, M. T., Surma, S., & Pakhomov, E. A. (2016). Southern Ocean Biological Iron Cycling in the Pre-Whaling and Present Ecosystems. *Philosophical Transactions. Series A, Mathematical, Physical, and Engineering Sciences*, *374*(2081), 20150292. doi:10.1098/rsta.2015.0292 PMID:29035257

Malin, S. A., Ryder, S., & Lyra, M. G. (2019). Environmental justice and natural resource extraction: Intersections of power, equity and access. *Environmental Sociology*, *5*(2), 109–116. doi:10.1080/23251042.2019.1608420

Manganda, I. E. (2004). The Deforestation of Rural Areas in the Lower Congo Province. *Environmental Monitoring and Assessment*, *99*(1–3), 245–250. PMID:15641386

Marcantonio, M., Rocchini, D., Geri, F., Bacaro, G., & Amici, V. (2013). Biodiversity, roads, & landscape fragmentation: Two Mediterranean cases. *Applied Geography (Sevenoaks, England)*, *42*, 63–72. doi:10.1016/j.apgeog.2013.05.001

Marchante, E., López-Núñez, F. A., Duarte, L. N., & Marchante, H. (2024). The role of citizen science in biodiversity monitoring: When invasive species and insects meet. In *Biological Invasions and Global Insect Decline* (pp. 291–314). Elsevier. https://www.sciencedirect.com/science/article/pii/B9780323999182000112 doi:10.1016/B978-0-323-99918-2.00011-2

Masih, S. C. (2021). Impact of Monocrotophos pesticide on serum biochemical profile in freshwater fish, Cirrhinus mrigala (Hamilton, 1822). *International Journal of Biological Innovations.*, *3*(2), 402–406. doi:10.46505/IJBI.2021.3222

McCarthy, T. S., Cooper, G. R. J., Tyson, P. D., & Ellery, W. N. (2000). Seasonal flooding in the Okavango Delta, Botswana—Recent history and future prospects. *South African Journal of Science*, *96*, 25–33.

McDonald, J., McCormack, P. C., Dunlop, M., Farrier, D., Feehely, J., Gilfedder, L., Hobday, A. J., & Reside, A. E. (2019). Adaptation pathways for conservation law and policy. *Wiley Interdisciplinary Reviews: Climate Change*, *10*(1), e555. doi:10.1002/wcc.555

McDonald, R. I., Mansur, A. V., Ascensão, F., Colbert, M. L., Crossman, K., Elmqvist, T., Gonzalez, A., Güneralp, B., Haase, D., Hamann, M., Hillel, O., Huang, K., Kahnt, B., Maddox, D., Pacheco, A., Pereira, H. M., Seto, K. C., Simkin, R., Walsh, B., & Ziter, C. (2020). Research gaps in knowledge of the impact of urban growth on biodiversity. *Nature Sustainability*, *3*(1), 16–24. doi:10.1038/s41893-019-0436-6

McGarigal, K., Romme, W. H., Crist, M., & Roworth, E. (2001). Cumulative effects of roads and logging on landscape structure in the San Juan Mountains, Colorado (USA). *Landscape Ecology*, *16*(4), 327–349. doi:10.1023/A:1011185409347

McHugh, P. A., Thompson, R. M., Greig, H. S., Warburton, H. J., & McIntosh, A. R. (2015). Habitat size influences food web structure in drying streams. *Ecography*, *38*(7), 700–712. doi:10.1111/ecog.01193

McIntosh, A. R., McHugh, P. A., Plank, M. J., Jellyman, P. G., Warburton, H. J., & Greig, H. S. (2018). Capacity to support predators scales with habitat size. *Science Advances*, *4*(7), eaap7523. doi:10.1126/sciadv.aap7523 PMID:29978038

McKay, S. F., & King, A. J. (2006). Potential ecological effects of water extraction in small, unregulated streams. *River Research and Applications*, *22*(9), 1023–1037. doi:10.1002/rra.958

McKinney, M. L. (2008). Effects of urbanization on species richness: A review of plants and animals. *Urban Ecosystems*, *11*(2), 161–176. doi:10.1007/s11252-007-0045-4

McLaughlin, J. F., Hellmann, J. J., Boggs, C. L., & Ehrlich, P. R. (2002). Climate change hastens population extinctions. *Proceedings of the National Academy of Sciences of the United States of America*, *99*(9), 6070–6074. doi:10.1073/pnas.052131199 PMID:11972020

Mehta, P. (2011) *An Assessment of Distribution, Status and Threats to Endemic Malabar Giant Squirrel (Ratufa indica) in Northern Western Ghats, India*. Rufford. https://www.rufford.org/projects/prachi-mehta/an-assessment-of-distribution-status-and-threats-to-endemic-malabar-giant-squirrel-ratufa-indica-in-northern-western-ghats-india/#:~:text=Although%20giant%20squirrels%20are%20widely,the%20squirrel%20by%20local%20people. 19.03.2024

Meijaard, E. (2018). Sumatran orangutan (Pongo abelii). In J. Kingdon & M. Hoffmann (Eds.), *The IUCN Red List of Threatened Species 2018: e.T121097935A123797042*. IUCN.

Meijide, A., Badu, C., Moyano, F., Tiralla, N., Gunawan, D., & Knohl, A. (2018). Impact of forest conversion to oil palm and rubber plantations on microclimate and the role of the 2015 ENSO event. *Agricultural and Forest Meteorology*, *252*, 208–219. doi:10.1016/j.agrformet.2018.01.013

Memmott, J., Gibson, R., Gigante, C. L., Henson, K., Huttel, H. R., Lopezaraiza, M. M., & Pearce, S. (2007). *The conservation of ecological interactions*. doi:10.1079/9781845932541.0226

Merciai, R., Molons-Sierra, C., Sabater, S., & García-Berthou, E. (2017). Water abstraction affects abundance, size-structure and growth of two threatened cyprinid fishes. *PLoS One*, *12*(4), e0175932. doi:10.1371/journal.pone.0175932 PMID:28414787

Meretsky, V. J., Atwell, J. W., & Hyman, J. B. (2011). Migration and conservation: Frameworks, gaps, and synergies in science, law, and management. *Environmental Law (Northwestern School of Law)*, *41*(2), 447. PMID:29332970

Mezőfi, L., Markó, V., Taranyi, D. Á., & Markó, G. (2023). Sex-specific life-history strategies among immature jumping spiders: Differences in body parameters and behavior. *Current Zoology*, *69*(5), 535–551. doi:10.1093/cz/zoac069 PMID:37637309

Micklin, P. P. (1988). Desiccation of the Aral Sea: A water management disaster in the Soviet Union. *Science*, *241*(4870), 1170–1176. doi:10.1126/science.241.4870.1170 PMID:17740781

Miettinen, J., Shi, C., & Liew, S. (2011). Deforestation rates in insular Southeast Asia between 2000 and 2010. *Global Change Biology*, *17*(7), 2261–2270. doi:10.1111/j.1365-2486.2011.02398.x

Millennium Ecosystem Assessment. (2005). *Ecosystems and Human Well-being: Biodiversity Synthesis*. World Resources Institute.

Millennium Ecosystem Assessment. (2005). *Ecosystems and human well-being: synthesis*. Island Press.

Millennium Ecosystem Assessment. (2005). *Ecosystems and human well-being: Synthesis*. Island Press.

Milliman, J. D. (1997). Blessed dams or damned dams? *Nature*, *386*(6623), 325–327. doi:10.1038/386325a0

Millner, N., Newport, B., Sandbrook, C., & Simlai, T. (2024). Between monitoring and surveillance: Geographies of emerging drone technologies in contemporary conservation. *Progress in Environmental Geography*, *3*(1), 17–39. doi:10.1177/27539687241229739

Mirski, P., Grosberg, J., Kull, T., Mellov, P., Tõnisalu, G., Väli, V., & Väli, Ü. (2024). Movement of avian predators points to biodiversity hotspots in agricultural landscape. *Royal Society Open Science*, *11*(1), 231543. doi:10.1098/rsos.231543 PMID:38204791

Mishra, S., Anuradha, J., Tripathi, S., & Kumar, S. (2016). In vitro antioxidant and antimicrobial efficacy of Triphala constituents: Emblica officinalis, Terminalia belerica and Terminalia chebula. *Journal of Pharmacognosy and Phytochemistry*, *5*(6), 273–277.

MoEFCC. (2014). *India's fifth national report to the convention on biological diversity 2014.* Ministry of Environment, Forests and Climate Change, New Delhi, India.

Monastersky, R. (2014). Life – A status report. *Nature, 516*(7530), 159–161. doi:10.1038/516158a

Montgomery, M. R., Stren, R., Cohen, B., & Reed, H. E. (2013). *Cities transformed: demographic change and its implications in the developing world.* Routledge. doi:10.4324/9781315065700

Moore, J., Kissinger, M., & Rees, W. E. (2013). An urban metabolism and ecological footprint assessment of Metro Vancouver. *Journal of Environmental Management, 124*, 51–61. doi:10.1016/j.jenvman.2013.03.009 PMID:23603775

Moreira Filho, O., & Buckup, P. A. (2005). A poorly known case of watershed transposition between the São Francisco and upper Paraná river basins. *Neotropical Ichthyology, 3*(3), 449–452. doi:10.1590/S1679-62252005000300015

Morgan, D., Strindberg, S., Winston, W., Stephens, C. R., Traub, C., Ayina, C. E., Ndolo Ebika, S. T., Mayoukou, W., Koni, D., Iyenguet, F., & Sanz, C. M. (2019). Impacts of Selective Logging and Associated Anthropogenic Disturbance on Intact Forest Landscapes and Apes of Northern Congo. *Frontiers in Forests and Global Change, 2*, 448311. doi:10.3389/ffgc.2019.00028

Murcia, C. (1995). Edge effects in fragmented forests: Implications for conservation. *Trends in Ecology & Evolution, 10*(2), 58–62. doi:10.1016/S0169-5347(00)88977-6 PMID:21236953

Mwale, V. D., To, L. S., Gogoda, C., Ngonda, T., & Nkhoma, R. (2024). Community resilience in Bondo community, Southern Malawi: Balancing energy, water and biodiversity. *International Journal of Energy Sector Management.* https://www.emerald.com/insight/content/doi/10.1108/IJESM-10-2023-0007/full/html

Naidoo, T., & Glassom, D. (2019). Decreased growth and survival in small juvenile fish, after chronic exposure to environmentally relevant concentrations of microplastic. *Marine Pollution Bulletin, 145*, 254–259. doi:10.1016/j.marpolbul.2019.02.037 PMID:31590784

NAS. (1995). *National Academies of Sciences, Engineering, and Medicine. Understanding Marine Biodiversity.* The National Academies Press. doi:10.17226/4923

Naughton, G. P., Hogan, Z. S., Campbell, T., Graf, P. J., Farwell, C., & Sukumasavin, N. (2021). Acoustic telemetry monitors movements of wild adult catfishes in the Mekong River, Thailand and Laos. *Water (Basel), 13*(5), 641. doi:10.3390/w13050641

Neal, M. (1998). *An investigation of altered wetland inflow partitioning due to landuse change in the catchment of Alfred Park, New Germany. BSc (Hons.) Research Report.* School of Life and Environmental Sciences, University of Natal.

Nepstad, D. C., Stickler, C. M., Filho, B. S., & Merry, F. (2008). Interactions among Amazon land use, forests and climate: Prospects for a near-term forest tipping point. *Philosophical Transactions of the Royal Society of London. Series B, Biological Sciences, 363*(1498), 1737–1746. doi:10.1098/rstb.2007.0036 PMID:18267897

Newbold, T., Hudson, L. N., Hill, S. L., Contu, S., Lysenko, I., Senior, R. A., ... Borger, L. (2015). Global effects of land use on local terrestrial biodiversity. *Nature*, *520*(7545), 45–50. doi:10.1038/nature14324 PMID:25832402

Nicholls, R. J., Marinova, N., Lowe, J. A., Brown, S., Vellinga, P., De Gusmao, D., Hinkel, J., & Tol, R. S. J. (2011). Sea-level rise and its possible impacts given a 'beyond 4_C 'world'in the twenty-first century. *Philosophical Transactions. Series A, Mathematical, Physical, and Engineering Sciences*, *369*(1934), 161–181. doi:10.1098/rsta.2010.0291 PMID:21115518

Nijmegen, R. U. (2017). Hunting accounts for 83 and 58 percent declines in tropical mammal and bird populations. *Science Daily*. www.sciencedaily.com/releases/2017/04/170413141120.htm

Nikiema, J., & Asiedu, Z. (2022). A review of the cost and effectiveness of solutions to address plastic pollution. *Environmental Science and Pollution Research International*, *29*(17), 24547–24573. doi:10.1007/s11356-021-18038-5 PMID:35066854

Nizzetto, L., Futter, M., & Langaas, S. (2016). Are agricultural soils dumps for microplastics of urban origin? *Environmental Science & Technology*, *50*(20), 10777–10779. doi:10.1021/acs.est.6b04140 PMID:27682621

NMFS (National Marine Fisheries Service). (2011). *Endangered species act status of west coast salmon and steelhead.* NMFS, Seattle, Washington. http://www.nwr.noaa.gov/ ESA-Salmon-Listings/upload/1-pgr-8--11.pdf

Norgaard, R. B. (1988). Sustainable development: A co-evolutionary view. *Futures*, *20*(6), 606–620. doi:10.1016/0016-3287(88)90003-1

Noss, R. F., & Cooperrider, A. Y. (1994). *Saving nature's legacy: protecting and restoring biodiversity*. Island Press.

Novelli, A., & Tarantino, E. (2015). Combining ad hoc spectral indices based on LANDSAT-8 OLI/TIRS sensor data for the detection of plastic cover vineyard. *Remote Sensing Letters*, *6*(12), 933–941. doi:10.1080/2150704X.2015.1093186

NPPC (Northwest Power Planning Council). (1989). *Smolt density model.* StreamNet, Portland. ftp://ftp.streamnet. org/pub/streamnet/projman_files/sdmdoc.pdf

Oliveira, G., & Oren, D. C. (2017). Exploring the long-term impacts of anthropogenic disturbances in the Amazon basin: A review of vegetation and carbon stock changes in the Brazilian state of Roraima. *Anthropocene*, *17*, 13–29.

Oliver, I., Dorrough, J., Doherty, H., & Andrew, N. R. (2016). Additive and synergistic effects of land cover, land use and climate on insect biodiversity. *Landscape Ecology*, *31*(10), 2415–2431. doi:10.1007/s10980-016-0411-9

Onoh, U. C., Ogunade, J., Owoeye, E., Awakessien, S., & Asomah, J. K. (n.d.-a). *Impact of Climate Change on Biodiversity and Ecosystems Services*. IIARD. https://www.iiardjournals.org/get/IJGEM/VOL.%2010%20NO.%201%202024/Impact%20of%20Climate%20Change%2077-93.pdf

Onoh, U. C., Ogunade, J., Owoeye, E., Awakessien, S., & Asomah, J. K. (n.d.-b). *Impact of Climate Change on Biodiversity and Ecosystems Services*. IIARD. https://www.iiardjournals.org/get/IJGEM/VOL.%2010%20NO.%201%202024/Impact%20of%20Climate%20Change%2077-93.pdf

Owens, A. C., & Lewis, S. M. (2018). The impact of artificial light at night on nocturnal insects: A review and synthesis. *Ecology and Evolution*, *8*(22), 11337–11358. doi:10.1002/ece3.4557 PMID:30519447

Pacheco, P., Mo, K., Dudley, N., Shapiro, A., Aguilar-Amuchastegui, N., Ling, P. Y., Anderson, C., & Marx, A. (2021). *Deforestation fronts: Drivers and responses in a changing world*. WWF.

Paerl, H. W., Hall, N. S., Peierls, B. L., Rossignol, K. L., & Joyner, A. R. (2014). Hydrologic variability and its control of phytoplankton community structure and function in two shallow, coastal, lagoonal ecosystems: The Neuse and New River Estuaries, North Carolina, USA. *Estuaries and Coasts*, *37*(S1), 31–45. doi:10.1007/s12237-013-9686-0

Palita, S. K., Panda, D., & Nayak, J. K. (2023). Indigenous Communities And Biodiversity Conservation: An Indian Perspective. *SCIENCE AND CULTURE*.

Palmer, M. A., Menninger, H. L., & Bernhardt, E. (2010). River restoration, habitat heterogeneity and biodiversity: A failure of theory or practice? *Freshwater Biology*, *55*(s1), 205–222. doi:10.1111/j.1365-2427.2009.02372.x

Pandey, K. (2014, January 25). Climate footprints in the Late Quaternary–Holocene landforms of Dun Valley, NW Himalaya, India. *Current Science*, *106*(2), 245–253.

Pandey, P., Dhiman, M., Kansal, A., & Subudhi, S. P. (2023). Plastic waste management for sustainable environment: Techniques and approaches. *Waste Disposal & Sustainable Energy*, *5*(2), 205–222. doi:10.1007/s42768-023-00134-6 PMID:37359812

Parendes, L. A., & Jones, J. A. (2000). Role of light availability and dispersal in exotic plant invasion along roads and streams in the HJ Andrews Experimental Forest, Oregon. *Conservation Biology*, *14*(1), 64–75. doi:10.1046/j.1523-1739.2000.99089.x

Parks, M., & Gardner, N. (2019). Pruning Practices for Canopy Health in Avenue Plantation. *Urban Forestry & Arboriculture*, *44*(2), 123–137.

Patarasuk, R., & Binford, M. W. (2012). Longitudinal analysis of the road network development and land-cover change in Lop Buri province, Thailand, 1989–2006. *Applied Geography (Sevenoaks, England)*, *32*(2), 228–239. doi:10.1016/j.apgeog.2011.05.009

Pathak, M. R., & Abido, M. S. (2014). *The role of biotechnology in the conservation of biodiversity*. Research Gate. https://www.researchgate.net/profile/Malabika_Roy_Pathak/publication/265126076_The_Role_of_Biotechnology_in_the_Conservation_of_Biodiversity/links/5721d93208ae5454b23210bd.pdf

Paustian, K., Lehmann, J., Ogle, S., Reay, D., Robertson, G. P., & Smith, P. (2016). Climate-smart soils. *Nature*, *532*(7597), 49–57. doi:10.1038/nature17174 PMID:27078564

Pejchar, L., & Mooney, H. A. (2009). Invasive species, ecosystem services and human wellbeing. *Trends in Ecology & Evolution*, *24*(9), 497–504. doi:10.1016/j.tree.2009.03.016 PMID:19577817

Peres, C. A., & Lake, I. R. (2003). Extent of nontimber resource extraction in tropical forests: Accessibility to game vertebrates by hunters in the Amazon basin. *Conservation Biology*, *17*(2), 521–535. doi:10.1046/j.1523-1739.2003.01413.x

Peters, G. P., Marland, G., Le Quéré, C., Boden, T., Canadell, J. G., & Raupach, M. R. (2011). Rapid growth in CO2 emissions after the 2008-2009 global financial crisis. *Nature Climate Change*, *2*(1), 2–4. doi:10.1038/nclimate1332

Pettorelli, N., Safi, K., & Turner, W. (2014). Satellite remote sensing, biodiversity research and conservation of the future. *Philosophical Transactions of the Royal Society of London. Series B, Biological Sciences*, *369*(1643), 20130190. doi:10.1098/rstb.2013.0190 PMID:24733945

Pimm, S. L., Jenkins, C. N., Abell, R., Brooks, T. M., Gittleman, J. L., Joppa, L. N., Raven, P. H., Roberts, C. M., & Sexton, J. O. (2014). The biodiversity of species and their rates of extinction, distribution, and protection. *Science*, *344*(6187), 1246752. doi:10.1126/science.1246752 PMID:24876501

Poff, N. L., Allan, J. D., Bain, M. B., Karr, J. R., Prestegaard, K. L., Richter, B. D., Sparks, R. E., & Stromberg, J. C. (1997). The natural flow regime. *Bioscience*, *47*(11), 769–784. doi:10.2307/1313099

Poff, N. L., & Zimmerman, J. H. (2010). Ecological responses to altered flow regimes: A literature review to inform the science and management of environmental flows. *Freshwater Biology*, *55*(1), 194–205. doi:10.1111/j.1365-2427.2009.02272.x

Polaina, E., Gonzalez-Suarez, M., & Revilla, E. (2015). Socioeconomic correlates of global mammalian conservation status. *Ecosphere*, *6*(9), 1–34. doi:10.1890/ES14-00505.1

Pongratz, J., Reick, C., Raddatz, T., & Claussen, M. (2008). A reconstruction of global agricultural areas and land cover for the last millennium. *Global Biogeochemical Cycles*, *22*(3), GB3018. Advance online publication. doi:10.1029/2007GB003153

Poole, G. C., & Berman, C. H. (2001). An ecological perspective on in-stream temperature: Natural heat dynamics and mechanisms of human-caused thermal degradation. *Environmental Management*, *27*(6), 787–802. doi:10.1007/s002670010188 PMID:11393314

Posey, D. A. (Ed.). (1999). *Cultural and spiritual values of biodiversity*. United Nations Environment Programme.

Postel, S. L., Daily, G. C., & Ehrlich, P. R. (1996). Human appropriation of renewable fresh water. *Science*, *271*(5250), 785–788. doi:10.1126/science.271.5250.785

Poudel, S., Funakawa, S., & Shinjo, H. (2017). Household Perceptions about the Impacts of Climate Change on Food Security in the Mountainous Region of Nepal. *Sustainability (Basel)*, *9*(4), 1–20. doi:10.3390/su9040641

Pounds, J. A. (2001). Climate and amphibian declines. *Nature*, *410*(6829), 639–640. doi:10.1038/35070683 PMID:11287931

Power, M. E., Sun, A., Parker, G., Dietrich, W. E., & Wootton, J. T. (1995). Hydraulic food-chain models. *Bioscience*, *45*(3), 159–167. doi:10.2307/1312555

Prakash, S. (2014). Global warming and its impact on ecosystem, agriculture and human health. *International Journal of Applied Research*, *1*(1), 292–296.

Prakash, S., & Srivastava, S. (2019). Impact of Climate Change on Biodiversity: An Overview. *International Journal of Biological Innovations.*, *1*(2), 60–65. doi:10.46505/IJBI.2019.1205

Prakash, S., & Verma, A. K. (2022). ANTHROPOGENIC ACTIVITIES AND BIODIVERSITY THREATS. *International Journal of Biological Innovations*, *04*(01), 94–103. doi:10.46505/IJBI.2022.4110

Prashantkumar, B. (2023). Impact of climate change and its importance on human performance. In Insights on Impact of Climate Change and Adaptation of Biodiversity (1st ed., pp. 1–9). KD Publication.

Prata, J. C., Silva, A. L. P., da Costa, J. P., Mouneyrac, C., Walker, T. R., Duarte, A. C., & Rocha-Santos, T. (2019). Solutions and integrated strategies for the control and mitigation of plastic and microplastic pollution. *International Journal of Environmental Research and Public Health*, *16*(13), 2411. doi:10.3390/ijerph16132411 PMID:31284627

Pretty, J., Benton, T. G., Bharucha, Z. P., Dicks, L. V., Flora, C. B., Godfray, H. C. J., & Wratten, S. (2018). Global assessment of agricultural system redesign for sustainable intensification. *Nature Sustainability*, *1*(8), 441–446. doi:10.1038/s41893-018-0114-0

Prokic, M. D., Radovanovic, T. B., Gavric, J. P., & Faggio, C. (2019). Ecotoxicological Effects of Microplastics: Examination of Biomarkers, Current State and Future Perspectives. *Trends in Analytical Chemistry*, *111*, 37–46. doi:10.1016/j.trac.2018.12.001

Qing, L., Usman, M., Radulescu, M., & Haseeb, M. (2024). Towards the vision of going green in South Asian region: The role of technological innovations, renewable energy and natural resources in ecological footprint during globalization mode. *Resources Policy*, *88*, 104506. doi:10.1016/j.resourpol.2023.104506

Qin, Y., Wen, Q., Ma, Y., Yang, C., & Liu, Z. (2018). Antibiotics pollution in Gonghu Bay in the period of water diversion from Yangtze River to Taihu Lake. *Environmental Earth Sciences*, *77*(11), 1–11. doi:10.1007/s12665-018-7558-4

Quintanilla, M., León, A. G., & Josse, C. (2022). *The Amazon against the clock: a Regional Assessment on Where and How to protect 80% by 2025*. WWF.

Ragusa, A., Svelato, A., Santacroce, C., Catalano, P., Notarstefano, V., Carnevali, O., Papa, F., Rongioletti, M. C. A., Baiocco, F., Draghi, S., D'Amore, E., Rinaldo, D., Matta, M., & Giorgini, E. (2021). Plasticenta: First evidence of microplastics in human placenta. *Environment International*, *146*, 106274. doi:10.1016/j.envint.2020.106274 PMID:33395930

Rajkhowa, R. (2014). Light pollution and impact of light pollution. [IJSR]. *International Journal of Scientific Research, 3*(10), 861–867.

Ramakrishnan, T., Mohan Gift, M. D., Chitradevi, S., Jegan, R., & Subha Hency Jose, P. (2022). Study of Numerous Resins Used in Polymer Matrix Composite Materials. Advances in Materials Science and Engineering. Hindawi. doi:10.1155/2022/1088926

Ramakrishnan, S., & Jayaraman, A. (2019). Global Warming and Pesticides in Water Bodies. In *Handbook of Research on the Adverse Effects of Pesticide Pollution in Aquatic Ecosystems* (pp. 421–436). IGI Global. https://www.igi-global.com/chapter/global-warming-and-pesticides-in-water-bodies/213519 doi:10.4018/978-1-5225-6111-8.ch023

Ranjan, H., Sanjeevi, R., Vardhini, S., Tripathi, S., & Anuradha, J. (n.d.). *Biogenic Production of Silver Nanoparticles Utilizing an Arid Weed (Saccharum munja Roxb.) and Evaluation of its Antioxidant and Antimicrobial Activities*. Research Gate.

Rao, K. S., Semwal, R. L., Maikhuri, R. K., Nautiyal, S., Sen, K. K., Singh, K., Chandrasekhar, K., & Saxena, K. G. (2003). *Indigenous ecological knowledge, biodiversity and sustainable development in the central Himalayas*. Academia. https://www.academia.edu/download/33916997/44110.pdf

Rathod, N. (2002). *Deforestation in great Aravalli Mountain Region of India. AARS (ACRS 2002) report*. Hazard Mitigation and Disaster Management.

Ray, R., & Jana, T. K. (2017). Carbon sequestration by mangrove forest: One approach for managing carbon dioxide emission from coal-based power plant. *Atmospheric Environment, 171*, 149–154. doi:10.1016/j.atmosenv.2017.10.019

Razak, F., Lua, W. Y., Abd Rasid, N. H., Aziz, N., Repin, I. M., Xue, X.-Z., Ashraf, A. R. M., Bachok, Z., Afiq-Firdaus, A., & Talaat, W. I. A. W. (2024). Adopting Marine Spatial Planning (MSP) using coral health assessment as indicator: A case study in Pulau Redang Marine Park, Malaysia. *Ocean and Coastal Management, 248*, 106943. doi:10.1016/j.ocecoaman.2023.106943

Rech, F., Narimanov, N., Bauer, T., & Schirmel, J. (2022). Urbanization increases fluctuating asymmetry and affects behavioral traits of a common grasshopper. *Ecology and Evolution, 12*(12), e9658. doi:10.1002/ece3.9658 PMID:36568863

Reed, E. S. (1996). *Encountering the world: Toward an ecological psychology*. Oxford University Press.

Reid, A. J., Carlson, A. K., Creed, I. F., Eliason, E. J., Gell, P. A., Johnson, P. T., Kidd, K. A., MacCormack, T. J., Olden, J. D., Ormerod, S. J., Smol, J. P., Taylor, W. W., Tockner, K., Vermaire, J. C., Dudgeon, D., & Cooke, S. J. (2019). Emerging threats and persistent conservation challenges for freshwater biodiversity. *Biological Reviews of the Cambridge Philosophical Society, 94*(3), 849–873. doi:10.1111/brv.12480 PMID:30467930

Rey, F., Dupire, S., & Berger, F. (2024). Forest-based solutions for reconciling natural hazard reduction with biodiversity benefits. *Nature-Based Solutions*, 100114.

Richardson, D. M., Pyšek, P., & Carlton, J. T. (2011). A compendium of essential concepts and terminology in invasion ecology. *Fifty years of invasion ecology: the legacy of Charles Elton, 1*, 409-420. Springer. doi:10.1002/9781444329988.ch30

Richardson, C. J., & McCarthy, E. J. (1994). Effect of land development and forest management on hydrologic response in south-eastern coastal wetlands: A review. *Wetlands*, *14*(1), 56–71. doi:10.1007/BF03160622

Richter, B. D., Mathews, R., Harrison, D. L., & Wigington, R. (2003). Ecologically sustainable water management: Managing river flows for ecological integrity. *Ecological Applications*, *13*(1), 206–224. doi:10.1890/1051-0761(2003)013[0206:ESWMMR]2.0.CO;2

Riitters, K. H., Coulston, J. W., & Wickham, J. D. (2012). Fragmentation of forest communities in the eastern United States. *Forest Ecology and Management*, *263*, 85–93. doi:10.1016/j.foreco.2011.09.022

Rist, L., Campbell, B. M., & Frost, P. (2013). Adaptive management: Where are we now? *Environmental Conservation*, *40*(1), 5–18. doi:10.1017/S0376892912000240

Rivera, K., Fidino, M., Lehrer, E. W., Torsey, H. R., Allen, S., Trouille, L., & Magle, S. B. (2024). Optimizing community science contributions in ecology: A case study on Zooniverse's 'Chicago wildlife watch.'. *Biological Conservation*, *292*, 110490. doi:10.1016/j.biocon.2024.110490

Rogers, K. H. (1995). Riparian wetlands. In G. I. Cowan (Ed.), *Wetlands of South Africa* (pp. 41–52). Department of Environmental Affairs and Tourism.

Rogers, K. H. (1997). Freshwater wetlands. In R. M. Cowling, D. M. Richardson, & S. M. Pierce (Eds.), *Vegetation of Southern Africa* (pp. 322–347). Cambridge University Press.

Rolsky, C., Kelkar, V., Driver, E., & Halden, R. U. (2020). Municipal sewage sludge as a source of microplastics in the environment. *Current Opinion in Environmental Science & Health*, *14*, 16–22. doi:10.1016/j.coesh.2019.12.001

Roodt, A. P., Röder, N., Pietz, S., Kolbenschlag, S., Manfrin, A., Schwenk, K., Bundschuh, M., & Schulz, R. (2022). Emerging midges transport pesticides from aquatic to terrestrial ecosystems: Importance of compound-and organism-specific parameters. *Environmental Science & Technology*, *56*(9), 5478–5488. doi:10.1021/acs.est.1c08079 PMID:35441504

Ruck, A., Van Der Wal, R., & Hood, S. C., A., L. Mauchline, A., G. Potts, S., F. Wallis DeVries, M., & Öckinger, E. (2024). Farmland biodiversity monitoring through citizen science: A review of existing approaches and future opportunities. *Ambio*, *53*(2), 257–275. doi:10.1007/s13280-023-01929-x PMID:37973702

Saba, N., & Balwan, W. K. (2023). Genetic Pollution: A Safe or Risky Bet? *Scholars Academic Journal of Biosciences*, *11*(04), 159–162. doi:10.36347/sajb.2023.v11i04.004

Sabater, S., Elosegi, A., & Ludwig, R. (Eds.). (2018). *Multiple stressors in river ecosystems: Status, impacts and prospects for the future*. Elsevier.

Sagarin, R. D. (2006). Recent studies improve understanding of population dynamics across species ranges. *Oikos*, *115*(2), 386–388. doi:10.1111/j.2006.0030-1299.14438.x

Sage, C. (2007). 7 - The scope for North–south co-operation. *Environmental Policy in an International Context*, *3*, 167–196. doi:10.1016/S1874-7043(03)80026-4

Sailu, G., & Unnisa, S. A. (2024). *Conserving precious urban freshwater bodies: A case study of Telangana, India*. Academic Journals. https://academicjournals.org/journal/IJBC/article-full-text-pdf/701C19771854.pdf

Sala, E., Mayorga, J., Bradley, D., Cabral, R. B., Atwood, T. B., Auber, A., Cheung, W., Costello, C., Ferretti, F., Friedlander, A. M., Gaines, S. D., Garilao, C., Goodell, W., Halpern, B. S., Hinson, A., Kaschner, K., Kesner-Reyes, K., Leprieur, F., McGowan, J., & Lubchenco, J. (2021). Protecting the Global Ocean for Biodiversity, Food and Climate. *Nature*, *592*(7854), 397–402. doi:10.1038/s41586-021-03371-z PMID:33731930

Sala, O. E., Chapin, F. S. III, Armesto, J. J., Berlow, E., Bloomfield, J., Dirzo, R., & Wall, D. H. (2000). Global biodiversity scenarios for the year 2100. *Science*, *287*(5459), 1770–1774. doi:10.1126/science.287.5459.1770 PMID:10710299

Salazar Loor, J., & Fdez-Arroyabe, P. (2019). Aerial and Satellite Imagery and Big Data: Blending Old Technologies with New Trends. In N. Dey, C. Bhatt, & A. S. Ashour (Eds.), *Big Data for Remote Sensing: Visualization, Analysis and Interpretation* (pp. 39–59). Springer International Publishing., doi:10.1007/978-3-319-89923-7_2

Saleh, M., Ashqar, H. I., Alary, R., Bouchareb, E. M., Bouchareb, R., Dizge, N., & Balakrishnan, D. (2024a). Biodiversity for ecosystem services and sustainable development goals. In *Biodiversity and Bioeconomy* (pp. 81–110). Elsevier. https://www.sciencedirect.com/science/article/pii/B9780323954822000055 doi:10.1016/B978-0-323-95482-2.00005-5

Sandbrook, C., Fisher, J. A., Holmes, G., Luque-Lora, R., & Keane, A. (2019). The global conservation movement is diverse but not divided. *Nature Sustainability*, *2*(4), 316–323. doi:10.1038/s41893-019-0267-5

Sanderson, E. W., Walston, J., & Robinson, J. G. (2018). From bottleneck to breakthrough: Urbanization and the future of biodiversity conservation. *Bioscience*, *68*(6), 412–426. doi:10.1093/biosci/biy039 PMID:29867252

Sandy, M. (2019). The Amazon Rain Forest Is Nearly Gone. We Went to the Front Lines to See If It Could Be Saved. *Time Magazine*.

Sanjeevi, R. (2011). *Studies on the treatment of low-strength wastewaters with upflow anaerobic sludge blanket (UASB) reactor: With emphasis on granulation studies*. Centre for Pollution Control and Environmental Engineering.

Sanjeevi, R., Haruna, M., Tripathi, S., Singh, B., & Jayaraman, A. (2017). Impacts of Global Carbon Foot Print on the Marine Environment. *International Journal of Engineering Research & Technology (Ahmedabad)*, *5*, 51–54.

Santana Marques, P., Resende Manna, L., Clara Frauendorf, T., Zandonà, E., Mazzoni, R., & El-Sabaawi, R. (2020). Urbanization can increase the invasive potential of alien species. *Journal of Animal Ecology*, *89*(10), 2345–2355. doi:10.1111/1365-2656.13293 PMID:32627190

Sathvara, P. B., Anuradha, J., Sanjeevi, R., Tripathi, S., & Rathod, A. B. (2023). Spatial Analysis of Carbon Sequestration Mapping Using Remote Sensing and Satellite Image Processing. In *Multimodal Biometric and Machine Learning Technologies* (pp. 71–83). John Wiley & Sons, Ltd. doi:10.1002/9781119785491.ch4

Scassa, T., & Taylor, F. (2017). Legal and ethical issues around incorporating Traditional Knowledge in polar data infrastructures. *Data Science Journal*, *16*, 3–3. doi:10.5334/dsj-2017-003

Scherr, S. J., & McNeely, J. A. (2008). Biodiversity conservation and agricultural sustainability: Towards a new paradigm of "ecoagriculture" landscapes. *Philosophical Transactions of the Royal Society of London. Series B, Biological Sciences*, *363*(1491), 477–494. doi:10.1098/rstb.2007.2165 PMID:17652072

Schreiber, E. S. G., Bearlin, A. R., Nicol, S. J., & Todd, C. R. (2004). Adaptive management: A synthesis of current understanding and effective application. *Ecological Management & Restoration*, *5*(3), 177–182. doi:10.1111/j.1442-8903.2004.00206.x

Secretariat of the Convention on Biological Diversity. (2010). *Global biodiversity outlook 3*. Secretariat of the Convention on Biological Diversity.

Secretariat. (2019, May 17). *Global Assessment Report on Biodiversity and Ecosystem Services*. IPBES secretariat. https://www.ipbes.net/node/35274

Seddon, P. J., Griffiths, C. J., Soorae, P. S., & Armstrong, D. P. (2014). Reversing defaunation: Restoring species in a changing world. *Science*, *345*(6195), 406–412. doi:10.1126/science.1251818 PMID:25061203

Sehgal, K. L. (1999). Coldwater fish and fisheries in the Indian Himalayas: rivers and streams. Fish and fisheries at higher altitudes: Asia. *Food and Agriculture Organization of the United Nations Technical Paper*, *385*, 41–63.

Şekercioğlu, Ç. H., Primack, R. B., & Wormworth, J. (2012). The effects of climate change on tropical birds. *Biological Conservation*, *148*(1), 1–18. doi:10.1016/j.biocon.2011.10.019

Selvakumar, P. (2024). Prediction of Earthquakes, Volcanic Eruptions, Tornadoes, Wildfires, and Droughts Through Artificial Intelligence. In D. Satishkumar, M. Sivaraja, & I. G. I. Global (Eds.), *Utilizing AI and Machine Learning for Natural Disaster Management* (pp. 133–145)., doi:10.4018/979-8-3693-3362-4.ch008

Senadheera, S. S., Withana, P. A., Dissanayake, P. D., Sarkar, B., Chopra, S. S., Rhee, J., & Ok, Y. (2021). Scoring environment pillar in environmental, social, and governance (ESG) assessment. *Sustainable Environment*, *7*(1), 1–7. doi:10.1080/27658511.2021.1960097

Sengupta, A., Bhan, M., Bhatia, S., Joshi, A., Kuriakose, S., & Seshadri, K. S. (2024). Realizing "30 × 30" in India: The potential, the challenges, and the way forward. *Conservation Letters*, *13004*(2), e13004. doi:10.1111/conl.13004

Sequeira, A. M. M., Bouchet, P. J., Yates, K. L., Mengersen, K., & Caley, M. J. (2018). Transferring biodiversity models for conservation: Opportunities and challenges. *Methods in Ecology and Evolution*, *9*(5), 1250–1264. doi:10.1111/2041-210X.12998

Seto, K. C., & Ramankutty, N. (2016). Hidden linkages between urbanization and food systems. *Science*, *352*(6288), 943–945. doi:10.1126/science.aaf7439 PMID:27199419

Shah, R. D. T., Sharma, S., & Bharati, L. (2020). Water diversion induced changes in aquatic biodiversity in monsoon-dominated rivers of Western Himalayas in Nepal: Implications for environmental flows. *Ecological Indicators*, *108*, 105735. doi:10.1016/j.ecolind.2019.105735

Shakil, M. H. (2020). Environmental, social and governance performance and stock price volatility: A moderating role of firm size. *Journal of Public Affairs*, *22*(3), e2574. doi:10.1002/pa.2574

Shankar, T., Sahoo, U., & Maitra, S. (2021). *Intensive Farming: It's Effect on the Environment*. Research Gate. https://www.researchgate.net/publication/357423421

Shik, J. Z., & Dussutour, A. (2020). Nutritional dimensions of invasive success. *Trends in Ecology & Evolution*, *35*(8), 691–703. doi:10.1016/j.tree.2020.03.009 PMID:32668214

Shi, T., Zheng, X., Zhang, H., Wang, Q., & Zhong, X. (2021). Coral reefs: Potential blue carbon sinks for climate change mitigation. [Chinese Version]. *Bulletin of Chinese Academy of Sciences*, *36*(3), 270–278.

Shivanna, K. R. (2022). Climate change and its impact on biodiversity and human welfare. *Proceedings of the Indian National Science Academy. Part A, Physical Sciences*, *88*(2), 160–171. doi:10.1007/s43538-022-00073-6

Shochat, E., Lerman, S. B., Anderies, J. M., Warren, P. S., Faeth, S. H., & Nilon, C. H. (2010). Invasion, competition, and biodiversity loss in urban ecosystems. *Bioscience*, *60*(3), 199–208. doi:10.1525/bio.2010.60.3.6

Siddique, R., Khatib, J., & Kaur, I. (2008). Use of recycled plastic in concrete: A review. *Waste Management (New York, N.Y.)*, *28*(10), 1835–1852. doi:10.1016/j.wasman.2007.09.011 PMID:17981022

Silcock, J. L., Fensham, R. J., & Lewis, T. (2020). Involving Indigenous Landowners in Feral Animal Management in Remote Northern Australia: A Case Study of Wunambal Gaambera Country, Kimberley, Western Australia. *Environmental Management*, *65*(5), 568–579.

Singh, A. K., & Akhtar, M. S. (2015). *Coldwater fish diversity of India and its sustainable development. Biodiversity for Sustainable development.* UP Biodiversity Board, Lucknow.

Singh, R. (n.d.). *Environmental policies in India.* Research Gate. https://www.researchgate.net/profile/Ravikant-Singh-2/publication/378590048_ENVIRONMENTAL_POLICIES_IN_INDIA/links/65e0ad05adf2362b635e43c3/ENVIRONMENTAL-POLICIES-IN-INDIA.pdf

Singh, G., & Agarwal, N. K. (2017). Impact of hydropower project (RoR) on the ichthyofaunal diversity of river Birahiganga in Central Himalaya (India). *Journal of Fisheries*, *5*(2), 507–512. doi:10.17017/j.fish.38

Singh, J. S. (2012). Biodiversity: An Overview. *Proceedings of the National Academy of Sciences. India. Section B, Biological Sciences*, *82*(2), 239–250. doi:10.1007/s40011-012-0112-3

Singh, S., Singh, P., Selvasembian, R., & Srivastava, K. K. (Eds.). (2021). *Global climate change*. Elsevier.

Smith, A. (2018). Global Perspectives on Compensatory Forestry: Challenges and Opportunities. *Environmental Science & Policy*, *63*, 45–52.

Smith, B. (2019). Native Species Selection in Avenue Plantation: Enhancing Biodiversity and Resilience. *Arboriculture & Urban Forestry*, *45*(3), 231–245.

Smith, B. T., Seeholzer, G. F., Harvey, M. G., Cuervo, A. M., & Brumfield, R. T. (2017). A latitudinal phylogeographic diversity gradient in birds. *PLoS Biology*, *15*(4), e2001073. doi:10.1371/journal.pbio.2001073 PMID:28406905

Sokół, M., & Łaska, G. (2024). Biodiversity and ecosystem services: Complementary approach for catchment protection and land use in northeastern Poland. *Ecological Indicators*, *159*, 111649. doi:10.1016/j.ecolind.2024.111649

Spotswood, E. N., Beller, E. E., Grossinger, R., Grenier, J. L., Heller, N. E., & Aronson, M. F. (2021). The biological deserts fallacy: Cities in their landscapes contribute more than we think to regional biodiversity. *Bioscience*, *71*(2), 148–160. doi:10.1093/biosci/biaa155 PMID:33613128

Srivastava, S., Shukla, S. N., & Singh, P. (2019). Climate change and Biodiversity management: A review. *International Journal of Environmental Sciences*, *10*(2), 71–75.

Sternberg, R. (2000). *Our cultural heritage: The role of biodiversity in preserving cultural landscapes. United Nations Educational, Scientific and Cultural Organization*. UNESCO.

Stolton, S., Shadie, P., & Dudley, N. (2013). *Guidelines for applying protected area management categories including IUCN WCPA best practice guidance on recognising protected areas and assigning management categories and governance types*. IUCN. https://portals.iucn.org/library/node/30018

Stromberg, J. C., Lite, S. J., Marler, R., Paradzick, C., Shafroth, P. B., Shorrock, D., White, J. M., & White, M. S. (2007). Altered stream-flow regimes and invasive plant species: The Tamarix case. *Global Ecology and Biogeography*, *16*(3), 381–393. doi:10.1111/j.1466-8238.2007.00297.x

Subasman, I. (2024). Evaluation of The Effectiveness of Group Cooperation In The Context of Problem Base Learning (PBL): A MIXED METHOD APPROACH. *Jurnal Konseling Pendidikan Islam*, *5*(1), 11–23. doi:10.32806/jkpi.v5i1.99

Sumaila, U. R., Cheung, W., Dyck, A., Gueye, K., Huang, L., Lam, V., & Zeller, D. (2012). Benefits of rebuilding global marine fisheries outweigh costs. *PLoS One*, *7*(7), e40542. doi:10.1371/journal.pone.0040542 PMID:22808187

Sumi, L., & Thomsen, S. (2001). *Mining in Remote Areas, Issues and Impacts. MinindWatch Canada/ Mines Alerte. Hudson, T. (2012).* Living with Earth, An Introduction to Environmental Geology. PHI Learning Private Limited.

Sunita, A., Daisy, R., & Singh, R. (n.d.). Sarus crane, biodiversity and pesticides: A review. *Fauna Journal.* https://www.faunajournal.com/archives/2024/vol11issue1/PartA/11-1-5-363.pdf

Sustainable Development Institute (SDI). (2022). *Guidelines for Effective Compensatory Forestry*. SDI. https://sustainabledevelopment.institute/

Sutton, R., & Morris, C. (2006). *Instream flow assessment upper Lemhi River and Canyon Creek, Idaho. U. S.* Department of the Interior, Bureau of Reclamation.

Swanson, S. S., & Ardoin, N. M. (2021). Communities behind the lens: A review and critical analysis of Visual Participatory Methods in biodiversity conservation. *Biological Conservation*, *262*, 109293. doi:10.1016/j.biocon.2021.109293

Synthesis, T. E. E. B. (2010). *The Economics of Ecosystems and Biodiversity: Mainstreaming the Economics of Nature: A synthesis of the approach, conclusions and recommendations of TEEB*. European Commission.

Tang, H. (2022). The Impact of urbanization on river channel pollution. *Highlights in Science. Engineering and Technology*, *17*, 51–57. doi:10.54097/hset.v17i.2435

TekmanM. B.WaltherB. A.PeterC.GutowL.BergmannM. (2022). Impacts of plastic pollution in the oceans on marine species, B*iodiversity and Ecosystems*, 1–221, WWF Germany, Berlin. Doi: doi:10.5281/zenodo.5898684

Tengö, M., Austin, B. J., Danielsen, F., & Fernández-Llamazares, Á. (2021). Creating synergies between citizen science and Indigenous and local knowledge. *Bioscience*, *71*(5), 503–518. doi:10.1093/biosci/biab023 PMID:33986633

Thatoi, H. N. (2021). Biodiversity Conservation and Livelihood Management. Daya Publishing House, a division of Astral International Pvt. Limited.

The New Nature Economy Report. (2020). Geneva: World Economic Forum.

The State of the Worlds Forest. (2020). *Food and Agriculture Organization of the United Nations*. FAO. https://www.fao.org/state-of-forests/en/

Thompson, K., Bakker, J. P., & Bekker, R. M. (1997). *The soil seed banks of North West Europe: methodology, density and longevity*. Cambridge university press.

Threlfall, C. G., Law, B., & Banks, P. B. (2012). Influence of landscape structure and human modifications on insect biomass and bat foraging activity in an urban landscape. *PLoS One*, *7*(6), e38800. doi:10.1371/journal.pone.0038800 PMID:22685608

Tiago, P., Evaristo, I., & Pinto, B. (2024). The role of BioBlitzes in citizen science: Insights from participants and experts. *Frontiers in Environmental Science*, *12*, 1347428. doi:10.3389/fenvs.2024.1347428

Tickner, D., Opperman, J. J., Abell, R., Acreman, M., Arthington, A. H., Bunn, S. E., Cooke, S. J., Dalton, J., Darwall, W., Edwards, G., Harrison, I., Hughes, K., Jones, T., Leclère, D., Lynch, A. J., Leonard, P., McClain, M. E., Muruven, D., Olden, J. D., & Young, L. (2020). Bending the curve of global freshwater biodiversity loss: An emergency recovery plan. *Bioscience*, *70*(4), 330–342. doi:10.1093/biosci/biaa002 PMID:32284631

Tilman, D., Fargione, J., Wolff, B., D'Antonio, C., Dobson, A., Howarth, R., & Swackhamer, D. (2001). Forecasting agriculturally driven global environmental change. *Science*, *292*(5515), 281–284. doi:10.1126/science.1057544 PMID:11303102

Tipping, J. M. (2011). Effect of juvenile length on Chinook Salmon survivals at four hatcheries in Washington state. *North American Journal of Aquaculture*, *73*(2), 164–167. doi:10.1080/15222055.2011.568862

Toivonen, T., Heikinheimo, V., Fink, C., Hausmann, A., Hiippala, T., Järv, O., Tenkanen, H., & Di Minin, E. (2019). Social media data for conservation science: A methodological overview. *Biological Conservation*, *233*, 298–315. doi:10.1016/j.biocon.2019.01.023

Tonkin, J. D., & Death, R. G. (2013). Macroinvertebrate drift-benthos trends in a regulated river. *Fundamental and Applied Limnology*, *182*(3), 231–245. doi:10.1127/1863-9135/2013/0404

Travis, J. M., Delgado, M., Bocedi, G., Baguette, M., Bartoń, K., Bonte, D., Boulangeat, I., Hodgson, J. A., Kubisch, A., Penteriani, V., Saastamoinen, M., Stevens, V. M., & Bullock, J. M. (2013). Dispersal and species' responses to climate change. *Oikos*, *122*(11), 1532–1540. doi:10.1111/j.1600-0706.2013.00399.x

Tyser, R. W., & Worley, C. A. (1992). Alien flora in grasslands adjacent to road and trail corridors in Glacier National Park, Montana (USA). *Conservation Biology*, *6*(2), 253–262. doi:10.1046/j.1523-1739.1992.620253.x

UNEP (United Nations Environment Programme). (2016). *Marine Plastic Debris and Microplastics – Global lessons and research to inspire action and guide policy change*. UNEP.

UNFCCC (2015). *Adoption of the Paris agreement, Proposal by the President, Draft decision CP.21*. UNFCC.

UNFCCC. (2015). 19th Session of the Conference of the Parties to the UNFCCC. *International Institute for Sustainable Development*. UNFCCC.

United Nations Environment Programme. (2009). *From Conflict to Peacebuilding: The Democratic Republic of the Congo*. Nairobi, Kenya: United Nations Environment Programme.

United Nations Environment Programme. (2010). *Global Forest Resources Assessment 2010: Main Report*. Food and Agriculture Organization of the United Nations.

United Nations Environment Programme. (2019). *Global environment outlook: Summary for policymakers*. United Nations Environment Programme.

United Nations. (2015). *Transforming our world: The 2030 agenda for sustainable development*. United Nations.

Valter, M. (2021). Plastic pollution: A focus on freshwater biodiversity. *Ambio*, *50*(7), 1313–1324. doi:10.1007/s13280-020-01496-5 PMID:33543362

Van der Plas, F., Manning, P., Allan, E., Scherer-Lorenzen, M., & Fischer, M. (2016). Biodiversity effect on ecosystem functioning stronger in more variable environments. *Ecology Letters*, *19*(4), 391–399.

Van Der Ree, R., Jaeger, J. A., van der Grift, E. A., & Clevenger, A. P. (2011). Effects of roads and traffic on wildlife populations and landscape function: Road ecology is moving toward larger scales. *Ecology and Society*, *16*(1), art48. doi:10.5751/ES-03982-160148

Vanneste, T., Depauw, L., De Lombaerde, E., Meeussen, C., Govaert, S., De Pauw, K., Sanczuk, P., Bollmann, K., Brunet, J., Calders, K., Cousins, S. A. O., Diekmann, M., Gasperini, C., Graae, B. J., Hedwall, P.-O., Iacopetti, G., Lenoir, J., Lindmo, S., Orczewska, A., & De Frenne, P. (2024). Trade-offs in biodiversity and ecosystem services between edges and interiors in European forests. *Nature Ecology & Evolution*, *8*(5), 1–8. doi:10.1038/s41559-024-02335-6 PMID:38424266

Vasilachi, I. C., Asiminicesei, D. M., Fertu, D. I., & Gavrilescu, M. (2021). Occurrence and fate of emerging pollutants in water environment and options for their removal. *Water (Basel)*, *13*(2), 181. doi:10.3390/w13020181

Vasiliev, D., Hazlett, R., Hutchinson, K. L., & Bornmalm, L. (2024). Light at the end of the tunnel: Innovative opportunities for saving tropical biodiversity. *Ambio*, *53*(5), 702–717. doi:10.1007/s13280-023-01970-w PMID:38353913

Velado-Alonso, E., Kleijn, D., & Bartomeus, I. (2024). Reassessing science communication for effective farmland biodiversity conservation. *Trends in Ecology & Evolution*. https://www.cell.com/trends/ecology-evolution/abstract/S0169-5347(24)00032-6

Velasquez-Camacho, L., Merontausta, E., Etxegarai, M., & de-Miguel, S. (2024). Assessing urban forest biodiversity through automatic taxonomic identification of street trees from citizen science applications and remote-sensing imagery. *International Journal of Applied Earth Observation and Geoinformation*, *128*, 103735. doi:10.1016/j.jag.2024.103735

Verma, R., Vinoda, K. S., Papireddy, M., & Gowda, A. N. S. (2016). Toxic pollutants from plastic waste-a review. *Procedia Environmental Sciences*, *35*, 701–708. doi:10.1016/j.proenv.2016.07.069

Vesk, P. A., Morris, W. K., Neal, W. C., Mokany, K., & Pollock, L. J. (2020). Transferability of trait-based species distribution models. *Ecography*, *44*(1), 134–147. doi:10.1111/ecog.05179

Vitousek, P. M., Mooney, H. A., Lubchenco, J., & Melillo, J. M. (1997). Human domination of Earth's ecosystems. *Science*, *277*(5325), 494–499. doi:10.1126/science.277.5325.494

von Schiller, D., Bernal, S., Dahm, C. N., & Martí, E. (2017). Nutrient and organic matter dynamics in intermittent rivers and ephemeral streams. In *Intermittent rivers and ephemeral streams* (pp. 135–160). Academic Press. doi:10.1016/B978-0-12-803835-2.00006-1

Vu, A. V., Baumgartner, L. J., Mallen-Cooper, M., Doran, G. S., Limburg, K. E., Gillanders, B. M., & Cowx, I. G. (2022). Diverse migration tactics of fishes within the large tropical Mekong River system. *Fisheries Management and Ecology*, *29*(5), 708–723. doi:10.1111/fme.12566

Wale, E., Drucker, A. G., & Zander, K. K. (2012). The economics of managing crop diversity On-Farm: Case studies from the genetic resources policy initiative. In The Economics of Managing Crop Diversity On-Farm: Case Studies from the Genetic Resources Policy Initiative. Taylor and Francis. doi:10.4324/9781849775649

Walker, K. F. (1985). A review of the ecological effects of river regulation in Australia. In *Perspectives in Southern Hemisphere Limnology: Proceedings of a Symposium, held in Wilderness,* (pp. 111-129). Springer Netherlands. 10.1007/978-94-009-5522-6_8

Wang, Y., Liu, Y., & Gu, B. (2022). *COP26: Progress, challenges, and outlook.*

Wang, K., Franklin, S. E., Guo, X., & Cattet, M. (2010). Remote sensing of ecology, biodiversity and conservation: A review from the perspective of remote sensing specialists. *Sensors (Basel)*, *10*(11), 9647–9667. doi:10.3390/s101109647 PMID:22163432

Wang, W., Feng, C., Liu, F., & Li, J. (2020). Biodiversity conservation in China: A review of recent studies and practices. *Environmental Science and Ecotechnology*, *2*, 100025. doi:10.1016/j.ese.2020.100025 PMID:36160928

Warren, R., Price, J., Graham, E., Forstenhaeusler, N., & VanDerWal, J. (2018). The projected effect on insects, vertebrates, and plants of limiting global warming to 1.5 C rather than 2 C. *Science*, *360*(6390), 791–795. doi:10.1126/science.aar3646 PMID:29773751

Waterman, C. (2021). Efficient Irrigation Systems for Sustainable Avenue Plantation. *Journal of Sustainable Urban Water Management*, *12*(4), 421–435.

Watkins, P. C., Woodward, K., Stone, T., & Kolts, R. L. (2003). Gratitude and happiness: Development of a measure of gratitude, and relationships with subjective well-being. *Social Behavior and Personality*, *31*(5), 431–451. doi:10.2224/sbp.2003.31.5.431

Watson, J. E., Evans, T., Venter, O., Williams, B., Tulloch, A., Stewart, C., & Lindenmayer, D. (2018). The exceptional value of intact forest ecosystems. *Nature Ecology & Evolution*, *2*(4), 599–610. doi:10.1038/s41559-018-0490-x PMID:29483681

Watts, J. (2018). Destruction of nature as dangerous as climate change, scientists warn. *The Guardian*. http://tinyurl.com/y56erzml

Watts, R. J., Richter, B. D., Opperman, J. J., & Bowmer, K. H. (2011). Dam reoperation in an era of climate change. *Marine and Freshwater Research*, *62*(3), 321–327. doi:10.1071/MF10047

Webster, K. L. (2014). *Sediment dispersal and accumulation in an insular sea: deltas of Puget Sound* [Doctoral dissertation, UW].

Wei, Y., & Hoganson, H. M. (2005). Landscape impacts from valuing core area in national forest planning. *Forest Ecology and Management*, *218*(1-3), 89–106. doi:10.1016/j.foreco.2005.07.002

Westley, F., & Vredenburg, H. (1997). Interorganizational Collaboration and the Preservation of Global Biodiversity. *Organization Science*, *8*(4), 381–403. doi:10.1287/orsc.8.4.381

Wilkie, D., Shaw, E., Rotberg, F., Morelli, G., & Auzel, P. (2000). Roads, development, and conservation in the Congo Basin. *Conservation Biology*, *14*(6), 1614–1622. doi:10.1111/j.1523-1739.2000.99102.x PMID:35701921

Williams, J. J., & Newbold, T. (2021). Vertebrate responses to human land use are influenced by their proximity to climatic tolerance limits. *Diversity & Distributions*, *27*(7), 1308–1323. doi:10.1111/ddi.13282

Williamson, P., & Gattuso, J. P. (2022). Carbon removal using coastal blue carbon ecosystems is uncertain and unreliable, with questionable climatic cost-effectiveness. *Frontiers in Climate*, *4*, 4. doi:10.3389/fclim.2022.853666

Wilson, K. (2019). Environmental Impact Assessment of Avenue Plantation Projects. *Journal of Environmental Science and Technology*, *48*(7), 4125–4133.

Witte, J. (1992). Deforestation in Zaire: Logging and Landlessness. *The Ecologist*, *22*(2), 58–64.

Wolfire, D., Brunner, J., & Sizer, N. (1998). *Forests and the Democratic Republic of Congo*. World Resources Institute.

Woodall, L. C., Sanchez-Vidal, A., Canals, M., Paterson, G. L., Coppock, R., Sleight, V., Calafat, A., Rogers, A. D., Narayanaswamy, B. E., & Thompson, R. C. (2014). The deep sea is a major sink for microplastic debris. *Royal Society Open Science*, *140317*(4), 140317. doi:10.1098/rsos.140317 PMID:26064573

Wright, S. L., Thompson, R. C., & Galloway, T. S. (2013). The physical impacts of microplastics on marine organisms: a review. *Environmental pollution (Barking, Essex: 1987)*, *178*, 483–492. doi:10.1016/j.envpol.2013.02.031

Wu, B., Liu, M., & Xu, Z. (2020). Accumulation of Microplastics in Soil and Their Potential Effects on Plant Growth. *The Science of the Total Environment*, *715*, 136952.

Wu, W. M., Yang, J., & Criddle, C. S. (2017). Microplastics pollution and reduction strategies. *Frontiers of Environmental Science & Engineering*, *11*(1), 6. doi:10.1007/s11783-017-0897-7

WWF's Living planet Report. (2022). WWF.

Xiao, Y., Zhou, G., Qiu, X., Liu, F., Chen, L., & Zhang, J. (2024). Biodiversity of network modules drives ecosystem functioning in biochar-amended paddy soil. *Frontiers in Microbiology*, *15*, 1341251. doi:10.3389/fmicb.2024.1341251 PMID:38328424

Xie, C., Liu, C., Liu, D., & Jim, C. Y. (2024). Charting the Research Terrain for Large Old Trees: Findings from a Quantitative Bibliometric Examination in the Twenty-First Century. *Forests*, *15*(2), 373. doi:10.3390/f15020373

Xu, B., Liu, F., Cryder, Z., Huang, D., Lu, Z., He, Y., Wang, H., Lu, Z., Brookes, P. C., Tang, C., Gan, J., & Xu, J. (2020). Microplastics in the soil envi ronment: Occurrence, risks, interactions and fate–a review. *Critical Reviews in Environmental Science and Technology*, *50*(21), 2175–2222. doi:10.1080/10643389.2019.1694822

Yanagi, M. (2024). Understanding the ecological significance of wheat: Roles, climate change impacts, and biodiversity conservation strategies. *Geographical Research Bulletin*, *3*, 6–9.

Yang, C., Yu, M., Li, Y., Hu, F., Jiang, Y., Liu, Q., Sha, D., Xu, M., & Gu, J. (2019). Big Earth data analytics: A survey. *Big Earth Data*, *3*(2), 83–107. doi:10.1080/20964471.2019.1611175

Yang, H., Suntornvongsagul, K., Wang, J., Wang, Q., & Zhang, W. (2024). Indigenous knowledge in the sustainable management of a biodiversity-based millennium-old tea forest in the highlands of Yunnan Province, southwestern China. *Biodiversity (Nepean)*, *25*(2), 1–11. doi:10.1080/14888386.2023.2290176

Yasin, E. H., Siddig, A. A., & Kornel, C. (2024). *Forests at the Crossroads: Biodiversity Conservation in the Era of Climate Change*. InTech Open. https://www.intechopen.com/online-first/1180035

Yokutkhon, R. (2024). EFFECTIVE STRATEGIES FOR TEACHING SECOND LANGUAGE ACQUISITION. *Ta'limning Zamonaviy Transformatsiyasi*, *5*(2), 107–111.

Zaidi, S. G. S., Singh, A. K., & Sarma, D. (2018). Breeding, seed production and rearing of coldwater ornamental fishes in Aquarium. *Published by ICAR-DCFR Bulletin*, (27), 43.

Zeng, N. (2008). Carbon sequestration via wood burial. *Carbon Balance and Management*, *3*(1), 1–12. doi:10.1186/1750-0680-3-1 PMID:18173850

Zhang, H., Jarić, I., Roberts, D. L., He, Y., Du, H., Wu, J., Wang, C., & Wei, Q. (2020). Extinction of one of the world's largest freshwater fishes: Lessons for conserving the endangered Yangtze fauna. *The Science of the Total Environment*, *710*, 136242. doi:10.1016/j.scitotenv.2019.136242 PMID:31911255

Zhang, J., Ren, S., Xu, W., Liang, C., Li, J., Zhang, H., Li, Y., Liu, X., Jones, D. L., Chadwick, D. R., Zhang, F., & Wang, K. (2022). Effects of plastic residues and microplastics on soil ecosystems: A global meta-analysis. *Journal of Hazardous Materials*, *435*, 129065. doi:10.1016/j.jhazmat.2022.129065 PMID:35650746

Zhang, W., Zhang, S., Zhao, Q., Qu, L., Ma, D., & Wang, J. (2020). Spatio-temporal distribution of plastic and microplastic debris in the surface water of the Bohai Sea, China. *Marine Pollution Bulletin, 158*, 111343. doi:10.1016/j.marpolbul.2020.111343 PMID:32753167

Zoker, M. E., Karim, S., Kargbo, B., Kemoh, R., & Yorpoi, L. D. (2022). Impacts of road constructions on ecological biodiversity and livelihood in Sierra Leone. *Resaerch Gate, 3*, 280–296. doi:10.54660/anfo.2022.3.3.18

About the Contributors

Ashok K. Rathoure brings an impressive and diverse background for environment protection. His extensive education, with a doctorate degree from Central University (HNBGU Srinagar UK), coupled with over 18 years of combined industry and teaching experience, establishes him as a seasoned expert in the environment field. Dr. Rathoure's expertise lies in Ecology & Biodiversity, Water Pollution Monitoring, Prevention & Control, and EIA (Environmental Impact Assessment) coordination for various sectors, aligning with NABET (QCI, New Delhi) guidelines. His experience encompasses a wide range of industrial sectors, including synthetic organic chemicals, mining, metallurgical industries, power generation, nuclear sector, coal, cement, pesticides, chlor-alkali, chemical fertilizers, paint, sugar, oil & gas, man-made fibers, distilleries, hazardous waste treatment, and more. In his academic and research pursuits, Dr. Rathoure has contributed significantly, with a remarkable record of 100 full-length research papers published in international and national journals. Additionally, he has authored over 55 books from reputed publishers in India and abroad. His dedication to advancing scientific knowledge is further evident through his role as a reviewer for more than 100 research manuscripts in international journals. Being a member of various international organizations such as APCBEES (Hong Kong), IACSIT (Singapore), EFB (Spain), and the Society for Conservation Biology (Washington), Dr. Rathoure has established a global network and stays abreast of international developments in his field. As a founder member of the Scientific Planet Society (Dehradun), he contributes to the promotion of scientific awareness and collaboration. Furthermore, Dr. Rathoure's commitment to the dissemination of knowledge is reflected in his editorial roles. He serves as the Editor-in-Chief for Octa Journal of Environmental Research, Managing Editor for Octa Journal of Biosciences, and Executive Editor for Scientific India. His significant publication record, including research papers, books, and book chapters, underscores his scholarly contributions to advancing knowledge in environmental biotechnology and related fields. Furthermore, his editorial roles for scientific journals demonstrate his dedication to fostering academic discourse and dissemination of research findings within the scientific community.

***

Arun Bhatt obtained his B.Sc.(Ag), M.Sc.(Ag) Genetics and Plant Breeding and Ph.D.(Ag) Genetics and Plant Breeding with specialization in Molecular Biology from GBPUA&T, Pantnagar. He is recipient of Young Scientist Award in field of Agriculture Sciences from Government of Uttarakhand. Dr. Bhatt has over more than fifteen years of teaching, research and extension experience in Genetics, Plant Breeding, Biotechnology, Molecular Diagnostic, Genetic Engineering, Environmental Biotechnology etc. subjects under various capacities. He has published more than fifty research papers, review papers in reputed scientific journals. Dr. Bhatt has served HNBGU Srinagar Garhwal (Uttarakhand) and VCSGUUHF Bharsar Pauri Garhwal as a faculty member. He has also delivered expert lectures in different institutions of repute. He has supervised more than thirty research scholars (UG, PG and PhD). Currently, he is working as Assistant Professor (Biotechnology) at G.B. Pant Institute of Engineering and Technology, Ghurdauri, Pauri Garhwal-246194 (Uttarakhand).

Naresh Chandra is presently working as Director (Education) B. K. Birla College, Kalyan (M.S.) with effect from 1st June 2018 after retiring as Principal from the same institute, bringing a wealth of experience in higher education administration. He joined as Principal, BKBCK in Oct 1988 and worked till 31st May 2018 (except when he was on lien). He has also been helping and guiding other educational institutions in the pursuit of Higher Education. He has been tirelessly working for the cause of quality education in HEIs and students. In the past, Dr. Chandra has served as Pro Vice-Chancellor of the University of Mumbai twice from April 1997 to May 2000 (first term) and from April 2012 to May 2015 (second term). As Pro Vice-Chancellor, he played a pivotal role in introducing innovative and need-based academic programs / courses, fostering a strong relationship between the University and its affiliated colleges, and creating a culture of mutual trust and appreciation. He has been instrumental in enhancing the university administration, improving campus development, streamlining the examination system, and fostering student welfare. He has been Chairman / Member of several Expert Committees of Government / UGC / NAAC / Universities and has visited several colleges in the country. He has also visited several Universities / Institutions in UK, USA, Australia, China, Sri Lanka and South Africa for exchange of academic ideas. He was invited to attend International Visitor Leadership Program on 'Higher Education – Administration and Curriculum Development: A Project for India', at Washington and New York in November 2008. He has attended a five-day Workshop on 'Multidisciplinary Approach to University Leaders Development' held at Harvard University, USA in August 2012. He is passionate about research. 19 students have been awarded Ph.D. Degree under his guidance. He has worked on two Major Research Projects sanctioned by UGC and NMPB, New Delhi respectively. He has published several research papers in National and International journals. He was the Organizing Secretary of 102nd session of Indian Science Congress Association, held at University of Mumbai from 3 - 7 January 2015. He is a life member of National Academy of Sciences, India. His rich experience in academic administration and his contribution to the cause has helped him carve a niche for himself in the field of higher education.

Ankita Chatterjee is a Research Scholar in the Department of Environmental Studies, Visva-Bharati, Santiniketan with UGC JRF fellowship.

Sharda Dhadse is currently working as Environment Scientist at CSIR-NEERI, Nagpur. She has completed his doctorate on Treatment of Industrial Wastewaters by Vermitechnology and Effects on Tissues and Cellular Organelles of Earthworms from Nagpur University. She has published more than 25 research articles and contributed more than 40 EIA projects.

Pravinsang P. Dodia boasts over three decades of academic and research excellence, with a remarkable 40 publications in prestigious national and international journals. His expertise spans the ecology and behavior of birds and mammals, emphasizing conservation biology, toxicology, and biodiversity monitoring. As a recipient of grants from esteemed agencies, He leads pioneering studies that address critical environmental challenges. His mentorship extends to nine PhD candidates and 23 MSc dissertation students, nurturing the next generation of conservation scientists. In administrative roles like course coordinator and as a member of reputable organizations, he shapes academic landscapes. His persuasive advocacy extends to expert lectures at conferences, inspiring action on pressing conservation issues. Dr. Dodia's enduring dedication and visionary leadership inspiring positive change and shaping the future of environmental stewardship.

Vandana Gupta is working as an Assistant Professor in the Department of Environmental Sciences at B. K. Birla College (Empowered Autonomous), Kalyan (MS), India. She has more than 14 years of teaching experience in degree college and over 10 years of research experience. She is also Quality Manager for NABL accredited Environmental Sciences Laboratory, B. K. Birla College. She has published 10 research papers and 5 chapters in reputed books. She has completed two research projects and guided more than 10 post graduate students for their academic dissertations. She is passionate about research and her field of interest is Environmental Biotechnology and Environmental Nanotechnology.

Anakha Mariya Jacob was born on the 16th of July 1999 at the Alappuzha district of Kerala. She has her graduation and postgraduation in botany from Sacred

Heart College, Thevara, Kochi. She has qualified for CSIR NET (Life Science) & GATE 2022, 2023 & 2024. Currently, she is pursuing a Bachelor's in Education at Sree Narayana Training College, Poochakkal, Alappuzha, Kerala.

J. Anuradha has received her doctoral degree in Pollution control and Environmental Technology from Pondicherry (Central) University. Her proficiency in field of Environmental Technology and Nanobiotechnology has been peer-recognized in the form of several international and national publications. She has thus far authored three patent abstracts; Twenty four research papers in high impact-factor international and national journals primarily in UGC approved Journals; Twenty eight of her papers have been taken part in international and nation scientific meet discussions. To add as an exemplary researcher, her work has been awarded with nine best paper's in scientific discussions, and does presided as a 'Resource person and Subject specialist' in (ANVESION - West Zone) annual students science meet (2017) in India. She has been 'nominated' for the 'Dr APJ Abdul Kalam Innovation award' in Indian Academy of Biomedical Sciences (2017) and had delivered 'Resource person' lecture on 'Facile synthesis of gold nanoparticles and its application on cancer diagnosis and treatment'.

J. Anuradha is working as Assistant Professor in Nims Institute of Allied Medical Science and Technology, Nims University Rajasthan, Jaipur. She received B.Sc. Botany (2003), M.Sc. Botany (2005), M.Phil Envi. Sci. & Engg. (2006), and Ph.D Environmental Technology (2014) from Pondicherry (Central) University. Her technical know-how in the field of Environmental Technology and Advanced Nanotechnology explored the cutting-edge R&D and has been peer-recognized in the form of several international and national publications. She has thus far published 3 patents, 55 publications in peer recognized international and national journals, 1 authored book and 9 book chapters. She has successfully guided three PhD scholars. To add as an exemplary researcher, she has been awarded with eight best papers in scientific discussions & 4 student best paper awards. She had presided as a 'Resource person and Subject Specialist' and Chairperson in several national/international scientific meets.

Ashwini Kardbhajne holds a M.Sc. degree in Botany from Rashtrasant Tukdoji Maharaj Nagpur University. Currently, she is working as a Project Assistant in EIA at CSIR National Environmental Engineering Research Institute Nagpur, with expertise in the fields of ecology and biodiversity.

Vineet Kumar is a renowned expert in the field of Malariology, Immunology, Animal physiology and anatomy. He has developed a fascination for the human body and his inquisitive mind and passion for learning led him to pursue a career in teaching and medical research. Dr. Vineet Kumar, has done M.Sc. (Hons.) M.Phil., Ph.D. form a premium institute Panjab University, Chandigarh, India and have a more than eight year of teaching experience in undergraduate and postgraduate level. He is presently serving as Assistant Professor and Head in department of Zoology at Dr. Yashwant Singh Parmar Govt. P.G. College Nahan, Distt. Sirmour, Himachal Pradesh. He researched on parasite immunology and antimalarial vaccines to get his Ph.D., from Panjab University, Chandigarh. His involvement in both teaching and research is active. His area of research is Protozoology (Malariology), Immunology, Entomology, Ecology Animal Physiology and Anatomy, contributing to advancements in medical understanding and technology. Dr. Vineet participated and presented papers at various national and international conferences. Numerous of his research articles have previously appeared in reputable journals on a national and worldwide level. He has awarded with Best faculty award and Young Scientist award. He is a member of Indian Science Congress Association, Vector and Vectors Borne Diseases, Indian Society of Parasitology, Microbiology Society of India and Editorial Board member of Journal of Infectious Disease Research and Guest Editor of Journal of Traditional Medicine Research, New Zealand. He continues to write and continually updates his knowledge in his profession through additional research and study.

Monika kumari Is currently a research scholar in Department of Zoology and Environmental Science (deemed to be) University , Haridwar. She has completed her M.Phil from Himalayan Garhwal University, Srinagar in year 2020. She has attended 6workshop , 5 international conference and1 national, 3 hands on trainings and 3 research paper and 1 book chapter. Her field of interest is impact of waste water on agriculture.

Poonam Nirmalkumar Kurve is a retired professor from the Department of Zoology. She is a recognized research guide for M.Sc. and Ph.D. Zoology and Environmental Science, University of Mumbai. She was the Ex-coordinator of the Department of Environmental Science, Dept of Biodiversity Wildlife Conservation & Management, VPM's B. N. Bandodkar College of Science, Thane 400 601, India. She has teaching experience of around 30 years and research experience of 37 years. She has 37 research publications in National and International Journals to her credit and has completed seven research projects in her career. She is an Hon. recipient of the UGC Travel Grant for the oral presentation of a research paper titled "Coastal Tourism in Alibaug; district Raigad: a menace or a bonus?" at the World Conference on Marine Biodiversity (26-30 September 2011) organized by the University of Aberdeen in collaboration with the University of St. Andrews, Scotland, UK .

Sangeeta Madan is currently an Associate Professor in Department of Environmental Science at Gurukul Kangri (deemed to be) University, Haridwar. She has published 52 papers in peer reviewd journals and 1 book. She attended 22 national , 14 International conferences and 20 workshops . She delivered 10 guest lectures. She has published 8 book chapters. 8 students have been awarded Ph.D degree under her supervision and 4 students are enrolled in Ph.D. Her field of interest is water , air pollution , phytoremedition and vermicomposting.

Vishal M. Makwana is currently pursuing his Ph.D. in Zoology at the Zoology Department, Sir P. P. Institute of Science, M. K. Bhavnagar University, specializing in the urban ecology of birds. He has made significant scholarly contributions, publishing five research articles and two book chapters in renowned national and international journals and books. His research interests span diverse and critical fields, including ornithology, entomology, and environmental toxicology. He is a sought-after speaker, having delivered 5 invited talks and presented his research at 6 prominent national and international conferences and workshops. He has reviewed for many International journals of repute.

Ritika Maurya Is a research scholar in Department of Environmental Science at Gurukul kangri (deemed to be) University, Haridwar, India. She has attended 2 workshops and 1 National conference. Her field of interest is modified biochar.

P. Selvakumar is an Associate Professor in the Department of Science and Humanities at Nehru Institute of Technology, Coimbatore, Tamilnadu, India-641105, where he has been since 2023. From 2023 he served as Department IQAC Coordinator. From 2011 to 2023 he served as Assistant Professor of Department of Science and Humanities in Info institute of Engineering and Dr.N.G.P Institute of Technology, Coimbatore, Tamilnadu, India. His area of interest is the field of phytochemistry. His Credit includes 30 Journal publications, 12 patents and 20 conference papers, conducted 5 seminars and conferences and honored with best teacher Awards.

Pradosh Kumar Parida, a PhD scholar at Tamil Nadu Agricultural University, Coimbatore is a dedicated agronomy researcher with extensive publications. He has completed his M.Sc. (Agronomy) at the College of Post-Graduate Studies in Agricultural Sciences, Central Agricultural University, Barapani, Meghalaya. His work focuses on sustainable agricultural practices, earning recognition in the academic community.

Foram P. Patel is an Assistant Professor of Zoology at D.K.V. Arts and Science College in Jamnagar, affiliated with Saurashtra University, Rajkot, India. She has completed her doctorate with a focus on the impact of anthropogenic activities on the ecology of the House sparrow (Passer domesticus). She has amassed over ten years of teaching experience at the college level, complemented by an equal period of research experience especially in the field of ornithology, biodiversity and conservation biology. She has published several research papers in recognized peer-reviewed journals, underscoring her significant contributions to the field. She is also dedicated to nature education, actively engaging students in field-based learning initiatives that emphasize ecological methodologies and environmental stewardship. Her efforts aim to develop conservation strategies for avian species especially in urban ecosystems, particularly through community involvement and educational outreach.

Pulak Kumar Patra is an Associate Professor at the Department of Environmental Studies, Visva-Bharati, Santiniketan. He has done his M.Sc. (Applied Geology) and PhD from IIT, Bombay. He has also obtained MTech degree from IIT, Kharagpur. Since 2022, he is teaching & doing research in Environment Science at Visva-Bharati. He has published more than 60 research papers and book chapters different aspects of environmental earth science. He is the author and editor of seven books. He has supervised nine PhD students.

R. Sanjeevi is an accomplished academician and researcher serving as the Associate Professor in Environmental Science of the Nims Institute of Allied Medical Science and Technology at NIMS University, located in Rajasthan, Jaipur. His extensive qualifications include M.Sc., M.Phil., Ph.D., and a Post-Doctoral Fellowship (CSIR-RA), along with a PG Diploma in Industrial Waste & Wastewater Treatment, all focused on Environmental Science and Engineering. With over 19 years of experience, He has thus far published 12 patents, 39 publications in peer recognized international and national journals, 1 authored book, 9 book chapters and 6 Awards. Dr. R. Sanjeevi's expertise encompasses a wide array of fields within environmental science, including water analysis, wastewater treatment, waste-to-energy conversion, renewable energy resources, air pollution monitoring and control, solid waste management, bioprocess technology, environmental biotechnology, and remote sensing & GIS. His multifaceted expertise and numerous accolades underscore his significant contributions to the domain of environmental science and his commitment to excellence in research and education.

Bharti Sadhiyan is a research scholar in Department of Zoology and Environmental Science at Gurukul Kangri (deemed to be) University. She has been awarded Inspire scholarship till her Masters. She has completed her Masters in Zoology in year 2020. She has attended 5 workshops and 2 National conference and 4 International conference. Her field of interest is phytoremediation in association with biochar , constructed wetland and fishery.

Prashantkumar B. Sathvara is currently employed at Nims Institute of Allied Medical Science and Technology, Nims University Rajasthan, Jaipur. He holds a diverse academic background, including a pursuing Ph.D. in Environmental Science (Thesis Submitted), M.Sc. in Environmental Science and Technology, P.G. Diploma in Geo-informatics System, and a B.Sc. in Safety and Fire Technology. His research contributions are primarily focused on environmental science, with a particular emphasis on carbon sequestration mapping, vegetation change detection, and the application of remote sensing and GIS techniques. He has published 3 publications in peer recognized international and national journals, 2 book chapter and 9 book chapter under reviewed and accepted He has also undergone extensive training programs, including projects at Space Application Center (ISRO), Ahmedabad and safety and fire courses, further augmenting his skills and knowledge. His computer skills include proficiency in advanced software such as ArcGIS, QGIS, ENVI, and Google Earth Pro, as well as familiarity with various operating systems and office tools.

Tari Vinaya Satyawan Savitri is working as a Postdoctoral fellow at, Universitas Airlangga, Indonesia. She formerly worked as a 'Grants Writer - Offshore' at Envirotech Education Pty. Ltd., Australia. She is an 'External Consultant' at GauEcoGram Agrovikas Producer Company Limited, Pune, Maharashtra, India (Part-time). She has completed her Ph.D. in Environmental Science from the University of Mumbai. She did undergraduate studies in Zoology, M.Sc., in Environmental Science, and a Diploma in Biotechnology (2 years) from the University of Mumbai and a B.Ed., from S.N.D.T. University, Mumbai. She has qualified State Level Eligibility Test (SLET) for Assistant Professorship valid in Maharashtra and Goa, India. She has been awarded the 'Young Scientist Award' from the Deccan Environmental Research Organization (DERO), India. Recently, she was awarded "Empowering Women" by the Gender Sensitization Cell, SIES College of Arts, Commerce and Science, Nerul, Navi Mumbai, Maharashtra, India. She has attended around 36 conferences/workshops/webinars/faculty development programs. She has published over 30 research items in reputed Journals, Books, Magazines/e-newsletters, and conferences. She has authored book chapters with several reputed publishers. She is an Associate Researcher at the Australian Centre for Sustainable Development, Research, and Innovation (ACSDRI), Australia. She is appointed a scientific reviewer and writer (Part-time) at 'The BrainAura', organization. She is appointed Managerial Board Member at IKIGAI, The Art of Living, NGO, Hyderabad. She has successfully acted as an organizing committee member/scientific committee member/conference treasurer/hostess for many national and international conferences. She is associated with various national and international scientific societies as a panel member, professional/official member/life member, etc., Her current research interests are environmental economics, sustainable crop protection practices, circular economy, environmental pollution, and biomedical applications of nanomaterials. She has also engaged as an Editor/Reviewer/Editorial board member/Young scientist board member for more than 20 reputed scientific journals. She is a certified mentor on 'Expertrons', for guiding students and professionals on their career path. She has collaborated with various overseas countries such as the USA, Australia, Qatar, Indonesia, Nigeria, Thailand, Malaysia, Turkey, etc.

Sandeep Tripathi is the Associate Professor of Biochemistry and Director, Institute of Public Health and Nutrition. His research degree was based on "Metal ion neurotoxicity" which was awarded as "Best PhD thesis" among the batch 2010 in the King George's Medical University, Lucknow, India. He has published 68 research articles and many book chapters. He has awarded Mrs. Abida Mehdi Award for Neuroscience Research by Indian Academy of Biomedical Sciences and Young Scientist Award by centre for education growth and research. He has established "National referral centre for fluoride poisoning (NRCFPI) in India", it was first fluoride referral centre in the country and continuously leading by Dr. Tripathi as Head, Research and Development and he is the recipientof Niloufer Chinoy Award- 2013 presented by the International Society for fluride Research. Dr. Tripathi is the member of editorial board of many journals. He is the member of various International and National Societies (IBRO, ISNR, SOT, IAN, ACBI, SFRR, SOPI, SPGHTN and InSLAR). He has attended more than 50 conferences and workshops as speaker/ chairperson / moderator or as delegate.

Index

D

E

F

G

H

I

L

M

P

R

S

U

W

www.ingramcontent.com/pod-product-compliance
Ingram Content Group UK Ltd.
Pitfield, Milton Keynes, MK11 3LW, UK
UKHW050948290926
15470UKWH00005B/1709

9 798369 369500